BALKAN CYBERIA

History of Computing
William Aspray and Thomas J. Misa, editors

A complete list of the titles in this series appears in the back of this book.

BALKAN CYBERIA

COLD WAR COMPUTING, BULGARIAN MODERNIZATION, AND THE INFORMATION AGE BEHIND THE IRON CURTAIN

VICTOR PETROV

THE MIT PRESS CAMBRIDGE, MASSACHUSETTS LONDON, ENGLAND

This book is freely available in an open access edition thanks to TOME (Toward an Open Monograph Ecosystem)—a collaboration of the Association of American Universities, the Association of University Presses, and the Association of Research Libraries—and the generous support of the University of Tennessee. Learn more at the TOME website, available at: openmonographs.org.

This book was set in Stone Serif by Westchester Publishing Services. Printed and bound in the United States of America.

Library of Congress Cataloging-in-Publication Data

Names: Petrov, Victor P., author.
Title: Balkan cyberia : Cold War computing, Bulgarian modernization, and the information age behind the Iron Curtain / Victor Petrov.
Other titles: Cold War computing, Bulgarian modernization, and the information age behind the Iron Curtain
Description: Cambridge, Massachusetts : The MIT Press, [2023] | Series: History of computing | Includes bibliographical references and index.
Identifiers: LCCN 2022030019 (print) | LCCN 2022030020 (ebook) | ISBN 9780262545129 (paperback) | ISBN 9780262373258 (epub) | ISBN 9780262373265 (pdf)
Subjects: LCSH: Electronic digital computers—History—20th century. | Computers—Bulgaria—History—20th century. | Computers—Soviet Union—History. | Computer industry—Bulgaria—History—20th century. | Cold War.
Classification: LCC QA76.17 .P46 2023 (print) | LCC QA76.17 (ebook) | DDC 004.09—dc23/eng/20220707
LC record available at https://lccn.loc.gov/2022030019
LC ebook record available at https://lccn.loc.gov/2022030020

10 9 8 7 6 5 4 3 2 1

To my parents, for everything

CONTENTS

LIST OF ABBREVIATIONS

ASU	Automated System of Governance
BAS	Bulgarian Academy of Sciences
BCP	Bulgarian Communist Party
BISA	Bulgarian Industrial Economic Association
BNB	Bulgarian National Bank
BSP	Bulgarian Socialist Party
CICT	Central Institute for Computer Technology
CoCom	Coordinating Committee for Multilateral Export Controls
Comecon	Council for Mutual Economic Assistance
CNC	computer-numerical controls
CPSU	Communist Party of the Soviet Union
CSTP	Committee for Scientific and Technical Progress (Bulgaria)
DKMS	Dimitrov Communist Youth Union
DS	Committee of State Security (Bulgaria)
DSO	State Economic Union
ES	Unified System (of Computers)
ESGRAON	System for Civilian Registration and Population Administration
ESSI	Unified System of Social Information
GKNT	Committee for Scientific and Technical Progress (USSR)
IIASA	International Institute for Applied Systems Analysis

ICCT	Intergovernmental Commission on Computer Technology
IMF	International Monetary Fund
ITCR	Institute for Technical Cybernetics and Robotics
IZOT	Computational, Recording and Organisational Technology
KESSI	Committee for the Unified System of Social Information
RB	Robot series (Bulgaria)
SM	System of Minicomputers
STI	Scientific-Technical Intelligence
SPC	State Planning Commission
TCC	Territorial Computer Center
TNMT	Technical and Scientific Creativity of Youth
UNIDO	UN Industrial Development Organization
VMEI	Higher Machine-Electrotechnical Institute
VTO	Foreign Trade Organisation

NOTE ON TRANSLITERATIONS AND ABBREVIATIONS

The book uses the Library of Congress (LoC) romanization system for Bulgarian and Russian, except for peoples' and place names that have a more common English spelling or an author who transliterates their name in their unique, non-LoC way.

Abbreviations of Bulgarian organizations come in two varieties: abbreviations of the English translation of the Bulgarian name or abbreviations using the LoC spelling of the Bulgarian organization. For ones that have appeared in literature, such as the Bulgarian Communist Party (BCP) or institutes' own publications (ITCR and CICT), the first convention is followed. For most others, the second has been followed. A note on IZOT: this transliteration is more commonly seen in secondary literature and is closer to the Bulgarian than "ISOT," which is sometimes seen on the company's own documents—hence, IZOT has been chosen.

Archive abbreviations are as follows:

AKRDOPBGDSRSNBA-M/R—Arkhiv na Komisii͡ata za Razkrivane na Dokumentite I Obi͡avi͡avane na Prinadlezhnost na Bŭlgarski Grazhdani kŭm Dŭrzhavna Sigurnost I Razuznavatelnite Sluzhbi na Bŭlgarskata Narodna Armii͡a–M/R (Archive of the Commission for the Uncovering of Documents and Notification of Belonging of Bulgarian Citizens to

State Security and the Intelligence Services of the Bulgarian People's Army, Sofia, Bulgaria)

AMVnR—Arkhiv na Ministerstvoto na Vŭnshnite Raboti (Archive of the Ministry of Foreign Affairs, Sofia)

ARAN—Arkhivĭ RAN (Archive of the Russian Academy of Sciences, Moscow, Russia)

BAN-NA—Bŭlgarska Akademii͡a na Naukite-Nauchen Arkhiv (Scientific Archive of the Bulgarian Academy of Sciences, Sofia, Bulgaria)

DA-V—Dŭrzhaven Arkhiv Varna (State Archive, Varna, Bulgaria)

NAI—National Archives of India, New Delhi, India

NMML—Nehru Memorial Museum and Library, New Delhi, India

TsDA—T͡Sentralen Dŭrzhaven Arkhiv (Central State Archive, Sofia, Bulgaria)

RGAE—Rossīyskiĭ Gosudarstvennyĭ Arkhiv Ėkonomiki (Russian State Archive for the Economy, Moscow, Russia)

PREFACE

"One of the things that hurts me the most is that we didn't save anything. There were machines, unique ones, that we built, that in the 1990s we housed in the basement of our institute, and then sold for scrap metal because we had no space anymore. Very few made it to a museum, if any. That hurts." So told me Petŭr Petrov, then in his 80s, over tea in his apartment in central Sofia. He had worked at the Institute of Technical Cybernetics and Robotics since its inception in the 1960s, had automated copper mines, and had made it a goal of his retirement to record his institute's history. I met him by chance in the archives of the Bulgarian Academy of Sciences, when, as it happened, we were both looking at the same files. Over the next couple of years, he became not just a source for this book, but a friend.

"I know that in the Soviet space program, there were Bulgarian computers, and you understand that they couldn't do their calculations without [them], of course." These were Koĭcho Dragostinov's words, spoken to me in his firm housed within the old Central Institute of Computing Technology, where he had worked during the socialist period. He took me around the premises, introduced me to many of the other surviving actors in the story of Bulgarian computing, and he, too, was invaluable for this book. Subsequently, much of this book was researched not just in the archives but at the tables and living rooms of multiple people who built, traded, and thought with Bulgarian computers.

All these people's stories have rarely been told. Petrov, Dragostinov, and the rest were engineers and scientists, involved in either fundamental or applied research in the electronics sector. In the socialist world, they fell under the umbrella of the "scientific-technical intelligentsia," which encompassed occupations that were both in the sectors of industrial production and management and in scientific research. They built the tools of modernity and often thought about the future horizons these tools opened up. Often these horizons are less explored than those advanced by the cultural intelligentsia that much of literature on the socialist period is interested in. These are the technical intellectuals this book is concerned with and whose stories it endeavors to bring to the front and center. These stories are global and full of pride in achievements that transcended political or parochial concerns. They are about the Bulgarian computer and its many lives, including its role at the very frontiers of human knowledge (such as in the Soviet space program mentioned in the quote above), but most importantly, the stories are about the people involved. In some ways, this book is a modest attempt to soothe Petrov's lament—if the machines never made it to a museum, I hope this book goes some way toward recovering the words and worlds of the people who made them.

ACKNOWLEDGMENTS

This book is a story born out of many stories I heard—as a child and then as a researcher. Some of these stories were told in dry reports, but the most interesting of them were told by people. And the big story, well, that came in conversations with so many people that these acknowledgments threaten to balloon beyond all proportion. Writing a book is never truly a solitary task. Thus, I want to repay some of my debts here.

I thank my department at the University of Tennessee, Knoxville, where this book took its final shape. It remains an incredibly hospitable place to write, think, and teach, and a shining example of the role that public universities and humanistic education have to play in our world today. All my colleagues here have made this a stimulating home, but I want to thank in particular Monica Black, Tore Olsson, Luke Harlow, Ernie Freeberg, Nikki Eggers, Matthew Gillis, Kristen Block, and Chad Black for their intellectual, social, and musical contributions to this work of prolix. I also thank my students who have made me a better teacher, and thus a better historian.

This book started its life as a dissertation under the supervision of Mark Mazower. It was in one of his classes that it was born as a paper, and it was in his office that the inspiration to make a Balkan story a global one was born. I will be eternally grateful for his push to make this story wider than originally envisioned, for his keen sense of style (which I hope has rubbed off on my own writing), and for the comments that made this book a sharper work. I thank Adam Tooze for not only chairing the defense but also for encouraging me to clarify every muddled issue and to draw out what my findings mean for European or economic history. His intellectual curiosity remains truly inspirational. Malgorzata Mazurek was an especially close reader and helped me place my story in the long durée of Eastern Europe. She continuously invited me to workshops, conferences, and other venues where her comments and questions have been part of the evolution of the thesis into a book. Matthew Jones's expertise in computer history was invaluable for this amateur. Without him, and his approach to my work, I would have never realized what kinds of arguments I was making—and his advice was always delivered with humor. I must also thank Elidor Mëhilli, whose guidance and support have been with me throughout the evolution of the project. It was his own dissertation, now an astounding book, read during that fateful class with Mark Mazower, that launched me on my topic and methodology.

This book has been influenced by conversations, classes, and musings with many others throughout the years. I thank Columbia University's history department for its global outlook and encouragement to follow the story wherever it takes me. Victoria de Grazia gave me my first chance to lecture in a university setting and spun the story of European history beautifully. Susan Pedersen's class on modern state formation is always in the back of my mind. Tarik Amar's insight into Soviet modernity and interest in socialist spying have been key. I also acknowledge the great privilege of being in a funded program, which is the sine qua non of doing global history—and I take this opportunity to call for more funding for graduate students to allow such work to be done. Many stories die at the planning stages; this book is a product of fortunate circumstances.

I thank my first university mentors, Dimitar Bechev and Robert J. W. Evans, for patiently teaching me how to be a historian and a writer, ever since my bachelor's degree. I thank the very first teachers who inspired me to do history, too, as such passions start early: the teaching staff at Gosford Hill School in Kidlington, England. I hope this work is a fitting culmination to all their efforts. Holly Case's continuous guidance and friendship has been an enormous source of energy throughout the years. Odd Arne Westad listened to my first ideas as they were taking shape and suggested India as a case study—for that I can't thank him enough. Larry Wolff's classes broadened my concept of the region immensely.

I also thank all my colleagues in Bulgarian studies who have shaped my thinking as it develops. Mary Neuburger and Theodora Dragostinova, especially, have read drafts and reviewed different versions, far too many times to count. Their work, methodologies, and insights into Bulgarian history have been absolutely invaluable in inspiring me and helping me find out what I wanted to say about the country.

At the European University Institute where I was lucky enough to spend a wonderful year, I thank Federico Romero for serving as my mentor; his and Corinna Unger's class was a strong vision of how to do global history. I especially thank my partner in crime, Veneta Ivanova, for not only arguments about history and socialism but also wine—so much wine—and all the amazing memories of Florence (and beyond).

Many people throughout too many conferences, workshops, and bars in Chicago, Boston, Philadelphia, San Francisco, New York, Leipzig, Geneva, Aarhus, London, Mobile, Florence, Laramie, Blagoevgrad, Sofia, Budapest, Cluj, Moscow, and places I have probably forgotten by now have contributed to this work. I also want to thank Betty Banks, Maria Baramova, Svetlana Borodina, Michel Christian, Filip Erdeljac, Yakov Feygin, Madigan Fichter, Falk Flade, Maria Galmarini, Konstantin Georgiev, Zoltan Ginelli, Mukaram Hhana, Ralitsa Konstantinova, Ana Luleva, James Mark and the whole "Socialism Goes Global" project; Robin Markwica, Johanna Mellis, John Paul Newman, Benjamin Peters, James Robertson, Miroslaw Sikora, Ksenia Tatarchenko, and Markus Wien.

The research and writing for this book have been supported by grants by the Leverhulme Trust and Social Science Research Council, as well as a Faculty Fellowship at the University of Tennessee Humanities Center, which I thank for allowing me the space and time to finish this book. Thank you to the American Research Center in Sofia (ARCS) for the fellowship that started this research, and especially Cynthia Lintz and Emil Nankov, who led the program when I was there. Diana Mishkova, Rumen Avramov, and Georgi Vladev at the Centre for Advanced Study in Sofia provided a warm and stimulating place during my affiliation. I thank Atanaska Stancheva for invitations to the Institute of Ethnography to present some of my results, but more importantly for the laughs and her warding off of snakes in the Rhodopes.

This research would have been impossible without the staff and archivists across numerous countries. I thank the staff of reading room 100 at the Central State Archive in Sofia, who must be tired of me, and Milena Milcheva for the help with images; the archivists at the Scientific Archive at the Bulgarian Academy of Sciences for their constant support; the staff at the Ministry of Foreign Affairs archive for allowing me to work there; those in the National Library, who never refused to find obscure technical journals for me. In Moscow, I thank the archivists at both the state archives and the Academy of Science archives, not only for their help but also for putting up with my terrible and accented Russian. In New Delhi, the staff at the National Archives and the Nehru Memorial Museum and Library patiently introduced me to a completely new world. I also am grateful to the staff at the Museum of Mosaics in Devnya for letting me rummage through their photo collection that apparently everyone else has forgotten exists.

Friends are what make this work possible. Its scope has left me with numerous groups all over the world to thank, and the bittersweet knowledge that such scattered distribution makes it rarer that I get to see them. Yet like nodes in a network, we are connected. I thank Gerard, Pat, Joe, Garriy, Kat, Derek, Kelli, and Betsabe for making Knoxville a home. My graduate school cohort remains an amazing group, and despite now being dispersed beyond New York, will always be the key to making my years there amazing. I thank Harun Buljina for Balkan solidarity and willingness to escape history into football or music discussions; Manuel Bautista for too many reasons to count or share in polite company; Sean O'Neill for endless back-and-forth insults; Hannah Elmer for laughs in at least two different countries for now; Chien Wen Kung for computer game and Liverpool v Arsenal tête-à-têtes; Dominic Vendell for Indian companionship and beer; him and Nishant Batsha for wit and a crash course in Indian history; Suzy Vuljevic for always making me laugh even if we see each other rarely; Ulug Kuzuoglu for mulled wine and walks by the Neva; Alana Hein for haircuts; Mary Freeman for skee-ball introductions and Robbie for cocktails; George Aumoithe for smiles; and Peter Walker and Melissa Morris for hurricane margaritas and my first trip to Wyoming.

I also thank Abigail Kret for all the discussions, food, coffee, and more importantly, wine, that we shared.

Thank you, ARCS buddies, for too many bad movies, trips, and laughs: Alli, Dragos, Eli, Francesco, Fraser, Ioana, Lucian, Manuela, Mikhail Thank you, Center for Advanced Studies in Sofia buddies for the late-night discussions-turned-marathon nightlife sessions: Elitza, Martin, Nadezhda, Veronika! I extend special thanks to Elitza Stanoeva, Tom Junes, and Jan Zofka for the always continuing friendship, wherever we are. Filip Lyapov, thank you talking about Bulgaria, football, music, and politics with me. In Moscow, I thank the whole of the Foreign Historians Conference for the social life while I was there, and especially Joe for agreeing to a beer right from day one. Also the Parque Ambue Ari cat-walkers: That experience was the right launchpad for it all.

In England, I thank Eddy, Roy, Paul, Alex, Kim, Joel, Boxy, and Ian for always being a home away from home (wherever home is). Thanks to all my friends from my undergraduate days: We see each other less and less, but you are more and more important. You have all put up with a lot from me! In Sofia, where a lot of this research and writing was carried out, the gang has always kept me wined, dined, laughed out, traveled out, and out and about after the archives close. I can never thank you enough, so cheers to Stunyo, Stela, Desi, Vanko, Stamen × 2, Vankata, Adamov, Nicole, Iva, Hristo, Geri, Yuliyana, Yoanna, and all the rest! Sasho: special thanks for not just Sofia, but now Rhode Island, too, and for everything since the age of seven! I also want to thank Yulkata for his help in setting up one of the interviews.

I thank Katie Helke at the MIT Press for seeing potential in this project and taking it on, as well as the series editors and Laura Keeler. Thank you so much for working with me, guiding me through this process, and putting up with missed deadlines. I also thank the anonymous reviewers whose comments and suggestions have been invaluable in making this a sharper, better book. I think my book has found exactly its right place.

Much of this research has been based on interviews with people who were part of this industry, and I thank all of them for their time. Two, however, stand out: thanks to Koĭcho Dragostinov for his tireless help and setting up interviews with colleagues; and Petŭr Petrov, whom I met by chance, but who became not only an invaluable font of knowledge but also a companion through this process. Sadly, he didn't live to see the publication of this book, but I hope in a small way it continues his legacy.

Another person who didn't get to see the end of this project, but was with it from its very early days, is Etien. We got there in the end, right? You are missed every day, and I can't thank you enough.

Finally, my family. I thank my grandparents, Pesho and Penka. And also uncles, aunts, and cousins: Dido, Megi, Petya, Stefcho, Joro, and big and little Ventzis. To my sister Lora: Thank you for always asking me what the hell I am doing, and then doing your best to show me the latest stupid video to distract me from writing. That's what siblings are for. My parents, Pavlina and Plamen, have forever supported me and cared about me and what I do. They dragged me to too many historical sites as a kid, so ultimately, they are to blame for all this. They lived through it, and were part of this story in more ways than one. This is why it is to them that I dedicate this work, with love.

All mistakes and attempts at humor remain, contrary to popular opinion, my own fault.

INTRODUCTION: THE WORLDS OF THE BULGARIAN COMPUTER

You can find the father of the computer at the center of Sofia, right in front of the imposing Telephone Palace. A solid bronze monument 6 meters high depicts a stylized tree—or is it an electronic schematic?—framing the face of a distinguished older man as a halo. Erected in 2003 to commemorate the centenary of John Vincent Atanasoff's birth, it embodies claims I had heard around the dinner table when I was a child—did you know that a Bulgarian invented the computer? Sure, I thought, his name sounded Bulgarian.

Born in Hamilton, New York, Atanasoff was a physicist and inventor now credited as being part of the team that developed the first electronic digital computer. His father had come to the new world as a teenager in the 1880s, his own father having been killed in the 1876 April Uprising. Atanasoff's father was a Bulgarian electrical engineer, his mother an Irish mathematics teacher. Virtually unknown in the country and largely in wider computer history until the 1970s, he was discovered for socialism and lauded by the young scientists of a nascent computer industry. By the time of his death in 1995, he had received various honors, including the US National Medal of Technology, and an annual prize in his name is awarded in Bulgaria to this day.

Atanasoff was a professor of physics and mathematics in Iowa State University before the Second World War. Tired of mechanical calculators,

I.1 John V. Atanasoff as a legend. (*Source*: Wikicommons.)

he started developing what would eventually become the Atanasoff-Berry Computer together with his graduate student Clifford Berry in 1939, completing it in 1942. The first electronic digital calculating device in the world, it had a mechanical rotating drum for memory usage. Atanasoff claimed that the inspiration had come in a flash, during a night drive in Illinois. Whatever the truth, the device lacked stored program capability and true programmability, distinguishing it from modern computing—and in any case, it was quickly forgotten due to the Second World War. It was during that conflagration that Atanasoff, like other early electronics pioneers, entered the service of American Big Science. While Norbert Wiener worked on antiaircraft guns, Atanasoff developed acoustic devices for the US Navy. During these years, he often met with John Mauchly, one of the creators of ENIAC, the first programmable, general-purpose, Turing-complete computer. By 1945, he was placed in charge of the Navy program to create a large computer, having been personally picked for this project by John von Neumann. However, the Navy had prioritized his acoustic duties, so by the time he could devote himself to his true passion, the ENIAC project had beaten him to the punch.

Between 1954 and 1973, however, Atanasoff became embroiled in the legal proceedings of Honeywell, Inc. v Sperry Rand, as older inventors fought to invalidate the patents given to the ENIAC team. By 1973, the judge proclaimed that J. Presper Eckert and John Mauchly did not themselves invent the automatic electronic computer, but instead derived the subject matter from one Dr. John Vincent Atanasoff.[1] Debates still rage about Atanasoff's place in computer history, but his and Berry's machine had been designated an IEEE Milestone by 1990. By then, thanks to the news of the case having reached Bulgarian science through specialized Western literature, he was a cause célèbre in his ancestral country. A young mathematician in the Bulgarian Academy of Sciences (BAS), Blagovest Sendov, had followed the legal proceedings beginning in the 1960s. In 1970, Sendov contacted Atanasoff, inviting him to visit Bulgaria, which he did in December. Atanasoff and his wife were official guests of the Academy, receiving a warm welcome, as well as the Order of St. Cyril and Methodius, First Class, Bulgaria's highest such honor. He also met distant relatives in his father's village of birth. In 1985, he visited again, delivering lectures at Sendov's conference in Varna on children in the computer world. Another medal—the Order of the People's Republic of Bulgaria First Class—and the keys to the city of Yambol, near which his father's village was situated, accompanied this. He had "become" Bulgarian.

Atanasoff was a providential gift to the regime but also to Bulgarian science. Here was proof that Bulgarians were uniquely gifted and tied to the birth of the machine that was being constructed in such cities as Stara Zagora, Sofia, and Plovdiv. In this narrative, here was the man who had set the whole thing in motion in the United States, visiting his father's rural cottage in southern Bulgaria. The future was meeting the past. His roots were in this land, the poverty of which his father had escaped. Now he was back to witness the leaps that "his" country had achieved. The Bulgarian seed of the computer was reaching its natural blossoming as the Bulgarian Communist Party (BCP) championed the information age.

When the communists seized power in September 1944, a generation before, they inherited an agricultural and rural country. Bulgaria was one of the perennial cases of underdevelopment that European economists pondered over before the Second World War, its people and resources largely tied to the land. Within 40 years, it was producing the high technology of the late twentieth century—computers, electronics, teleprocessing systems, and even space exploration instruments. This achievement seemed, and indeed was, incredible. By the end of the regime, despite having undergone Stalinist breakneck industrialization (as well as other hallmarks of the takeover of Eastern Europe, such as camps, repression, and the party-state), Bulgarian socialism had a different language and image than those of its allies.

This self-image was most evident around many tables and dinners shared with family and family friends when I was younger. Talk among contemporary Bulgarians often turns back to tropes, and sometimes jokes, about "what we lost," a particular case of *Ostalgie*. A commonly heard narrative continues to be that Bulgaria had been a technologically advanced nation, an industrial exporter with its own specialties and achievements. They would throw out the names "Pravetz" (a Bulgarian PC) or "Atanasoff" as proof, and only later would I find out a bit more about the provenance of both words. At the same time, a childhood in the first post-socialist decade was replete with a sense of something shutting down: shuttered factories, buildings, and institutes. I do not claim that I understood what that meant then, but the impression of the past that "we had lost" was all around. In my home city of Varna itself, a radio-electronics factory that my father had worked in stood vacant and imposing on a lot at the edge of the city. Many of his stories centered on it, but there was a keen sense for me that this time was irreversibly gone.

At the same time, I could play games such as *Prince of Persia, Duke Nukem*, or *Karateka*, on those fabled computers as I found them in the offices my father worked in at the technical university in Varna, and then in the software company he set up. Computers were intriguing, they were entertaining, and they were mysterious. I didn't want to program them, but I did want to use them. Years later, when I searched for a topic for a dissertation, these strands coalesced as I discovered that this story was largely untold but concerned numerous questions I had about history. How did countries modernize under socialism? Why did factories shutter in the way they did after 1989, yet the technologies and skills persist so that my father could create his own software firm? And to what extent were the stories that my childhood ears heard true—how Bulgarian was the computer, and what did we lose?

This book tells the story of the socialist bloc's biggest electronics producer. In 1944, when the Red Army crossed the Danube and the communist party took power in Sofia, the country was a garden, one of its golden exports being the tobacco that calmed Axis troops in occupied Europe. By the 1980s, when the Interkosmos 22 satellite beep-beeped its way through the heavens, Bulgaria held a 45 percent market share of electronic exports within the Eastern Bloc.[2] Its population, just under 9 million by the end of the regime, was heavily engaged in the industry. The exact number of workers in the electronics industry depends on whom you ask and who is counting: In the official annual statistics for 1987, out of just over 1.4 million workers engaged in some sort of industry or extraction, 161,302 worked in "electrical and electronics industry"—or just over 11 percent of all industrial workers. If we take the important statistical sector of "machine-building manufacturing," which employed around 419,000 people, the proportion was over 38 percent, a gargantuan number for the small state.[3] Kiril Boi͡anov, a computer specialist closely involved with the industry, estimates that 181,000 individuals were at work in it by 1990,[4] Milena Dimitrova, in her compendium of interviews with luminaries from the sector, pushes the figure to as high as 215,000.[5] The exact number is difficult to gauge—some in the official statistics worked in power engineering, while the number does omit scientific workers, technological experts, data and software specialists, and other workers that can be seen as working with computers or in the electronics sector. What is clear from these numbers, however, is that the sector was huge and extremely important for a country presenting itself as an advanced industrial nation that created complex machines.[6]

The surprising existence of this industry in the supposedly gray, small, orthodox, and backward satellite is alone worth telling this story, but uniqueness does not necessarily bring relevance. The Bulgarian computer is a prism, however, through which one can tell a story of the late twentieth century's multiple trajectories of technological change, socio-economic shifts, and socialism's modernity. The unique perspective of a small state on the periphery of its alliance ("which I knew wasn't an-all-that-important place," as the US ambassador in the 1970s saw it)[7] allows multiple stories to be intertwined: the maturation and fall of the socialist regimes, but also the rise of the information economy, the nature of technological innovation, the relationship between politics and technology, and the impact of the computer on society. In this respect, Bulgaria was just as an important place as anywhere else, and it entered the computer age with all its hopes and anxieties, just like the rest of the world. But the size of the sector relative to the rest of the economy, the prestige that the party wanted to derive from it, and the international openings it entailed for a socialist economy mean that the Bulgarian case had its own unique characteristics, too: cybernetics as a tool to create a new human, or the fertile interaction of economic shortages and widely proclaimed dreaming of a new world that ensnared the last socialist generation.

This book treats the computer as both a commodity and a tool. This approach helps keep the disparate threads united through the materiality of the product as it was created, circulated, and imbued with meaning. Other innovative work on the region's commodities, especially tobacco, has successfully told a deep and broad story through one locus.[8] The electronics industry was created primarily as a means of profit and prestige, but its construction required not just factories but thousands of trained specialists. Thus late socialist Bulgaria became a veritable Cyberia:

supposedly locked away behind the Iron Curtain, isolated, part of the socialist camp but in fact richly engaged with the meaning of the information age and how it could help build socialism, create a new type of persona, and eventually pit it against the regime's failed promises. Its promise of automation in labor, as well as its abilities of modeling and prognosis based on the huge quantities of data that a modern economy produced, captured the imagination of intellectuals, party members, and society as well. Thus, this commodity history of the computer allows us to trace the circulation of the ideas, technologies, and money that created the item, but also the vistas of intellectual history and cultural ramifications of this tool of the future.

Thus this book is also about people. Utilizing oral-history methods, it uncovers the world of those who managed the industry and worked in both its scientific and managerial positions. Some are high-ranking members of the party and state, the captains of industry and candidate members of the Politburo, while others are pitched at what I call the middle level—scientific workers in institutes in academia and industry, and in the gargantuan electronics conglomerate IZOT (the Bulgarian abbreviation for Computational, Recording and Organizational Technology). To those we can add a motley crew of philosophers, pedagogues, sociologists, psychologists, and novelists, who grappled with the information age, its tools, and its implication for the future of socialism, Bulgaria, and humanity. Interviews with such actors add detail to how these people lived, what they thought about what they were doing, and how they made decisions and forged links—information that is sometimes absent from official sources. A rich culture of debate emerges around the computer as a "trading zone" where different scientific fields can meet and exchange knowledge.[9] However, moving beyond this view, I include other cultural and artistic fields in which participants both dreamed and were anxious about the roles played by the computer and cybernetics. It was not just scientists who thought about the computer, but increasingly the whole of society. This book is not a bottom-up history of the industry, as its source base and methodology do not allow a full study of those hundreds of thousands of workers who toiled in factories or in offices and automation centers equipped with computers. Yet by using popular magazines and oral interviews, it does posit that there was wide engagement with the computer in a society that is usually seen as not having entered the information age. The actions of workers facing automation, or of children who had to take computer classes or flocked to clubs, do speak of an engagement—on their own terms—with the ideas circulating in both the regime and specialist literature at the time. This book thus argues that the history from below can be uncovered by also looking at the interaction between action and plan, which often also meant that some of those from "below" assumed middling positions in the information economy of late socialism as it sought to mobilize creative powers beyond the institutes.

TWO PRISMS: COMPUTERS AND SMALL STATES

Words that abound in this book include "prism," "small," "dream," and "anxiety." They speak to both methodological and intellectual concerns. One is the computer as an organizing principle of socialist politics and a zone for fertile discussion. Zubok and Pleshakov called late socialism "the senile cold war." However, it was anything but: The relative security and power that the Eastern Bloc achieved in the Brezhnev era was just as important as the identified need to intensify the economy in creating the space for the computer and cybernetics to become powerful symbols of the future.[10] But this development was also playing out in a small state. The limits but also possibilities that both the Cold War and the need for technological progress offered, however, allow a small state to show us new ways of how geopolitics operated. High technology allowed Bulgaria to carve out niches and defend its own economic interests not only in Eastern Europe but also in the global market. Thus, dreams and anxieties were always intertwined. The BCP dreamed of profiting from this industry and of solving its economic problems without giving up its political monopoly. At the same time, it was anxious about its hard-won gains in the socialist international economy, as well as whether its bet on the computer and cybernetic governance would actually enhance the economy. The technical intellectuals dreamed of professional advancement, modern technology, and its successful application. But they were also anxious about the shortages, the difficulty in implementing computers into everyday life, the uneven access to Western knowledge, and their own position in the party–state hierarchy. For ordinary people, it was a dream of less arduous work or of a creative future, but also a cause for anxiety about job loss, surveillance, and new burdens.

There is a temptation to tell stories like this as ones of failure. This book is not a prehistory of the fall of socialism, with 1989 looming over all. It argues that the technological and political projects of the Bulgarian computer should be seen on its own terms and within the contours of its own logic. People under socialism didn't live with the constant expectation of its imminent failure, even if they welcomed its eventual demise—in Alexei Yurchak's immortal phrase, everything was forever until it was no more.[11] Well, while it was forever, it was vital, creative, and constructive. The Bulgarian computer industry had its successes—it allowed the party to function as a nonagricultural periphery in the socialist world system. Moreover, it became a pervasive language and framework for the development of the economy and society. It was an increasing part of everyday life in the factories, offices, schools, and even homes of the socialist citizens. And it was also becoming a wide-ranging framework for thinking about the economy, society, psychology, and the future itself. The Bulgarian computer is part of the universal story of the late twentieth century's rise of an information society where knowledge work became increasingly important. It was a part of the socialist order's concept of the scientific-technical revolution as the next step in boosting economic production and bringing the communist future. But it was also Bulgarian—it served that regime's aims, often contra the interests of Moscow and others, and it intersected with a cultural landscape that was its own. This book brings together the local, socialist, and universal dimensions of these predicaments.

As soon as the regimes fell, specialists rushed to explain why, after being supposedly forever, the socialist order was no more. Economic reasons were tantamount in many explanations of 1989, with the socialist bloc's heavy focus on such industries as steel being cited as an obvious reason—they failed to enter the information age. Written as the regimes fell, Francis Fukuyama's by-now infamous "end of history" thesis is a solid summary of the myriad texts that can be cited here: "the failure of central planning in the final analysis is related to the problem of technological innovation."[12] Manuel Castells' powerful trilogy on the information age also makes it clear: the USSR and the socialist world system as a whole never made the jump from industrial and Fordist to postindustrial, post-Fordist, and informational organizations. Imperfectly reformed and nascently innovative, these societies were doomed.[13] This is the conventional narrative of the computer revolution's failure in the East.

A still-influential thesis is that of Charles Maier, who argued that the collapse of the German Democratic Republic (GDR) is at least partly tied to its investment in computers in the 1980s: "a race between computers and collapse," as he called it.[14] The East Germans tried to master the construction of such systems and expand them to mass production, leading to increasing costs and investments that took away from the rest of the economy while giving few results. But that is a country- and time-specific thesis. With a lens on the late 1980s, it ignores the longer history of the computing industry during socialism, and even tacitly admits it: Bulgaria had already cornered the low-cost market. It was, as this book argues, cornering other markets, too, much to East German chagrin. Maier's view is true in one sense—Bulgaria also incurred debt as it tried to modernize its industry while still failing to generate convertible Western currency from sales abroad, and this trap did contribute to the collapse. But to see the whole industry's history as failure misses both its chronology and logic. Bulgaria's high-technology story starts in the 1960s, and its computer production—with all its problems and costs—was a successful sector right up until the end. This success can only be explained through the logic of the socialist international economic order. Bulgaria had maneuvered itself into primacy in the electronics sector years ahead of the GDR. East German computer failure is thus in some ways the story of a latecomer. Moreover, this book shows that the failures in computer policy were instrumentalized by party and technical elites to call for reform, but they cannot be disentangled from a general internal diagnosis of total socioeconomic problems, not just sectoral problems. The Bulgarian computer industry was created for a specific local purpose in the logic of Comecon, and within that framework, it succeeded. If different political choices had been made in the late 1980s in Moscow, Berlin, Sofia, and elsewhere, this industry could have survived. This is not a triumphalist nationalist slogan, but a recognition that even while falling ever further behind economically, if Comecon had continued, it would have perpetuated the niches where the logic of Bulgarian production could have continued (of course, comparison with the ideological enemy was unavoidable and was the kernel of reformism within the Bloc). Taking the Second World—the world of the socialist states that opposed the Western First World—seriously as a market, shows us a way to tell the story that is different from that of Maier and other authors.

Through this prism, the Cold War appears as a much freer space for smaller states, where possibilities for independent policies and paths existed. Bulgaria has often been peripheral to these stories, sidelined by the other Balkan mavericks: Yugoslavia with its nonaligned path and charismatic marshall; Romania, which opposed Soviet policies obstinately and eventually became a basket case of shortages and oppression, while its neighbors seemed to liberalize; or Albania, which slalomed between Moscow and Beijing. Bulgaria, by contrast, was led at the start by the Stalinist mouthpiece of Georgi Dimitrov and eventually by the wily but seemingly bootlicking Todor Zhivkov. Thus the story is often told as one of loyalty.[15] This view was also widespread among the US diplomatic corps. Contrasting Zhivkov to the Romanians, the political officer at the US embassy in the late 1960s, Donald Tice, said "he [Romanian leader Georghiu Dej] seemed actually to think for himself, whereas the Bulgarians all just 'hewed' to the Soviet line. I could read the opening paragraphs of a speech by Todor Zhivkov, put the newspaper aside and write the rest of it, because they were all the same."[16] Yet political orthodoxy, trumpeting socialist fraternity and Soviet primacy, masks real economic and intellectual divergences. The computer as a focus for Bulgarian efforts was not because Moscow wanted it so—in fact, as my book shows, the Soviets and others increasingly bristled at Bulgarian practices—but because Sofia did, for domestic reasons. Once the industry was set up, the party and intellectuals dreamed big of a cybernetic future, which did speak a language similar to that of Moscow and its scientific-technical revolution but was never identical.

Moreover, loyalty can prove beneficial. The maverick story in the Cold War is eye opening, and attractive, but if we take the lens of integrating into alliances, we can see the contours of an alternative modernity arising even more clearly. The socialist bloc integrated economically, not just militarily, in order to win the Cold War. Close links at the personal level (Zhivkov was a successful wooer of both Khrushchev and Brezhnev) enabled both the financial and technological help for setting up high-technology industries and access to huge markets that made these industries viable. At the same time, the troubled but real road of socialist economic integration created a Second World, juxtaposed to the Western First World, where experts and their languages circulated widely, creating a tangible material and intellectual culture that was distinct. Small states could thus both exploit their superpower backers and show locally the image of the world that was being built. The Bulgarian computer was part of a wider world of socialist technology, that of Comecon and its unified system of ES (mainframe computers) and SM (minicomputers) series machines. Industrial and scientific cooperation created a common experience from the inner German border to Vladivostok, but also with outposts in Havana, Hanoi, and Maputo. The circulation of the Bulgarian computer, which was a part of the material integration of this space, shows how the socialist world order made itself into a project distinct from capitalism as both a geopolitical force and a technological system.[17]

The Iron Curtain was porous, however. Metaphors have already been advanced of replacing the term "iron" with "nylon" and the term "curtain" with "membrane."[18] The computer industry was dependent on knowing about and learning from the centers of the information age, which of course lay in the West. The state not only trained thousands of engineers to leapfrog into the computer age but also financed a huge intelligence operation to acquire the items and expertise denied to them by the West's restrictions placed on this high-value good, which was key to the arms and space race. My book argues that this massive know-how transfer needs to be integrated into our stories of technology, where there were many channels of engaging with ideas, especially in authoritarian states. Spies transferred not just computers but also models, and they were involved in a very complex symbiosis with the civilian sector. The Bulgarian case shows that the distinction between licit and illicit exchange becomes meaningless, as the Cold War logic defined these terms. Instead, we should concentrate on how the dreaded socialist security services were a key part of the story of technological exchange in the twentieth century. The feared Bulgarian State Security was violent—against its internal opponents but also in infamous cases, such as Georgi Markov's assassination in 1978 in London—and none of my work aims to minimize that. However, intelligence was also a channel without which the Bulgarian computer industry could not have existed.[19]

The computer was also a pathway to the new world that the late twentieth century had wrought. The Cold War is a key part of my story; but it can't and shouldn't be privileged, especially in the period when it intersected with decolonization and newly liberated states' projects. The independent states of the Global South sought allies in their own quests to build modern institutions and economies. Bulgaria was a self-avowed friend of these states, an anti-imperialist state, and it reached out to this new world. Bulgaria's opening to the global market was thus played out as much in Africa, Asia, and Latin America as anywhere else. But this book sees the Third World also as a place where the Second met the First—in technological terms, markets such as India were open to all computers, and the restrictions that prevented socialist specialists from operating freely in the West were mostly absent. India and other states were places to learn but also to sell. Often the Global South has been presented as a space where development models competed directly.[20] However, they were also markets where socialists had to sell, in much more open markets than Comecon, often competing with Western firms that practiced modern advertising and management techniques. As such, the Bulgarian computer was a commodity to sell but also one for which Bulgarians had to learn to do things differently. What did socialists learn, and how did they change when they acted like capitalists? The experience of the Bulgarian computer in India thus goes some way toward answering questions about how nonsocialist ideas can enter the regime's mindset and its professionals' practices.

When the regime ran into trouble and fell, it was the people who were politically involved who made the transition to the post-socialist world—not so much the computer industry, which having lost its market and raison d'être, quickly laid off workers and closed down factories. This book doesn't consider the industry as merely the enterprises, machines, and tools that it created and sold throughout the world, but much more importantly, as the workers and intellectuals it spawned, fostered, and developed. To say that socialism ended in 1989 would be reductive when looking through this prism. At the highest levels, some of the elite transformed their positions and continued to exercise political and economic power well into the democratic period. At the level of the engineer and the technical intelligentsia, the end didn't signal a cessation of their professional activities but a change of context. Communism collapsed as a framework for science, with particular goals and investment, but as this book shows, the world in which these intellectuals had been operating for decades wasn't colored just by the regime but also by their professional concerns. To build machines for a socialist regime didn't mean that every day you thought about how it would bring about Marx's dream. Instead, you pondered over chips, blueprints, and development plans. If we thus take the lens of that middle strata, engaged in creating the tools of tomorrow, the endpoints of this story do not coincide with the fall of the party that kickstarted it all. Conventional chronologies driven by pure political events obscure the richness of life, as it continues despite these changes. Even more importantly, it is impossible to explain the political, technological, and economic landscapes of post-socialism if 1989 is taken as a tabula rasa.

WHAT WORLDS?

The subtitle of this book boldly claims that this narrative is an engagement with the world: both geographic and geopolitical worlds, as well as intellectual and cultural ones. The Bulgarian computer circulated throughout the socialist and decolonizing world, while it also invaded the horizons of the party, the intelligentsia, and the ordinary person. The computer changed the economic, cultural, and intellectual horizons of the late twentieth century, and Bulgaria was no different. Yet within this universal stream is a local story, in which the machine interacted and changed discourse and possibilities in particular ways.

It is Bulgaria's engagement with the whole world that is of note. The call to write a pericentric history of the Cold War, with multiple polarities, is now more than 20 years old, but it still resonates.[21] My book offers this perspective as part of an increasing literature on the subject. Seen from the sidelines, both Cold War and global developments take on a new hue—a space to engage with the world, learn from it, and participate in it.[22] I argue that the need to build a computer in a state with no previous techno-economic base for this industry made the regime more open to global developments in ways hitherto hidden. The world was a dangerous place in the conditions of superpower confrontation, but it was not uniformly so.

Bulgaria's desire to create a technologically advanced economy pushed the elites to create the conditions for an engagement with global circulation of knowledge, expertise, material, and even capital, in order to achieve the computer dream. Thus, the pericentric perspective I offer combined with the history of technology opens up vistas that are rarely explored in the history of socialist states. It builds on work that has shown how Bulgaria opened up culturally or politically, demonstrating that it was often technology that created the conditions for a wider interaction and an "opening to the West" as the regime (mistakenly) often saw technology as a value-neutral sphere after Stalinism's end. Whatever the political reservations the party may have had about its global engagements, it granted much wider freedom to its technological elites than to the general public.

The transnational circulation of electronics know-how, however, concerned not just the party's managers or its technological elites but also spies. My book shows that we must incorporate hitherto underappreciated stories into our understanding of the global information age. In our age of increased concerns about surveillance, this approach is maybe not so controversial. The world of expertise was not created just by those who created the machines but also by those who acquired them. This book shows how often the lines blurred, and how the civilian economy controlled the concerns of the spying agencies. Moreover, I argue that in the conditions of not just the Cold War but also in any technological competition, we must jettison distinctions such as "legal/illegal" when talking about knowledge acquisition. Often the actors in this book received knowledge from their colleagues through the simple chit-chat over a conference coffee, or by visiting one another in their workplaces, wherever they may be in the world. Espionage and expertise combined and melded in the global information age, as they did at many other times. This book thus argues that we should expand our source base and purview when discussing how and why knowledge systems were created, building on the work of scholars who have shown the close interrelation of the military and the dawn of computing. What constitutes Big Science in the socialist world was slightly different from that of the Western world, where it is usually studied.[23]

The computer was part of both opening and closing the worlds available to a society. It has been seen as constituting a closed world—both in itself and in helping model the Cold War as such. Born as part of the British and American military-industrial complexes, it also fostered a mentality of "closing the world" as it became a self-contained set of logic, models, and techniques. Of course, the story is similar in the USSR: From its earliest days, the computer was a tool for the nuclear and space confrontation between Moscow and Washington.[24] However, as I argue in this book, the computer operated differently in Bulgaria. It is true that it became a tool for closing the Cold War: The embargo of goods imposed by CoCom (Coordinating Committee for Multilateral Export Controls) created a closed world of socialist computing, which in fact enabled the Bulgarians to win their markets. However, although the Bulgarian military undoubtedly used the computer for its own Cold War ends, the "seepage" of models and knowledge between the military and civilian sectors was much less in the socialist world than in the West. The need to constantly update the technology meant that the computer allowed Bulgarians to open up to the world; at the same time, the closed world of socialism was a boon and a positive, rather than a negative, as it created the international economic conditions for the Bulgarian computer to flourish. Moreover, as my book shows, the overwhelmingly civilian nature of the computer industry in socialist Bulgaria created different intellectual horizons centered on new forms of creativity and seamless production control: the right steps toward utopia. The world of ideas, of reform, of a future without work, allowed Bulgarian intellectuals to open up discussions beyond the world of the party's mantras.

What of that socialist world, however? My book shows that this world existed and was vital in its exchange. I treat Comecon seriously as a project that self-consciously constituted an alternative economic order that was aimed at winning both the ideological and economic conflict with the West. Facing an enemy that constituted itself into a powerful economic bloc in the form of the European Economic Community, in the 1960s, Comecon increasingly sought a unified division of labor, too. This was the world in which the Bulgarian computer could be born and make its claim to fame. Recent scholarship has pushed us to see this economic space as a geopolitical order that allowed countries to leverage their own national positions in the hierarchy supposedly dictated by Moscow.[25] Western contacts on a national level—which Bulgaria fostered as well as any country—were then instrumentalized to gain comparative advantages. Comecon was not a straitjacket but an opportunity for the Bulgarian computer industry. But it was not just a world in which to generate a profit, but a world in which to compare, learn, trade, and proclaim successes. Mëhilli has excellently argued that tensions existed between internationalism and conflicting nationalist interests, but illiberal regimes' interactions and constitution into a system of institutional arrangements—Comecon and the Warsaw Pact—had tangible, real results.[26] My book similarly illustrates this vitality of socialist economic alliances and how states hitherto relegated to the southern and underdeveloped tier maneuvered this world to show themselves as just as (if not more) advanced as their East German or Czechoslovak allies. The socialist world was thus a place of possibility and widespread economic-technological circulation, rather than a moribund space doomed to fail. But it wasn't just an alternative economic order but an alternative modernity, shaping and being shaped by a socialist political project that resulted in similar approaches to politics, philosophy, and culture.

The computer also allowed Bulgaria to operate as an agent and participant in globalization. The world was framed not only by the Cold War but also by decolonization, and the story of interconnectivity in this new age should not and cannot be told just as Westernization. The increasingly rich literature on socialist globalization has pushed many observers to consider the myriad political, economic, and cultural ways that the Soviet's supposed satellites hewed their own roads into the Global South. Eastern European allies, the Bulgarians among them, were important agents of modernization in political, economic, and military terms, as they trained engineers, agriculture specialists, doctors, and armies throughout the newly independent states—or at home.[27] This book continues on this path to look beyond Moscow as the agent of socialist globalization and to show the rich technological engagement between East and South. Recent works have shown the circulation of architects, women's organizations, and cultural diplomacy as part of this Bulgarian engagement with the decolonized world.[28] The computer contributed not only to the circulation of expertise but also to the creation of different geographies of exchange: Bulgaria possessed unique advantages, even if its technology was not as modern as that offered by the Americans and Japanese. States in the Global South desired this tool of the future and were willing to look for it in the socialist world for both political and economic reasons. The computer allowed Bulgaria to present an image to its newfound allies—that of a modern country that had leapfrogged backwardness in a single generation, from the apple orchard to the computer age.

My book is also an argument about the limits of globalization and in particular, the socialist kind. Through the prism of the computer, we can see how important capitalist globalization remained for the development of the socialist world's technology, and more importantly—the priestly class that created it and used it. Trying to avoid using the hindsight of knowing that this system failed, my book investigates how the limits of socialist globalization in terms of technical expertise fostered a strata of transnational experts that developed reformist ideas about politics in Bulgaria. Moreover, their very existence and circulation are also evidence of other globalizing limits, just as they are of the limits of the party's electronics dream. Both fostered the exchange of ideas between Bulgarian experts and global institutes or markets, but they often distanced those same experts from the shop floors and villages that surrounded them, the latter being objects of these forces, to be acted on by the more confident and technically capable state. At the same time, the workers in those villages and shop floors were also political subjects—but ones with different ideas, different anxieties, and different experiences from those who were most mobile in socialist globalization.

My book looks beyond the Global South as a space where socialists tried out their ideas and presented development projects as proof of their modernity's success, however. The biggest Asian market that Bulgaria targeted was India, a country that held a unique place in Cold War history. Odd Arne Westad posits it as an almost "anti–Cold War" nation that aimed to create an alternative to the superpower ideological confrontation.[29] India was a key world meeting ground during the Cold War, a place that both capitalist and socialist globalization tried to invade and sway to their respective models, and a locality that is a great prism for seeing the contours of this exchange and clash. Although much of the literature focuses on the contracts between center and periphery—Washington and Moscow as the centers around which models orbited—there was more to contact with the Global South than specific development projects.[30] This book shows that the computer, despite being a tool to develop and modernize, was also a commodity that states such as India desired without necessarily wanting to harness them to the same projects that the superpowers desired. I argue that the exchange between East and South must also be seen through the lens of business history rather than just international development. Moreover, Bulgaria could meet the world's technology and practices without the same restrictions it faced in the West. India, like other places in the Global South, was a world of business transactions and competition. Bulgarians had to cut their teeth on models that were absent in their trade and business dealings within the logic of the Second World, as they competed with the West for Indian contracts. My book shows that through engagement with India's electronics landscape, Bulgarian specialists also met and learned from the world of advertising, customer relations, and modern marketing. The Global South was thus a space to learn and not just to sell or manipulate. The computer was a channel through which new, nonsocialist techniques could filter back into the Second World's economies, with important consequences in creating a "worldly" professional class.

However, the computer also opened up other worlds beyond geography. The computing age opened new vistas that were technological, intellectual, and political, touching on every part of life in a modern society from the ideas about how one should govern to what kind of literature people read. The late twentieth century was also the age of cybernetics, even when the term lost its popularity in the West. In Bulgaria, it was part of political discourse and the popular imagination for much longer, a key ingredient of the scientific-technical revolution that the communist parties imagined they were harnessing. Chapter 1 delves into the definition of the term more fully, but it suffices to say that in socialism, cybernetics became a driving force of planning and economic thinking from the 1960s on. Robert Kline has highlighted the "disunity of cybernetics," which as a science took on many different forms, depending on national contexts, despite its birth in America.[31] My book shows how it meshed with the party's dream of nonmarket reform, its burgeoning industry, and particular national concerns. The story of the Bulgarian computer is also that of engaging with the world of cybernetics, a field that spread octopus-like across disciplines as well as nations. But development was not simple diffusion, as despite reading Wiener or Shannon, Bulgarian cyberneticians developed their own concerns born out of the state's attempts to reform the socialist workspace into the office or factory of the future.

The history of cybernetics in a non-Western context is rich and growing. Slava Gerovitch's work on Soviet cybernetics has been highly successful in showing how the discipline emerged from its tainted position as "bourgeois science" in the Stalinist period to become an exciting, exact language in the 1960s. Soviet scientists latched onto it to defend their positions and autonomy versus the Soviet state, before it became subsumed within the official language as little more than a language of rent-seeking by Soviet science, losing its earlier vitality. My book, however, shows that the specifics of the Bulgarian example are different. Most importantly, Bulgarian cybernetics was a late comer—it enters at the "official" stage, when the party enshrines it in its development plan, in service to a state-led industrial effort to build the computer into the Comecon structures. Thus from the very start, it allowed Bulgarian science to frame its projects in an official language. However, the BCP's obsession with cybernetics as the key to their dream of intensifying the moribund economy meant that qualified engineers of the new ilk were much more likely to have strong positions in the power structures. In effect, Bulgarian science—to a larger degree than its Soviet counterpart—used cyberspeak to get a position at the high table. Unlike the Soviet case, this late-coming status meant that the chronology is reversed: Cybernetics, and the failure of the centralized party to win the wager it had placed on this new technology and attendant science, became mobilized against it toward the end as a language of reform.[32]

The harnessing of cybernetics to different socialist projects has also been explored in the Eastern Bloc context, where the vision of a network of computers was attractive to parties that sought to find the optimal way to run their command economies. This appeal resulted in gargantuan, ambitious projects such as OGAS (the Russian acronym for National Automated System for Computation and Information Processing) in the USSR, which never came to fruition. Bulgaria was no different, with its own dreams of a networked economy. The story of the Bulgarian computer thus allows us to look into the world of socialist governance and its utopian aims, building on work such as that by Benjamin Peters, who shows how competing interests in a centralized party ultimately stymied the project.[33] Following his lead, I also trace the Bulgarian techno-utopian dream of reforming both society and economy through the possibilities of the computer. However, once again, the local matters as much as the global. It is true that the computer was part of the alternative modernity of Comecon, but it also assumed a larger relative weight in Bulgarian policy thinking due to the industry's predominance; as a result, the actors who championed it were often more successful than their Soviet counterparts, even if ultimately they, too, were defeated. Moreover, this book shows how the social networks of socialist society mattered as much as the computer networks. The computer allowed automation, and the computer-controlled machine entered the workforce alongside the socialist worker. By meditating not just on the rival political factions that used the computer to advance communism, or the scientists who created it and wrote papers on it, my book looks at how its successes and failures were also embedded in the deeper structures of Bulgarian society, especially labor practices. The computer gave birth to the dream of automation, and little research has been done on how this march of the machines impacted the workers themselves. What did it mean to be a laborer with a robotic coworker? Socialist governance had real results, despite the ultimate failure of the project in both the USSR and Bulgaria. This book explores the world it wrought not just in the Politburo meetings or the cybernetic institutes, but also in the anxious minds of ordinary people.

Computing at the periphery thus complicates our notions of what this tool could be used for. Technology was embedded in political visions, and those visions could change the brief of the design. Not every computer was harnessed to creating the Western information age or the socialist command network. Other worlds existed. Historians have looked at particularly illuminating projects, such as the Chilean Cybersyn, a vision of a participatory society through the technology of the computer. Such a view from the periphery has shown us that technology circulation was not unidirectional; nor did it necessarily carry predetermined political baggage.[34] Similarly, the Bulgarian computer engaged with the wider world—technology, ideas, and discourses often came from the West or from Moscow. But they also existed in a local Bulgarian world. The political project of Bulgarian socialism was different, tied to its own concerns.

Cultural politics in Bulgaria demanded a new type of human, driven by the politics of Liudmila Zhivkova, and Bulgarian computing discourse dutifully answered: How could the new machine and its network liberate humanity and foster creativity? My book contributes to our understanding of how local conditions impacted the ways technology was used. However, by locating the story in a country that had put such a premium on the industry relative to its exports and economy, I also show how it was actually at the periphery that the computer and cybernetics could become a political cipher that filled political ideology with meaning. All socialist states proclaimed that the "scientific-technical revolution" was the driving force of their policies and that only they could harness it properly. The industrial policy of Bulgaria, however, made this notion increasingly computerized, right down to the regime's end. Late socialist ideology, with its focus on science as productive force, was not just empty words—or at least not in the mouths of all party members. It is precisely from the Balkan periphery, where the ruling elite and much of society had placed such hopes in the computer as both a source of currency and a solution to slowing productivity, that we can see how cybernetics was a vital discourse that nested right at the heart of the regime's utopian utterances. Because the factories existed, and because they were the future, elements of the Bulgarian elite could sincerely believe that Wiener's science could give meaning to the chosen vanguard party, fusing the two into a cyber-socialism that claimed it had the tools to reach the utopian horizon. But precisely because this story starts with the profit motive, it also made it thinkable that the regime's politico-economic framework had to be discarded if the industry were to truly succeed and be modern. The Bulgarian computer is thus a path to both the socialist and capitalist worlds, dependent on both the period and subjects involved.

The computer connected various intellectual and dream worlds in Bulgaria itself. As a trading zone, it was itself a world, a black box, which fascinated and promised solutions to myriad problems. A true "cyborg science" existed—computer science, operations research, and game theory were all discussed in socialist countries as they were in capitalist ones. Information was a paradigmatic key to all kinds of problems faced by an increasingly complex society, stretching from economic planning to pedagogy to self-perfection. The computer and its terminology thus allow us to blur the lines between different professions and spheres of life. Computer specialists were in demand in many different aspects of Bulgarian life—from the automation of factories to discussions of workplace psychology. The social life of the technology thus meant that the worlds of science and politics intertwined and informed each other. The desire to automate more and more aspects of life also impacted the personal worlds of many people, as well as the social webs that they used to make sense of their position. Women were workers and also mothers, who had to contend with the computer as it invaded their offices and also their duties raising the next generation. That very next generation, too, was connected to politics and the economy through the computer—subjected to computer education, they were not just observers but also participants in the increasingly obvious slippage between the promise and the reality of cyber-socialism. Children were supposed to be the future cyborgs, but by the end of the regime, the computer had already made them cyborgs, as they entered the world of work through the software they created. Cyborg science invaded more and more areas of life as the computer was applied to every dilemma posed by modern industrial society. By the 1970s, within only a few years of the creation of the industry, a cyborg culture had emerged, too—in the workplaces that were to be automated, but even in literature, where writers in fact modeled the future that computers were supposed to bring. My book shows how if we follow the thread of computing as both a commodity and intellectual tool, we can weave together multiple strands of how an entire society reacted to the information age, to socialism, to the West, and to its own anxieties.[35]

If the computer was to change everyone's life, it had to be everywhere; thus each social world had to react to the new age. The technical intellectuals who were its priestly class were a serious force in politics and increasingly even in culture, popularizing the terms "governance," "information," and "automation." Used by the party, these terms also confronted ordinary people. This book does not claim that the history of late socialism is dominated only by computing, because life is much more than a screen. What I do ultimately argue is that due to the particular choices made in Bulgaria to create this important industry, which dominated economic discourse and party proclamations, the computer became a solution to be applied to multiple problems and thus also a vessel to be imbued with multiple expectations and fears. The particular content of those expectations and fears depended on whether you were a high-ranking bureaucrat, economic tsar, trader, cybernetic specialist, science fiction writer, or a teacher, student, woman, or worker. Thus, by focusing on the electronics world in Bulgaria, we glimpse the multiple worlds of social, cultural, and political life in the country.

STRUCTURE

Why Bulgaria chose computing as its niche is of course the most pressing question, and chapter 1 places this emergence in both the longer-term history of perceived Bulgarian backwardness and the challenges and opportunities that the BCP faced in the early 1960s. These local developments not only made sense because of the Bulgarian entanglement in debt but also because of a new desire to create an alternative modernity in the socialist bloc. Chapter 2 takes up this story and follows the developments that helped this industry grow into the behemoth it became by the 1980s. By doing so, the industry took advantage of both the logic of cooperation and competition within Comecon, and the openness of countries like Japan to trade, creating a vibrant and growing scientific community and the actual factories that made up the industry. These two chapters constitute an argument for the existence of a real, viable Second World of material and intellectual exchange.

But the majority of cutting-edge knowledge remained in the West, and chapter 3 thus explores the intelligence services' role as an industrial research arm of the civilian economy, a real conduit of information into the supposedly cut-off technical community. This engagement with the wider world through illicit means is followed by the engagement through licit means, as Bulgarian computers sought new markets in the Global South. Chapter 4 follows the attempts to break into the Indian market. The chapter argues that socialists could become capitalists when they needed to, which fed back into thinking about business practices behind the Iron Curtain. Chapters 3 and 4 are thus an exploration of the global outlook of this sector and the various ways it learned from and interacted with both the capitalist and newly liberated worlds.

Taken together, the first four chapters discuss the global outlook and positioning of the Bulgarian computer, creating either a global socialist community or chasing markets. Then the next two chapters zoom back into the country, right down to the shop floor. The computer created fascinating vistas for the BCP and for intellectuals, but it also provoked resistance and anxiety among workers. Chapter 5 juxtaposes the political program computing was harnessed to with some labor responses to it, while chapter 6 follows the story to the academy and schools, and finally to science fiction. Taken together, these two chapters argue for a local response to the information age, which only makes sense within the political and cultural debates and forces that shaped the BCP rather than the whole Eastern Bloc.

Chapter 7 ties these stories together by showing how the forces set in motion by these elite decisions in the 1960s created a new managerial class that was increasingly transnational and at odds with the older BCP generation. This chapter follows the globetrotting technocrats as well as the global dreams of the new electronics generation to argue that the human developments created by this industry complicate our view of both the fall of socialism and 1989 as a convenient end point to our stories.

In the following pages, the reader will meet old party functionaries, young computer specialists, charmed philosophers, worldly trade representatives, anxious writers, confused spies, ordinary workers, and fascinated children. Although the computer is the trading zone and organizing principle of these multiple networks, this story is above all about people and the worlds they wrought or met through the machine. In this corner of the world, the electronics choice made by one party opened up a Pandora's box of possibilities and anxieties, many of which are still with us today. Bulgaria was not gray and loyal as it is often presented, a backwater that may as well have been in Siberia. Just like the real Siberia, the Balkan Cyberia was its own world, with its own concerns—vital, real, and in this case cybernetic.

1

THE CONJUNCTURE: THE ROAD TO THE BULGARIAN ELECTRONICS INDUSTRY

At the end of the 1950s, socialist Bulgaria didn't look too different from most countries that had undertaken the Stalinist path of development—a communist party sat atop the political peaks, having eliminated all its opponents, and the government presided over a rapidly urbanizing and industrializing society. In the sphere of technology and economic planning, gigantism ruled, as well as such industries as iron, coal, and power generation. But these years were also a conjuncture that paved the way for the take-off of Bulgarian electronics in the following decade, as three main events combined—a victory, a crisis, and a possibility.

As Khrushchev's Secret Speech blew open the horizons of communist orthodoxy in 1956, the Bulgarian Communist Party (BCP) started taking stock of the achievements of its first few years in power. By 1958, at the end of the second five-year plan (1953–1958), it proclaimed its seventh Congress that of "victorious socialism," a politically important watershed. At the same time, the BCP finally looked at its accounts and realized that it was facing its first and very serious debt crisis. The confidence of the victory was thus combined with the doubts of financial emergency, and the party's future had to be put on a new footing. This new path, however, would lie at least in part in the new world that the Thaw was opening up before the Eastern Bloc—a softening rhetoric toward capitalism and the possibility of contacts with the West. Closer to home, the autarkic nature of Stalinism was making way for a move toward true cooperation within Comecon itself, as the possibilities of an international division of labor were discussed and specializations in technological areas loomed. Historians have recently finally begun to focus on the long-neglected Comecon as a field of mutual cooperation, but they largely concentrate on the later 1960s and 1970s, when it formed its main intergovernmental bodies.[1] Yet by the late 1950s and early 1960s, it was clear that this period was coming, at least to some party officials. It was at the juncture of these three factors that the surprising electronics revolution in the following decade was made thinkable.

This chapter thus highlights the 1956–1965 period as the culmination of several medium and long-term trends in Bulgarian development that constitute the prehistory of the country's computer industry. Having exhausted the benefits of Stalinist-type industrialization, the BCP faced the specter of continual peripheral status within the framework of a socialist economic world order that beckoned, as its allies sought to delegate it to the position of agricultural producer. The chapter shows how such factors combined to drive the party into an identity crisis, to which the technological solution of electronics proved an attractive solution. In this context, particular individuals could enact structural changes, and this chapter highlights the role of the father of the electronics industry in Bulgaria: Ivan Popov.

ALWAYS THE PERIPHERY?

Some of the first theories of international development used the region of Eastern Europe as their case studies of backwardness, with the Balkans figuring prominently. Paul Rosenstein-Rodan and Kurt Mandelbaum dubbed it an area of disguised rural unemployment, lacking in structural investment, and in dire need of basic infrastructure. One of the potential paths for unlocking this was the big push model which Rosenstein-Rodan advocated, stating that bit-by-bit investment, especially by the private sector, would never suffice to save South-Eastern Europe from the low-level equilibrium trap that it was in. The state would decisively have to step in and uplift multiple sectors of the economy at once. While neither economist pushed for Stalinist-style economics, in effect they had already identified some of the key characteristics of the path Bulgaria and others took after 1945 to tackle their underdevelopment.[2] Moreover, these debates overlapped with internal Bulgarian ones on the state of the country's development, which also identified the lack of necessary investment capabilities in the local private sector and the problem of the massive surplus labor force locked to the land. Segments of the Bulgarian intelligentsia itself thus understood their country to be backward, perpetually playing catch-up but lacking the right tools to develop at the necessary pace. It saw the problems as the weak development of the market-oriented sector, too much reliance on a state (which was weak anyway), and the predominance of the small producer as the main figure in industry.[3]

In the memorable phrase of economic historian Ilii͡ana Marcheva, Bulgaria's history of modernization after 1878 swung between industrialization and agrarianism. Liberation from the Ottoman Empire brought political independence but an economic loss—the nascent textile industry depended on state contracts from the Sultan's army, and Bulgarian traders had made their money thanks to access to the large markets of Constantinople but also those of Syria and Egypt. The paucity of available capital or raw resources, as well as the periodic losses of foreign markets, made independent Bulgaria's capitalism dependent on the state. Through both an oversized state sector and its economic policies of étatism and protectionism, the Bulgarian state was a major force in development, a path not too unlike many other newly independent countries in Europe.[4] It is too much to call the developments in Bulgaria in those years "evolution without development," as Michael Palairet did, but growth was lopsided, and industry was slow to appear.[5] By the 1940s, all results so far achieved still left Bulgaria a lagging agrarian state on the European periphery.[6] The securing of a large foreign market remained key, too—after the Ottomans, Bulgaria had fallen for a time in the sphere of German trade, but the Nazi defeat foreclosed that access as well.

During the late 1930s and the 1940s, only around 8 percent of national income was produced by industry, of which over half was in the food sector, which included the profitable tobacco industry.[7] Sectors such as metallurgy or power generation were negligible, each supplying less than 5 percent of an already meager total industrial output.[8] The sector was characterized by an almost artisanal nature in its scale and agglomeration: In 1939, there were 3,355 private enterprises with more than ten workers or output of energy higher than 10 horsepower, accounting for 10 percent of all production in industry.[9] Hampered by the weak investment power of the Bulgarian bourgeoisie, with only around 500 joint-stock companies in the whole country, the state established some of the very high protectionist barriers, ensuring a captive market.[10] Similar to its neighbors in terms of low industrial development, the country was lagging by a factor of 10 to 30 behind its Central and Western European counterparts, to which it was aspiring, in this indicator.[11] By 1946, the rural population was still more than 80 percent of the total; less than 9 percent of people were employed in industry, and even then, around 2–3 percent of them were employed in the heavier sectors.[12] This was the proletariat that the BCP inherited.

Jan Gross was right to point out that socialist industrialization was a continuation of already existing tendencies of state economic intervention in the region, amplified by the Second World War,[13] but the transformation of agricultural Bulgaria into a modern and industrial country was the explicit aim of the newly installed BCP from the very start. The first Economic Declaration, of September 1945, stated that its aim was "accelerating all aspects of economic development in Bulgaria in such a way as to turn it, in the shortest amount of time possible, into a modern industrial and agriculturally prosperous country."[14] The first five-year plan of the BCP, started in 1949, put this into practice, earmarking over 80 percent of investment for the heavy industrial sector, with the aim of leaving the agricultural past behind. This went hand in hand with widespread nationalization, and by 1951, around 86 percent of industry was in state hands. Together with the problems of this industrialization, such as falling real wages, concrete truths were created, such as 26 power stations, the first sizeable metallurgical factories, and reservoirs. Bulgaria got its own planned city as a monument to Stalinist modernity—Dimitrovgrad—complete with gargantuan chemical works.[15]

The following five-year plan tried to rectify some of the shortfalls of this gargantuan industrial effort, mainly in goods for the wider population and agriculture. The death of Stalin in March 1953 let loose new policies but also some tremors that were caused by the intense drive of the plan—in May, tobacco workers in Plovdiv rose up against low wages and high norms, and were violently repressed by the militia. Yet the Party's lower rates of investment in areas such as metallurgy or extraction didn't signal a retreat from the general path taken in the late 1940s, and uprisings such as those in Plovdiv did not directly threaten its power. Instead, the BCP felt confident enough to tackle the big task of building socialism—the problem of agriculture and collectivization. The peaks of this in Bulgaria came in 1955–1956, and by 1958, over 92 percent of arable land was united in the local collective farm form TKZS. This was also the year that Bulgarian agriculture finally reached and exceeded its prewar (1939) levels of production, thanks to lower fertilizer and seed costs, and higher prices paid for grains by the state. This victory was key for the party, as it seemed to prove that it had tackled successfully the agricultural issue in a country whose prewar politics often revolved around the land issue.[16] Moreover, the growth and delivery of consumer goods did not materialize as promised, and despite the temporary growth in unemployment during the mid-1950s, during that decade Bulgaria maintained one of the highest rates of economic growth in the world, at 14.8 percent (also higher than the Comecon average of 12.1 percent).[17]

Alongside the appearance of smokestacks in hitherto nonindustrial cities, or dams in remote mountain areas, there was another visual clue to the transformation that Bulgaria was going through: the streaming of people into the towns and cities. Between 1953 and 1956 alone, more than 410,000 people moved from villages to towns, accounting for two thirds of internal migrations during the period. For the 1955–1959 period, just under 69,000 people per year moved from villages to towns. Work opportunities in the towns and agricultural collectivization meant that between 1947 and 1967, a staggering 1.3 million people left the villages (with a further 440,000 leaving by 1972), completely changing the demographic landscape of bucolic Bulgaria—a process that was accelerated to gigantic proportions precisely in the 1950s.[18] Bulgaria became one of the fastest urbanizing countries in Europe, prompting the state to widen the regime of address registrations that were applied by the Tsarist government to Sofia citizenship in 1942, expanding them to other major cities in 1955. Eventually, most towns in Bulgaria would be subject to address registrations tied to workplace, which was an attempt to direct these massive migration flows. Despite this fraught urbanization, with its often shoddy housing, this country was on the move—in temporal and spatial terms. The road from the village to the town had a sociopolitical goal, which was simple: the creation of the proletariat that a country based on its rule sorely lacked. But this was also a move to the future, as socialist modernity was the city, and its civilization could never be based on the farmhand, in ideological terms. This was social engineering on a gargantuan and crude scale, assuming that class consciousness would eventually be formed when you work in a factory and live in a city apartment (even if it wasn't yet built). The space of the factory floor and the home was the site of the future—modern industrial labor, the very thing that the electronics industry would eventually aim to modernize or even replace. The result was that between 1948 and 1960, largely during the first two five-year plans, around 63,000 people per year joined the working class by virtue of their employment. This hyper-proletarization, in human terms, was the social flipside of the hyper-industrialization of the economy during these years.

If in the 1940s the Bulgaria that BCP took over was agricultural, nonindustrial, and rural, then its early years of autarkic-minded industrialization left a very different landscape by the late 1950s. The statistical almanacs of the state, inflated and massaged as they were, still reflected a real change: In 1960, agriculture was down to contributing only 24 percent of national income, while industry was at 58 percent (with construction adding a further 9 percent). Nearly 22 percent of people worked in industry, 5 percent in construction, 4 percent in transport and a further 4 percent in trade—leaving just over 55 percent to agriculture, down from 82 percent 12 years earlier. By 1957, there were over 800,000 people classed as "material sphere workers"—the nascent proletariat of Bulgaria.[19] The growth was uneven, lopsided, lumpy; problems remained in shortages, housing, real wages. Yet there was also a sense of progress, reflected not just in the panegyrics of socialist realist art but also in perceived improvements in life. Most importantly, however, the BCP could now feel that it represented a significant part of the population and was building a country that was succeeding in demonstrating the socialist path of development. The physical landscape was transformed by cities and industries, while the political landscape was now cleared of any last vestiges of potential enemies.

THE CONGRESS OF VICTORY

Celebrating these achievements but also using them as a springboard for the future thus dominated the Seventh Congress of the BCP, held June 2–7, 1958, and named "The Congress of the Victorious Socialist Order." Virtually 100 percent of industry was state-owned, and over 92 percent of land was collectivized, so there was no longer any road back to capitalism. Todor Zhivkov's closing speech stated that the congress "notes the undeniable fact that in the People's Republic of Bulgaria socialism has won and is paramount in all areas of social-political, economic and ideological life."[20] Stating the fact of political victory was important, as despite the repressive apparatus in place or the backing of the USSR, the party had been ideologically rocked by de-Stalinization. Zhivkov himself had emerged the victor of the new "April Line" (named after the April Plenum of 1956, the Bulgarian fallout of the Secret Speech) over the Stalinist figure of Vulko Chervenkov, and the seventh Congress was key in showing that the new line was the natural heir to socialist achievement. Now that the politics were solidified, the BCP could turn toward a future of a better material-technical base and increased socialist consciousness among the populace, with the two going hand in hand.[21] The victory, however, had not been total: The Congress noted discrepancies between its programs and actual social phenomena. For example, it criticized the banning of private plots in many TKZS as being the reason for expected yields not being achieved.[22] The proclamation of victory also changed the framework of expectations—if until now shortcomings or deviations could be explained away by the struggle to solidify the new order or the presence of insidious internal enemies, 1958 was signalling a new phase where such excuses could not fly. No longer would the ends justify the means, as Kandilarov puts it, as the new system would now have to be proven to be superior to the old.[23] This was the start of "real existing socialism"—a self-proclaimed end to its revolutionary maturation and the start of attempts to square the promises with the realities.

The Congress's tone also determined the goals of the third five-year plan. The State Planning Commission was warning that the developments of the 1950s, focused as they were on agriculture and primary industries, was leaving the country further behind its Central European allies. Machine-building was being neglected, as were any kind of high-value-added goods, condemning the country to a continued Balkan style of a natural resource exporter. At the same time, the economic experts were enamored with the Maoist model unfolding concurrently in the fraternal Chinese nation. The plan was thus colored by a "Great Leap Forward" mentality, aiming at fulfilling the industrial goals in three years and agricultural goals in four years, rather than five.[24] The mark of voluntarism was more pronounced at the November plenum, which set the goal as not mere quantitative improvements, but a qualitative developmental jump into the future.[25] Over two-thirds of investments were earmarked for heavy industries, a sector that was supposed to increase by 77 percent (against an average of 62 percent), with machine-building supposed to be a priority. Yet the gigantism continued—the giant steelworks at Kremikovtzi near Sofia (a site where iron ore proved to be of poor quality), the oil refinery at Burgas, the zinc works at Plovdiv.[26] By 1962, this Leap was officially complete. Yet the Sino-Soviet split, combined with the shortcomings of many of its goals, pushed the party to abandon such Maoist experiments and adopt the more Soviet-influenced long-term development programs that aimed to increase machine-building by a factor of 17 by 1980 (among other sectors—chemical industry output was supposed to increase 25 times!).[27] In many ways, this redirection was the real start of the Bulgarian machine-building industry after its relative neglect in the 1950s. It would be the basis for the construction of communism.

This change also raised a key problem for the BCP—how to move from extensive to intensive growth, now that the expansion of the urban labor force was plateauing and the economy had built up the basic industrial and construction projects that accounted for the hitherto impressive growth numbers. Now individual workers had to become more productive, rather than abstract sectors. Throughout the following decade, various attempts at economic reform would be made, the particulars of which are beyond this book. However, it is their minimization over the 1960s that also make the choice of electronics more understandable, as a "surrogate" sector that could offset the shortcomings of structural reform failures. The period between 1963 and 1968 was marked by attempts to introduce a new system of planning, away from the "leap" and toward steady acceleration. In April 1964, some enterprises started applying the principles of profitability and sales, and they achieved good results—the Bulgarian reforms were, in some views, a testbed for the Soviet Premier Kosygin's reforms in the USSR itself the following year.[28] These attempts, however, met with serious opposition among members of the Central Committee at plenums in both 1965 and 1966, so that the final decisions made at the July Plenum of 1968 contained a minimalist version of the reform. This was not a radical break with planning, and remained largely within the central-administrative confines of the party-state logic. In many ways it was was stillborn even before it was implemented, and yet it remained the formative economic document over the next 20 years.[29] There were thus heated debates in the party about the very structure of the economy, and the choice of electronics should be seen as an attempt to find a politically neutral solution, or surrogate, for the inability to reach a consensus on the future of economic reform. Of course, this technology and the sector it spawned would have a bearing on subsequent debates and could never have remained "neutral."

But just as the party was turning toward these problems and announcing its victories in socialist construction, it had to face a concrete problem: financing. For unsurprisingly, this solution cost money, and much of that came from outside the country.

BANKRUPTCY

Rapid industrialization demanded machines as well as know-how from abroad. Much of this trade expansion was not only within Comecon, but also with the West: Trade with Western countries increased from $45 million to $200 million between 1954 and 1959.[30] But the export profile of 1950s Bulgaria was poor, and after 1956, its trade balances with both East and West were in the red. Tobacco, grain, vegetables, fruits, seeds, some ores—this was what Bulgaria could export. Agricultural growth was still relatively sluggish, not enough to finance technological imports. Yet import reductions were out of the question, because industrial expansion was paramount and was resource-hungry. Comecon allies often fell short of providing the needed technology, leading Bulgarian enterprises to make up the shortfall in Western markets in pursuit of fulfilling the plan at all costs—and besides, the best-quality goods and machines were available in the Western market.[31]

The regime's foreign trade organizations (VTOs) were poor at their job, too.[32] They underestimated the importance of aesthetic design and were woefully ignorant of local markets. Quota fulfilments also meant that often there were rushes of production toward the end of the year, leading to the dumping of Bulgarian goods onto markets with no regard for local needs. Conversely, the VTOs often bought the wrong technology or ones that enterprises could not implement for years—machines gathered dust in warehouses until they were obsolete. VTOs ended every year with large numbers of unfulfilled aims in both the import and export lines. The years 1956 and 1959 were particularly bad, contributing to a debt in Western currency amounting to 872 million Bulgarian levs by 1959. This equaled $115 million according to official conversion rates, a sum that was significant, given the poor export prognosis.[33]

These loans were Western in a strange way—as they were in fact Soviet. Yet Bulgaria owed Soviet banks based in Paris and London, both branches of Gosbank—the Banque Commerciale pour l'Europe de Nord and Moscow Narodny Bank. Those banks were thus following Western banking laws, and besides, political solutions were less likely, because the USSR was growing impatient after providing further loans in 1957, staving off the worst of the crisis just before the Congress of Victory. By 1959, the financial situation was dire once again, and no more loans were forthcoming.

In early 1960, a delegation made up of the Bulgarian trade representative in Paris, the deputy director of BNB, and the head of a section in the bank, undertook a whirlwind tour of the UK, France, the Federal Republic of Germany, and Italy in a desperate bid for new extensions or credits.[34] Societe Generale, Midland Bank, Westminster Bank, Bank of England, Deutsche Bundesbank, Banca Nazionale del Lavoro: Everywhere the reception was frosty. Even the sympathetic Soviet banks made it clear that the country should build up currency reserves rather than relying on short-term credits. Yet by the middle of the year, the trade balance was a negative 100 million levs and falling rapidly, with no end in sight. Bulgaria was simply producing goods that were uncompetitive even in the Soviet market. In such desperate times, desperate measures were proposed by Kiril Nestorov, the director of national bank BNB: the selling off of Bulgaria's gold reserve of about 21 tons. On May 7, 1960, Nestorov wrote to the President of Gosbank, Alexander Korovushkin, raising the issue as a possibility.[35] The Soviet replied on the same day, saying it was an option. In fact, the gold itself was already in the USSR, sent there in the 1950s because the BNB did not yet have a nuclear-proof vault, and the gold had been reduced by 14 kilograms after refining in Novosibirsk.[36] The sale was, of course, an extreme step, delaying the ministerial decision until 1961, when BNB had to declare that it had other gold reserves in order to circumvent the law that protected the sale of the state reserve. By the end of the year, more than 20 million levs had been raised by these deposits, with a further delivery of nearly 4 tons of gold in bars and 2 tons in coins to the Moscow Narodny Bank in London in 1963, guaranteeing a further credit of $6 million that year.[37] These were still envisioned as deposits for further loans, but by 1964, Nesterov wrote to Zhivkov asking for the sale of at least 4 tons—in fact, 9 tons were sold on the Zurich gold market by the end of the year, according to Khristo Khristov.[38] Over 30 years later, Zhivkov's memoirs would deny that any such sales happened and stated that he had actually increased the gold reserves to tens of tons.[39]

This drastic step would not be the ultimate solution that the regime sought, and political decisions had to be applied: In 1965, Moscow agreed to forgive Bulgarian debts to its two Gosbank branches, as well as to delivering goods that Bulgaria would otherwise have to buy on the world market. Five thousand tons of cotton, hundreds of tons of key chemicals (such as phenol), 150 tons of nickel—all on top of the normal Soviet contingents for the year—were delivered, easing the problems of Bulgarian industry. Flowing the other way were thousands of tons of sugar, cheese, poultry, and more than 20 million eggs, which the Soviets were to buy at world prices.[40] Despite its industrialization, first following the precepts of orthodox Stalinism and then attempting to emulate Maoism, Bulgaria was still dependent on agriculture in its exports and political negotiation in its finances for the solution of its economic problems. The financial crisis had laid bare to both the BNB and the Politburo the shortcomings of its economy, which needed to radically change its profile. No quantity of eggs or canned tomatoes would ever be able to provide the convertible currency needed to finance the machine-building factories or consumer goods that were part of the long-term development plan (to say nothing of the humiliation of selling off the family gold).

The debt crisis can also be seen as a symptom of increasing participation in world trade. The deep changes in Bulgarian economic structures during this and subsequent periods were realized with the help of outside resources, whether Soviet or Western credits.[41] Despite being part of a longer history of Bulgarian debt, in which loans were always preferred to foreign investment in the post-1878 period, the scale of economic change and thus indebtedness during the 1950s was unprecedented. Simultaneously enabling modernization and disturbing the state, this entanglement with the international market brought into sharp relief the need for a different structure for Bulgarian exports. What that nature would be was heavily dependent on Comecon, where countries with an industrial pedigree (such as Czechoslovakia) were positioning themselves as suppliers of the latest technologies. If Bulgaria were to avoid remaining the perennial basket-case and breadbasket, it had to utilize the emerging socialist division of labor within the Eastern Bloc to its own advantage.

TOGETHER AGAINST CAPITALISM

Comecon was formed as a reaction to the Marshall Plan in 1949, yet during Stalin's life, it had existed more on paper than in reality. The organization always had a primarily political goal of shoring up Soviet influence in the East and orienting the countries toward Moscow through economic ties. Developmental goals to bring agricultural states like Bulgaria to the same level as advanced industrial countries, such as the GDR, were secondary.[42] Yet the concept of a division of labor was inbuilt in this community from the very start, with member states expected to coordinate on the basis of a general economic plan, which would ensure that the states would complement one another rather than compete in the same sectors. But until the later 1950s, its main success had been in doing its part in solidifying the division of Europe into two competing politico-economic blocs. The blanket application of Stalinist autarkic planning in the early years of industrialization also ran counter to socialist unity, as it encouraged parallel rather than complementary development. Trade continued to be done on a bilateral basis rather than through a common framework.

Stalin's death was a watershed for Comecon as much as it was for most other things. Intra-bloc trade would now be a part of the wider program of finding an economic victory over capitalism.[43] East of the Iron Curtain, the 1955 Warsaw Pact Treaty had already committed the countries to closer integration in geopolitical matters; to the West, the 1957 Treaty of Rome gave the real start to a European integration that Comecon would both compete with and wish to emulate in some ways.[44] Comecon got its charter in 1959, setting out an organizational structure as well as annual council sessions, an executive council, and permanent commissions on various economic issues. As early as 1956, however, the issues of socialist nations' specialization had been raised, with 600 products earmarked for such treatment. The vast majority of these went to the developed industrial states, contradicting the interests of states such as Bulgaria and highlighting the problem of uneven development.[45] Despite getting some pan-bloc responsibilities in the fields of copper, cement, and certain chemicals (to be developed with allies' machines and plans), the main sectors that fell to Bulgaria were grain and other primary resources.[46] More advanced states argued strongly against Bulgaria getting any machine-building specializations, correctly pointing out its low technological capacities. As Marcheva has stated, these biases remained throughout the late 50s and up until at least 1965.[47] Khrushchev himself supported such a focus for the country and provided loans earmarked for Bulgarian agriculture, light industry, and extractive sectors, after 1955.[48] All these were sectors where profits were low and adding value was difficult. Developmental leaps were not easily built on such exports, and neither were loan repayments.

Yet Comecon itself was changing to become a real multilateral body. In 1962, a Central Dispatching Board was created to unify electrical power systems; in 1963, an International Bank for Economic Cooperation (IBEC) was set up, followed in 1964 by a Bulgarian Foreign Trade Bank, facilitating financial exchange and settlement among member states.[49] Joint institutes were created, such as the Dubna Institute of Nuclear Research (although established in 1956, it became much more active in the early 1960s). This encouraged growth among member states, helping Bulgarian foreign trade blossom to be 2.5 times bigger in the 1958–1962 period, largely within Comecon.[50] Slowly but surely, the organization was growing more ambitious, culminating in the fifteenth Council Session in 1962, where the Basic Principles of the International Socialist Division of Labour were adopted. Bulgaria was one of the countries that saw the dangers of this document, which would concentrate production in developed countries, such as Czechoslovakia. Khrushchev waxed lyrical about a "socialist commonwealth" under the auspices of a central Comecon planning committee, but this only increased the fears of the poorer members: Unless they could quickly demonstrate machine-building capacities in this new world, they would be doomed to perennial catch-up. Which was precisely what the BCP had been trying to avoid throughout the 1950s.

Zhivkov made promises that Western licenses and know-how would be purchased, and the country would reach new levels in structure-defining sectors. The national jostling within the organization meant that as the GDR and Czechoslovakia defended their positions, Bulgaria was pushed closer toward the USSR in seeking technical assistance. In the short term, this would deliver the industrial capacities it sorely needed but would tie the country to an increasingly less innovative economy in the long run.[51] Throughout 1964, the country cooperated closely with the Soviets, ensuring support for the Kremikovtt͡si steelworks; chemical plants; and new specializations, such as electrocars. Most importantly, it secured a 400 million rouble loan for its economic plan, larger than all the credits envisioned for the period up to 1970. Moreover, Bulgaria was playing a wily political game to garner Soviet favor in distributing the Comecon specialization pie. The most controversial but politically useful move was to suggest to Khrushchev that Bulgaria could become the USSR's sixteenth republic. Patently infeasible, not least due to the international implications, the suggestion demonstrated to Moscow that Sofia was doggedly loyal and interested in closer integration. The benefits, economically, were real, as Bulgaria pulled away from its agricultural role within the socialist family thanks also to Soviet aid.

However, the move toward dividing labor among countries was opposed more vocally by Romania. The 1962 Principle had already been protested by Bucharest on grounds of national sovereignty, which it held to be a key pillar of Comecon. Gheorgiu-Dej railed against the plans, insisting that every country had the right to determine its own road. As countries such as the GDR raised the issue that the 1962 Principles were not being acted on as they required unanimous agreement even when projects did not concern all countries,[52] there were calls for institutional reform to allow groups of countries to move ahead and cooperate. In April 1964, the Romanian party issued a declaration stating that talks of economic integration were "withdrawing the economic activity and decision-making from under the national authority."[53] The declaration had the desired effect, torpedoing the reforms and ensuring that post-1964 talks were not of integration but of coordination of plans. While Romania was the most vocal, the ideas faced passive resistance from others, including the Bulgarians. A compromise institution—the Bureau for Integrated Planning—limped on as an advisor to the Executive Committee.

The fall of Khrushchev in that year focused the USSR on internal matters, while Hungary and Poland pushed for a convertible currency within Comecon to allow some market relations to emerge in inter-country trade. The "transferable rouble" had already been set up in 1963 with the creation of IBEC, but it was meant for inter-country trade accounts and was not freely convertible into national currencies. The Polish-Hungarian proposals would further the creation of a true supranational credit system and transfer some of the market liberalizations of "goulash communism" to Comecon as a whole.[54] Despite conservativism on the part of many parties, debate continued in specialized journals, where the champions of market mechanisms clashed with those who preferred supranational solutions to socialist development. These debates were helped by increasing East-West meetings, as econometrics, linear programming, and other ideas unified economists on both sides.[55] Although the issue was clearly not being resolved, any intelligent observer could see that the discussions meant that Comecon was perceived to be a vital part of the future, and when Moscow focused on it again, some sort of reorganization would follow.

A less astute observer would have still noted that even in its current form, the organization offered immense possibilities and lifelines. As Randall Stone noted critically, Soviet Bloc trade operated according to complex calculations that underpriced commodities, especially oil, and overpriced machines, which were considered to meet Western standards. Any trade would incur a cost either for the seller or buyer, as world prices could be obtained in Zurich or London. East European satellites tailored their negotiation positions accordingly, taking advantage of highly distorted prices.[56] Stone's influential analysis rings true for the realities of Comecon dealings, where satellites minimized contributions, defended national interests, and extracted the maximum possible from a Soviet partner hampered by its own bureaucratic intransigence that prevented it from enforcing trade commitments. Moscow was politically unwilling to translate its obvious preponderance into a real integrative project, allowing the socialist division of labor to be hampered by weaker states, such as Romania. At the same time, it was increasing its subsidies to satellites every year, sold them oil at below-market prices, and took more and more goods from them at inflated values. The satellites did not become a real burden, as they offset Soviet costs by other contributions—not least in a military sense, where a full third of European-theater forces were non-Soviet.[57] And so trade was both mismanaged and unbalanced. The satellites could look at the USSR and see a captive market ready to be tapped. If you found a niche in Comecon, your machines would equip enterprises from Berlin to Vladivostok. But that niche had to be new, as it would be suicidal to compete with the GDR in optics or Czechoslovaks in cars, for example. A clever country would find a new sector in which to get a head start, with an eye to the obviously upcoming decisions over the ideas of socialist integration.

Indirectly, Bulgaria benefited from a geopolitical reality that helped it garner more favor and resources. It was the only Warsaw Pact state bordered by two NATO members, and increasingly it was the only reliable member on the Southern Front. The original maverick was Yugoslavia, and Albania became one by 1961. Romania's obstinacy in Comecon was reflected in similar moves in the military sphere—by 1964, it had adopted the policy of nonintervention in other countries' affairs and became the first country to remove KGB supervisors from its intelligence.[58] In 1966, it blocked pretty much every structural change aimed at creating a multilateral military council, culminating famously in their condemnation of the crushing of the Prague Spring.[59] By the 1960s, Bulgaria had assumed an oversized importance to the Warsaw Pact, despite the secondary importance of the Southern Front. Zhivkov used this fact to offset military costs by securing gifts worth hundreds of millions in rubles by both Khrushchev in 1963 and Brezhnev in 1965.[60] Domestic investment could thus be focused on the civilian economy, while tying Moscow to even more aid commitments. Sofia had many tentacles wrapped around its giant patron.

FATHERING BULGARIAN ELECTRONICS

But of all the goods a country could specialize in, why choose electronics? This question can only be answered by the addition of historical contingency. Bulgaria's turn toward the sphere was driven by a highly connected actor at the peak of political power and the man who every veteran of the future computer industry would name if asked about its genesis. Professor Ivan Popov was, in the words of Vasil Nedev, "[Bulgaria's] biggest scientific industrialist in its whole history . . . the patriarch of its modern industry."[61] Popov is indispensable to the story of Bulgarian socialist modernization as a whole and of electronics in particular. His rare combination of international education, experience, political connections, clout, managerial skills, and personal contact came together in one of those actors who forms one of the sides of historical conjuncture, allowing a single individual to become the main conduit of ideas that could shape entire structures and put them on new paths. As Stoi͡an Markov notes, "he knew that Bulgaria was poor in resources [and] . . . he knew electronics was a profitable area that did not depend on raw resources that Bulgaria lacked."[62] Popov was thus the man who offered solutions to the problems and possibilities outlined in this chapter, championing electronics as the way out of the regime's predicaments and the best way to capture the Comecon and Soviet markets.

Ivan Popov was born in 1907 in the medieval capital of Veliko Tŭrnovo.[63] His parents were socialist schoolteachers who encouraged his studies, which he continued in Sofia in 1921, where he also became a member of the Communist Youth Union, aligned with the "narrow socialists." This was followed by his arrest in the wide anti-communist sweeps after the 1925 Sveta Nedeli͡a terrorist act, for which he was sentenced to two years in prison. After eight months, he was amnestied and continued his studies in the Mathematical Faculty of Sofia University. He showed great aptitude as a student, graduating with distinction and working as an assistant in the Faculty of Higher Analysis in 1930–1931. His first scientific work dates from this time, helping him secure a stipend to Toulouse University in France, from which he graduated in 1933 with a gold medal, specializing in electrical technology and hydrology. He stayed on in Paris to work on the practical applications of his thesis on neutral currents, which he managed to patent. In 1934, he returned to Bulgaria, opening Electrotherma, a private firm that produced heating elements and medical instruments, which proved to be sufficiently successful in the local market to expand in 1939–1941. Political events, however, caught up with him. He was not a member of the workers' party at this time, but his brother and son-in-law were involved in some capacity, leading to their arrest in 1941 and subsequent execution by firing squad. Understanding his position to be precarious, he left for Budapest, where he worked as a researcher in the Agrolux factory up to 1943 and then as a designer for the German electric equipment producer AEG up to 1945. The end of the war found him as the director of the factory, where he worked hard to prevent its technical equipment from being carted off by the retreating Germans, and he resumed its production lines under Soviet occupation. During this period in Hungary, he traveled widely in Germany, Austria, Czechoslovakia, and France, forging business and personal links with people in the electrical industry throughout Europe. In 1949–1950, he came back to Bulgaria, becoming the director of the power engineering factory "Kliment Voroshilov," which was to become a key school for Bulgarian engineers. During these postwar years, he also bolstered his professional profile with political memberships in line with his youthful convictions—a member of the Hungarian Workers Party between 1945 and 1949, he joined the BCP in 1950. Here, his history as a repressed communist youth combined with his technical experience—in short supply among party members—to facilitate his quick rise through the ranks: head of District Committee, and then member of the Central Committee from 1961. His final position would be the highest—a Politburo membership between 1966 and 1976—concurrent with his apogee as the strategist of the Bulgarian economy.

His economic clout grew more gradually—director of the newly created State Union Elprom, putting him in charge of the growing Bulgarian power industry. At the same time, since 1949, he had resumed his academic career as the head of the Faculty of Electrical Engineering at the State Polytechnic (later the Higher Machine Electrical Institute "Lenin"—VMEI—the premier technical university in the country). His style of work was often authoritarian, and people remember him as an exacting, workaholic, somewhat humorless but always extremely professional, competent, and fair boss.[64] He was always demanding, expecting quick and accurate work by his subordinates, and in return, he championed them in ministries and the party. Because of such methods, he was the subject of a 1952 article in *Rabotnichesko Delo,* titled "Short Circuits," which accused him of authoritarian and dictatorial work in Elprom.[65] Despite a subsequent rebuttal in the same pages and a disciplinary action against the article's author, in 1952 Popov was moved to a permanent position in the State Polytechnic and dismissed from his managerial positions. In his academic capacity, he developed new programs in engineering education as well as designing electrical engines and regulators that found applications in the industry. His clout meant that between 1954 and 1958, he was deputy rector of VMEI, during which he worked on more than 20 scientific projects and monographs, some published in both Germanies, the USSR, and France. His academic star was shining bright, and after 1958, he spent four years at the prestigious Scientific Research Institute of Electrical Technology Testing in East Berlin, where he was made the head of the section dealing with transformers. Every year he would spend up to four months lecturing in Bulgaria. He was still, however, a relative political unknown. In the apocryphal story, it was during a Zhivkov visit to the GDR that Walter Ulbricht joked that he was thinking of appointing a Bulgarian scientist to the post of deputy minister of the electrical industry—Popov was indeed a member of SED, GDR's ruling Socialist Unity Party, since 1958, continuing his astuteness for the political climate. The more prosaic and likely story is that he came to the attention of Zhivkov in 1961, when he won a prize and doctorate from the Higher Technical School in Ilmenau, and he was recalled to Bulgaria, to become a member of the Central Committee and rector of VMEI in 1962, as well as a member-correspondent of the Bulgarian Academy of Sciences (BAS). This post, however, lasted for only four months, as he was being groomed for the much higher position—head of the newly founded State Committee of Science and Technical Progress (CSTP), the successor to the Technical Progress Committee founded in 1959. This organization and position, the importance of which will be seen in chapter 2, gave Popov the commanding heights over the strategic direction in Bulgaria of research, innovation and its implementation into industry, and power over the universities and BAS. In party economic terms, he was now one of the most powerful people in the country; in terms of party science policy, he was unquestionably the most dominant. It was during this quick rise that he also became one of Zhivkov's favorites, who saw in him a capable and innovative professional.

At the start of this pivotal moment for Bulgarian policy in the early 1960s, Popov was almost unique among high-ranking BCP members—he had an internationally tested and recognized expertise in a technological field. He was also fluent in Hungarian, Russian, German, and French (skills bolstered, as he joked, by marriages to both a French and a German wife), unheard of among the mostly monolingual BCP functionaries and, together with his contacts cultivated in the 1930s and 1940s, giving him unprecedented access to foreign ideas and industrial trends. His own expertise in power and electrical engineering kept him interested in the latest global trends in this field, and it was only logical that he noted the ideas in the parallel field of electronics that arose during the Second World War. He also experienced the GDR's scientific climate, which convinced him that if Bulgaria tried to compete with such countries in established fields of industry, it would inevitably lose. He expressed these ideas in the many personal meetings he had with Zhivkov in the first months after his return to Bulgaria. In touch with the first Bulgarian doctoral students who studied in the nascent field of electronics in the USSR and the GDR, he advised Zhivkov that "cybernetics, computer technology, fine mechanics. Here is our strength."[66]

Popov is the father of Bulgarian electronics not because of his innovations or his scientific work, as he was an academic in a very different field. But he had the intellectual tools to recognize the importance and possibilities of the nascent sector, as well as to understand the general trends of specific research and production, and what would be the avant-garde in the sector. His scientific network was wide, both beyond and behind the Iron Curtain, bolstered especially by the time he spent as the head of a laboratory in East Berlin. But his key characteristic was his managerial style, which contemporaries described as "American-style." The 1970s electronics minister Ĭordan Mladenov describes him as "more like an organiser in the American sense of the word 'manager'": finding and mobilizing financial resources for projects, organizing cohesive design teams, attracting the best cadres, and having a general awareness of the industry and market.[67] His iron working discipline, often from 6 a.m. to midnight, helped his productivity during these years. Once he moved away from his academic work, he became a supreme organizer of science, utilizing his languages, experience as an academic administrator, and political connections (which stretched to Moscow, where he was colleagues with similar party-engineering cadres in the radio industry). Unlike many other sectors of the

1.1 Ivan Popov (right) meeting Konstantin Rudnev, the Soviet Minister of Automation, 1972. (*Source*: Central State Archive, Sofia.)

socialist economy, his personal clout and desire for accurate reports helped instil a more internally accountable, if stressful, working atmosphere. He kept abreast of electronics by reading the newest dissertations published by the emerging cohorts of students. An anecdote illustrates this—a student, sure his committee would not read the thesis in full,[68] promised the reader a full case of beer if he had reached this particular page. Months after his defence, Popov called him at home, asking for his beer.

Popov's claim to "fathering" this field in Bulgaria is in championing electronics as a profitable area and organizing a productive, well-financed environment that would allow an emerging group of scientific and engineering cadres to make this field a possibility. After 1962, he was in the halls of power that he needed to push through his project of a high-technology, low-resource but high-yield sector and make it a success. But one man alone, responding to the problems of Bulgarian industrialization and the possibilities of the Comecon market, could not be enough to create something from scratch. He could develop, however, some existing capacities and emerging scientific potential.

THE EXISTING CAPACITIES

This chapter has made it clear that the country had little tradition in most high-technology sectors, but there were cores of expertise available from which to start. These were tied to the preceding decade's industrialization, with the biggest school of many of the new specialists being the electrical factory Kliment Voroshilov in Sofia, specializing in communications equipment and since 1949, uniting all smaller companies and enterprises of the sector in the capital. Built with Soviet help, it blossomed into a large site for telephone and radio production but also a veritable school for engineers.[69] It was also Popov's first industrial appointment once he was back in Bulgaria. In the early 1950s, Soviet engineers used the plans of a Rostov factory to organize the shop floor and also delivered numerous manuals and blueprints. Within a few years, the factory was producing serial runs of telephones and phone exchanges, ultra high frequency stations for civilian and military use, and it was developing the first Bulgarian TV, the Opera-1. Groups of up to 40 engineers at a time were sent to the GDR and Czechoslovakia to train, quickly building up a core of experienced technicians.[70] To make sure this knowledge diffused to those who stayed behind, the factory also became a field school. Those who knew English, German, or French were assigned foreign journals to follow, with the task of submitting commentaries on two articles at the end of each month. These were collated together in publishable form, so that the factory quickly built up a library of the latest Western developments.[71]

Such moves allowed the Bulgarian workers to start their own research and development efforts, rather than slavishly following the Soviet licenses (such as that for a military field radio, which was proving unreliable). A team lead by a military engineer, Stoĩan Dzhamiĭkov, set out to produce a better radio for the Bulgarian and also the Warsaw Pact armies. Together with the development of the Opera TV, this project became the factory's first foray into its own research and development, as well as a testing ground for the next engineering generation. The improved radio would enter serial production in 1964 and equip the Bulgarian, Hungarian, and Polish armies, securing the factory's reputation.[72] At the same time, the production of TVs would make it the country's first factory with some degree of automation, installing two mechanical conveyor belts in 1962.[73] Throughout these years, the factory possessed the best-trained engineers in electronics and communications, and its staff were used to set up the newly differentiated factories that split off. The Kliment Voroshilov factory would retain radio relay and long-distance communication duties, having made its mark on Bulgarian industry.

Future electronics specialists got their first taste of modern technology here. Lîubomir Antonov, part of the team that designed the first electronic calculator (as we will see in chapter 2), started off in the television laboratory there.[74] When given the chance to join BAS's Institute of Communication, he decided to stay, as the factory remained the best-equipped place in the field into the early 1960s.[75] Often the work still had to be improvised and depended on the young engineers' creativity, as blueprints were lacking. Antonov created the first Bulgarian digital measurement instrument, scrapped together from sensors given to him by friends in various institutes, when a Romanian delegation visited in 1958 and talked to him about the future of electronics. At other times, all he had to go on was the basic idea that a particular machine existed, for example, an analog computer after reading an English book in Russian translation. To create it, he had to go back to basics, speaking with his university lecturers in mathematics, or emulating transistor technology he had seen in a Phillips catalog.[76] This machine was delayed by the lack of silicon transistors, leading him to replace them with radio lamps, which made the machine obsolete the instant it was created (the elements base for socialist electronics would plague its whole history). Yet the Voroshilov factory was an invaluable school for young engineers, pushing them to find solutions in a veritable information desert.

The bloc's cutting-edge electronics research in those years remained outside Bulgaria, in the USSR and the GDR. In 1956, as Soviet science "thawed," and the maligned field of cybernetics was rehabilitated,[77] the first theoretical work on computational machines in Sofia was done by Prof. Bozhorov and Prof. Nedi͡alkov. Around same time, Prof. Li͡ubomir Iliev of Sofia University attended a Moscow conference on the "Development of Soviet Mathematical Machine-Building."[78] On his return, he pushed the university to send students to Moscow to complete undergraduate degrees in the field. By 1957–1958, there was at least one student studying electronic engineering there (Stoĭcho Chamarov),[79] while the first larger group of students and teaching assistants was dispatched from Sofia University's "Digital Methods" course in 1959—among them future luminaries of the field, such as Blagovest Sendov. The first doctoral student in the field, Racho Danchev, also started work at Moscow State University that year.[80] Others were sent to the GDR, the other preeminent center, with Antonov specializing in Berlin in 1960[81] and Petŭr Petrov (who would go on to work at the Institute of Technical Cybernetics and Robotics, ITCR) specializing in electronic automation there in 1962.[82] Others, such as the future director of the ITCR, Angel Angelov, started off with semiconductor specializations in Moscow in 1956 and continued to work on joint East German–Bulgarian projects on the bloc's first digital telephone exchanges in 1960.[83]

A critical mass of intellectual interest and cadres was thus being created before Popov came to head Bulgarian science in 1962. A Council of Ministers order from April 1961 created the country's first electronic Calculation Centre at the Institute of Mathematics at BAS, as well as the Faculty of Higher Analysis at Sofia University.[84] Iliev became the deputy-director of the Mathematical Institute (renamed Mathematical Institute with Calculation Centre), under the director Academician Nikola Obreshkov; the main engineer was Ilko I͡Ulzari. This center would be the core that took the first steps in domestic computing development, as I describe in chapter 2. Iliev, the true champion of the field at the time, organized a summer school for his most promising mathematics students at Dubna, the bloc's Joint Institute for Nuclear Research and home to a powerful computer center.[85]

There were other nuclei of potential, situated in industry itself. The regime's awareness of the coming need for more specialists in research and

1.2 Radio production in the Kliment Voroshilov factory, 1962. (*Source*: Sandacite.bg).

development than academia could provide led to the creation of a network of Bases for Technical Development (BTD) in various enterprises in 1961. Some would eventually become independent institutes due to their importance, the paramount being the "Instrumental Industry" BTD.[86] ITCR's predecessor, building on a small foundation in 1959, also became a fully fledged electronics and automation site in those years.[87] The first batch of engineers also took existing institutes in new directions after the completion of their studies abroad, such as Angel Angelov, who returned from specialization in Berlin in 1963 to set up an Industrial Electronics section in the Research Institute for Electrical Industry.[88]

At the start of the 1960s, Bulgaria found itself with various sites of expertise in radio, communication, and television (the Voroshilov factory), and electronics and automation due to the first generation trained in the USSR and the GDR. This potential was recognized by the BCP, which supported both BAS and industrial sectors in their development of research and development centers in the field. Popov's emergence would serve to unite these many areas under one coordinated and centralized vision, which was sorely needed in a field where other states were also taking their first steps.

THE STATE OF SOCIALIST COMPUTING AND CYBERNETICS

Computing in the Eastern Bloc was a perpetual game of catch-up throughout its history. In 1950, the head of the Institute of Precise Mechanics in Moscow stated that the country was fifteen years behind the US in the field and would have to make this good in five to keep the USSR from losing the arms race.[89] At the same time, this was to be done without copying any of the philosophical ideas that came with the field, which were seen as idealistic and metaphysical deviancy. Cybernetics was a reactionary, bourgeois pseudo-science, as articles in 1952–1953 dubbed it. Under the pen name of "Materialist" (the author's pseudonym), criticisms of Norbert Wiener and Claude Shannon abounded, taking them to task for repeating eighteenth-century routes that were anachronistic and had failed.[90] In 1954, the *Short Philosophical Dictionary* solidified this official reputation, pushing Soviet science even further behind the trends.[91] Yet this inertia from the Stalinist period was about to be displaced as restrictions eased, and by 1955, the existence of Soviet computers was declassified. Cybernetics, too, was rehabilitated.

This famous word denoted a multidisciplinary field that explores systems' structures and restraints, applicable to computing and maths but also increasingly to social engineering. It posited that a system with a goal can take action to achieve that goal, and in the process can also be self-correcting through "feedback" (a concept that originates in cybernetics) at all levels of the system. This is applicable not just to simple organisms but also to the whole universe. Norbert Wiener, who is credited as cybernetics' originator through his 1948 book *Cybernetics: Or Control and Communication in the Animal and Machine,* defined it as "the scientific study of control and communication in the animal and the machine." He expanded on its social implications in his 1950 work *The Human Use of Human Beings.* Since its inception, the discipline has lost its preeminence as a standalone field, but it lies at the core of or has informed multiple important fields of study, such as game theory, system theory, neuroscience and cognitive psychology, and organizational theory in business management. In the Eastern Bloc, it held people's fascination for much longer than it did in the West.

By 1958 the first book aimed at the general public, Igor Poletaev's *Signal,* encapsulated the discipline's promise:

> The laws of existence and transformation of information are objective and accessible for study. The determination of these laws, their precise description, and the use of information-processing algorithms, especially control algorithms, together constitute the content of cybernetics.[92]

The *Short Bulgarian Encyclopedia,* a five-volume set that could be found in many Bulgarian homes, was published in 1966 just as the industry was being born in the country. Quoting the field's founder, Wiener, the encyclopedia defines cybernetics as "the science of connections, governance, and control in the animal and machine." It highlights the importance of feedback loops (*obratna vrŭzka*), where a system receives information about the results of its processes and can thus regulate its behavior. The short article states that this science affects a wide variety of areas, from machine translation to mathematics, but thanks to the advancement of computing, it also impacts the automatic control and governance in the economy and beyond.[93] Unleashed, cybernetics became a dominant language for Soviet scientists, who sought in it everything from a confirmation of Marxism to the "cyberspeak" of Slava Gerovitch's argument: a precise and objective language of science and methodology, where the precision of the algorithm was opposed to the regime's unverifiable slogans.[94] The computer and cybernetics combined became the superstar of Soviet science, which was searching for a "panacea for Soviet economic woes."[95] By December 1957, the Soviet Academy was telling the Politburo that the use of computers in planning was of exceptional importance to efficiency, inaugurating a long-lasting effect of cybernetics on the bloc's economic thinking. Planning was a cybernetic feedback system of control of enormous proportions, and the Soviet economy was potentially a fully controllable system with multiple information flows.[96] Mathematical modeling of the economy through computing was discussed in 1960, and the cybernetic utopian mantra was incorporated in the CPSU's congress of 1962. The party had embraced the new world and was ready even to incorporate Western management techniques if they proved useful. Khrushchev's words of November 1962 envisioned society functioning like an automated assembly line, just as Sofia's Voroshilov factory was installing its first such line:

> In our time, the time of the atom, electronics, cybernetics, automation, and assembly lines, what is needed is clarity, ideal coordination and organization of all links in the social system both in material production and in spiritual life.[97]

The imperial center's embracement of the field changed the trajectories of the satellites' science, but it also called into question the material base. Computers did of course exist in the bloc by the late 1950s, spurred on by the military arms race and then economic needs. While Stalinism rejected Western cybernetics, it fully understood the need for computers in the arms and nuclear race. The first Soviet stored-program computer appeared in Kiev in December 1951, the MESM (from the Russian abbreviation for "Small Electronic Calculating Machine," which was of course anything but small), which was also the first such machine in Europe. The Automatic Computing Machine M-1 appeared early in 1952, built by a Moscow team. Both computers were quickly harnessed to the needs of nuclear physics, jet propulsion, radio location, and aviation.[98] In 1955, the BESM (Large Electronic Calculating Machine, the name now closer to the reality) equipped the first purpose-created computer center in the Soviet Academy, yet again harnessed for military matters. It remained the fastest machine in Europe for at least two years and spawned a full range of machines, culminating in the BESM-6 in the mid-1960s, a mainstay of Soviet computer centers.[99] The first serially produced computer, the Strela series, started appearing in 1953, and most of the machine time went to missile defence. The Strela's design team was also tasked with creating the heart of the first antiballistic missile defence of Moscow, the specialized M-40 and M-50 in 1958–1959.[100]

The Soviets were not alone. The GDR's tech prowess produced the first non-Soviet machine in 1955. The Oprema was created by the Carl Zeiss firm and became a testing ground for the ZRA-1 in 1958, a much more advanced machine. Dresden Technical University developed its own machine in parallel, the D-1, in 1956.[101] In Czechoslovakia, the pioneering Antonin Svoboda had already designed his first machine in 1951 but could only complete the project in 1957, when his SAPO started calculating in the Academy of Sciences. It burned down in 1959, but by then Svoboda's team had produced the EPOS-1 and EPOS-2.[102] In Romania, Victor Toma created the first digital computer in the form of the CIFA-1, at the Institute of Atomic Physics.[103] It was his visit to Sofia that inspired Antonov to turn toward digital machines. Next door, the Hungarians built the MESZ-1, an experimental machine, in Budapest's Technical University.[104] The Polish ODRA series was probably the most innovative, starting in 1959 and inspired and compatible with the famous British ICL series.[105] Only two socialist countries in Europe had no domestic machine by 1960—Albania and Bulgaria.

Yet these machines were almost unique—none were ever placed in serial production. Even the Strela existed in a run of only seven machines, while the BESM would enter true serial production 10 years after the first machine. The early ODRAs were single models, for training purposes; the SAPO and CIFA were literally unique. Even the GDR managed to produce only 32 ZRA-1s, over a number of years.[106] Soviet machines were used for military, nuclear, and space-race calculations; those for satellites were products of scientific visionaries who waited years for institutional backing. Almost all of these machines emerged from the national academies or universities, as not even the GDR or the USSR had the necessary infrastructure and organization to mass produce them. These were artisanal rather than industrial computers, unable to solve the grand tasks Khrushchev had set them rhetorically by 1962. The Cold War tensions had locked them behind blast-proof doors, with military-specific software, a far cry from being applicable to civilian life. Comecon had a market gap, and specialization contracts were looming.

In Bulgaria, there was a party scrambling for a golden export, an emerging cadre of bright engineers, and a patron in the figure of Ivan Popov. Starting far behind its allies-cum-competitors, the country could now take advantage of one of the boons of economic backwardness: borrowing and adapting the latest technology developed elsewhere, under the auspices of a centrally led, capital investment process.[107] Electronics demanded little in the way of nonexistent natural resources, was high-profit, and existed nowhere in the East in a serial fashion. Whispering in Zhivkov's ear, Popov could drive this point home to the Politburo, paving the way for the creation of a true computer industry that could reap the enormous benefits that the BCP wanted.

2

THE CAPTIVE MARKET: THE RISE AND APOGEE OF AN INDUSTRY

The first Bulgarian computer spelled out two phrases—"Bulgaria builds socialism!" and "Only Levski!"[1]—when it was turned on in July 1963 at a Moscow exhibit dedicated to the country's technological prowess.[2] Within 20 years of these small steps, the country would be responsible for 45 percent of the Eastern Bloc's electronic trade,[3] producing the full gamut of computing technology: large mainframes, PCs, hard discs and magnetic tapes, industrial robots, tele-processing systems, and automated systems for manufacturing and administrative uses. It would employ thousands of scientists and engineers and hundreds of thousands of factory workers. From inauspicious beginnings, it would leapfrog its Comecon allies to capture a disproportionate amount of the market. In a feedback loop the cyberneticists would have been proud of, the more the country sold, the more the technology permeated political and social thinking, demanding even more production, more sales, and more implementation. But how the Balkan state rose so meteorically also illuminates the Second World's internal workings as well as the state-led nature of computing's development on both sides of the Iron Curtain.

This leap only makes sense if we conceive of the Eastern Bloc as a real attempt at an alternative and unified modernity, rather than as a purely imperial-ideological Moscow venture.[4] The industry needs to be approached as both a project and a failure, for as Mëhilli argues, it created a shared material and mental culture without political unity.[5] For this industry to flourish, the homogenizing ideology of socialist fraternity, allied with the economic need to outpace capitalism, created the market that Bulgarian industry under Popov could capture. The failure to unify all for a common goal also meant that both state and substate actors could use the vaunted ideas to gain positions in the world. The Bulgarian state backed a winning card and supported it to the hilt, garnering vast profits and prestige by utilizing the language of fraternity, cooperation, and alliance but also planning, quotas, trade agreements, and promises. The small state wagged the tail of the Soviet dog. This success story is thus intimately tied to seeing Comecon as a real attempt at integration, with logic that allowed the satellite states to defend their own interests.

Zhivkov was a venture capitalist, owning an entire state's resources and willing to bet it on ten different ideas as long as one came through. He was quoted by a professor who had been a member of the Central Committee as stating that "For a country like Bulgaria 10 million levs is nothing. But even if one of these ten crazy ideas to become successful, to create production on a world level with which we can win new markets—that's enough."[6] Popov delivered just such an idea. The sector became both an engine of ideas and of billions in profit due to the logic of Comecon that was intimately tied to fighting the Cold War, a striking illustration of the embeddedness of economics in politics. At the same time, the attempt at socialist integration that followed the bloc's ideas of divisions of labor, and their failure, show that the logic of economic benefit and market capture operated differently, depending on whether you were in Sofia or Moscow. The Cold War shouldn't be the only prism through which to view this, even though it is what made possible the dominance of Bulgarian electronics in the socialist world. Political unity was the not the order of the day, on either side—the very story of Japan's role in Bulgaria's rise is proof that the capitalist world also often put economics above politics.

This chapter picks up where the last one ended and traces the apogee of Bulgaria's "golden factories" under a concerted state-led investment policy. It shows how Comecon created the conditions of a unified material culture from the Inner German Border to the Pacific, as well as how it was possible for small states to carve out and defend their own interests. Bulgarian electronics thus operated under the sign of the "market" in a market with very specific rules that could be used to the country's benefit. At the same time, under this geopolitical aegis grew a thriving scientific-technological community, with the chapter showing how scientific teams were formed and flourished in the last two decades of socialist rule. By situating this industry and its technical intelligentsia within the alternative modernity context, the chapter also shows the tensions and promises inherent in forming the socialist bloc. Following the growth and development of the technology, we see how important a concerted state policy was, how key the capture of just a few technological niches was to the profitability of this industry, and the ways it broke through the Iron Curtain from its very inception.

THE FIRST BULGARIAN COMPUTER

Starting the 1960s with no domestic computer, the emerging champions of the discipline in Bulgaria that we saw in chapter 1 ran up against the older generation's conservativism. Professor Tagmatliĭski of the Mathematics Institute at the Bulgarian Academy of Sciences (BAS) asked Sendov to calculate 2+2 on an analog MH-7 machine when he heard that the newly founded Computing Center was given the task of designing Bulgaria's first digital computer in June 1961. The machine came back with 3.95 as it was not used for adding sums, prompting the professor to joke that the new machine will get to 3.98 at a much higher financial cost.[7] Li͡ubomir Iliev, however, wouldn't budge, accelerating the development process. The year 1962 saw multiple seminars on Boolean algebra and programming as well as the division of tasks—Kiril Boi͡anov, who recalls the anecdote above, was responsible for the power bloc. Teams visited Romania and the GDR to learn from Toma's CIFA-1 and Dresden's D-1. The creation of a digital machine was made number one on the list of technological goals for the years 1962–1963, an important part of the state plan for technological progress under the leadership of the Committee for Scientific and Technical Progresss (CSTP). The institute wanted to finish it by the party's eighth Congress, which was held in November 1962, but the hardware was completed in early 1963.

The newborn machine consisted of radio lamps with an expected life of about 10,000 hours each, and it was formidable in size—4 by 2 meters.

2.1 The Vitosha. (*Source*: Sandacite.bg.)

The goliath needed a special ventilating and cooling system, built from scratch, and it sported a command panel and an electric typewriter as an output device. It required around 70 square meters of specially ventilated space.[8] It also needed a name, and the scientists settled on "Vitosha," the name of the mountain that could be seen from the institute's windows.

The creation of such a machine was no easy task for a team that had no practical experience in the field apart from theory and observations of foreign examples, and they often proceeded through trial and error. Some of the elements had to be created from scratch, as there was no ready availability of components in Comecon. The precision needed to create the magnetic memory lathes was available only in a military factory in Sopot. The outer layer of the disc required manufacture in Dubna, USSR. Boi͡anov also recalls mishaps—a *kollergang* was ordered from the USSR: an industrial edge mill to create the ferrites needed for the memory devices. He had never ordered such a thing, so he had ended up ordering one suited for factory rather than scientific work, leading to a scramble, as there was little financing or time for a new mill. Fortunately, a factory in Stara Zagora needed such a machine and accepted it, allowing a laboratory-grade one to be purchased from Hungary.[9]

Ivan Popov understandably took personal interest and visited the institute in early 1963, urging the team to not miss a more important deadline than the Eighth Congress—the autumn exhibit "Bulgaria Builds Socialism!" in Moscow. The team decided that the best bet was to go straight to Moscow and finish the montage on the pavilion grounds rather than setting up in Sofia and then dismantling the giant. Problems at customs had to be overcome, as well as the lack of enough space in the Moscow exhibit. The latter was often solved by Ilich I͡Ulzari and his cache of "Sunny Beach" brandy, a tried and tested method of Bulgarians visiting the USSR. Another problem appeared, however, as the electricity cable for the pavilion was not sufficient to feed the hungry computer. A Soviet engineer, bribed by yet more brandy, solved the problem by leaving the Indian exhibition space next door without power in the interests of fraternal friendship.[10] By late July, fueled by Bulgarian engineering and alcohol, the machine was operational.

The exhibit at Sokolniki Park was visited by both Khrushchev and, importantly, the man who would oust him a year later: Leonid Brezhnev. Zhivkov was there, too, extolling the Bulgarian Communist Party's (BCP) achievements in technology. All were duly impressed, including local administrators who listened in awe to the team's exaggerations of the machine's capabilities and ordered five of them after the sales pitch. This was of course impossible, as the Vitosha was in essence a demonstration. After the exhibit, it was crated back to Bulgaria, calibrated again, and tinkered with by the team. However, in a story common to many early computers, it was disabled in late 1964 by a small flood in the operations room.[11]

The deluge didn't matter as the aim had been achieved—a team of Bulgarian scientists had created a computer in an environment of embargoes and fairly scant information. Mistakes had been made; unanticipated problems overcome; and in solving them, invaluable experience had been built up. Importantly, the Moscow unveiling had put the country on the map in terms of computing and had showed that the previously agrarian country could now produce the tools of the new age domestically. The Soviets did seem sufficiently impressed, acknowledging that Bulgarian computing now existed, and after negotiations delivered a MINSK-2 for the needs of the Mathematics Institute, as of course the Vitosha was only a demonstration model and was dead by 1964.

The first Bulgarian computer was a testing ground, a school, and a statement. It proved to both Comecon and the Bulgarian Politburo that the country had the scientific capacity to see through a highly complicated technological project. Popov had insisted on Moscow as the demonstration with both Zhivkov and the Soviets in mind—the Bulgarian leader saw the machine for the first time alongside his CPSU backers. While in technological terms, the Vitosha gave Bulgarian scientists the recognition of their Soviet partners and secured them a more advanced machine with which to continue their work, in political terms, it was the first step to securing the BCP's support for organizing a new sector of the economy.

CALCULATING SUCCESS

As the MINSK-2 was being installed, Ivan Popov was entering his second year as the head of the CSTP, a body that was increasing in importance as both the CPSU and BCP adopted a political language that trumpeted science. The term "scientific-technical revolution" would become an all-encompassing phrase in the following years, and the CSTP's strength only grew alongside it as economic growth demanded more and more technological management and improvement. This revolution would grow in importance during the years, but its seeds were already there with Khrushchev's proclamations, the rehabilitation of such fields as cybernetics and genetics, and the emergence of machines like the Vitosha. These things were clear to Popov, but not so much to the Politburo members, who lacked the education to fully grasp the winds of change. Together with Academician Li͡ubomir Krŭstanov and the head of the Committee on Machine-Building, Mariĭ Ivanov, Popov thus compiled a report titled "On the Development of Computer Technology." On June 24, 1964, the political elite was first lectured to about the latest developments in the field, its benefits to society, and its possible application in the economy.

This report hammered on what Popov already had noted to Zhivkov: The sector required few raw materials or rare metals and could use women's labor, which was just beginning to be harnessed. Importantly, it is a sector that would return its investments quickly and many times over.[12] It taught the Politburo the difference between digital, analog, and hybrid machines, accentuating that the first were the future. Importantly it planted a seed that would grow into a veritable obsession for the party as it stated that the machines were "the heart of automation" in production and administration, something that would only grow beyond the use of statistical calculation that the party grasped.[13] It also pointed out that states such as the USSR, Poland, and the GDR had well-developed research in the area but were still far from production.

The Politburo might not have understood fully the difference between types of discs or processors, but it grasped intuitively the importance of money. Two weeks later, on July 7, it made its first decision on the development of Bulgarian electronics and computing. Recognizing Popov's lead, all powers were vested in the CSTP, which was to head a Coordinating Council on the question. It would coordinate all fundamental research across the universities and BAS, would make decisions on cadre training numbers and distribution as well as on financial questions, with the aim of using the least amount of resources in the shortest amount of time.[14] The ultimate path was to be set out by this body in a plan on computer development up to 1970, which was to be approved by the Council of Ministers by the end of the year. A new body was also set up to organize the industrial production of computers and automation components, on the basis of a Unified Industrial Enterprise for Automatics. Fast cash was made available for the existing computer center's upgrade (200,000 levs in 1964, half of it in scarce Western currency). The CSTP was to coordinate with the Foreign Trade Ministry for the purchase of at least four foreign digital computers. The ministry was also to see what possibilities lay in utilizing existing UN programs, such as the Technical Aid Program and UNESCO initiatives in computing. In conjunction with the Ministry of Finance, CSTP was to initiate membership in the International Federation of Information Processing in 1965. It was also to invite foreign specialists to both lecture on the latest developments and advise on the organization of institutes in these new conditions.[15]

A particularly important international contributor throughout the socialist period was the UN Industrial Development Organization (UNIDO). Created in 1966, it favored industrial planning and ran technical aid programs in some Eastern European countries. Eight of the 33 assistance projects that UNIDO ran in Bulgaria from the 1960s to the 1980s concerned information technologies, automation, and electronics. These projects helped key institutes, such as the cybernetics institute, and the numerical control machine tools center. Project managers lived in Sofia and helped deliver instruments and English courses in such institutes, as well as providing fellowships in both Western and Eastern European countries. As Michel Christian shows, this was beneficial for both Bulgaria, which received Western expertise through these international programs, and for UNIDO itself, which sought to influence development and politics. By the latter part of the socialist period, experts from Bulgarian institutes ran UNIDO electronics development projects in places such as Cairo, showing how this was both a key source of knowledge and eventually, proof of growing technical confidence and competence. As Christian argues, for UNIDO, the line of "development" ran through Eastern Europe itself—with Bulgaria being in the less developed area; something that the Bulgarians shrewdly used to access all avenues in building up the electronic sector.[16]

The cadre question was solved by the creation of new university courses that started in late 1964, only two months after the decision was made. "Computer Machines" was added as a specialization in the "Semiconductors and Industrial Electronics" course in VMEI-Sofia. Concerted state policy in the field of science also created new technical universities outside the capital, such VMEI-Varna in 1963 and VMEI-Gabrovo in 1964. The first students in electronics at Varna were admitted in 1964,[17] the first at Gabrovo in 1967.[18] As higher education expanded to provide specialists throughout the country, so did school education. The technical high school "A. Popov" in Sofia created classes in the production and fundamentals of calculating machines to ensure a school-to-university pipeline in the coming years. The fine mechanics high school of Sofia created classes in mechanical devices used in calculating machines.[19] Over the next few years, other technical schools in the country also created such classes. This was the final step in the general offensive that the Politburo envisioned, alongside international cooperation, the setting up of industrial production, and technological upgrades. The creation of universities and courses ensured the long-term provision of specialists spread throughout the country.

By April 1965, the CSTP had also organized the first high-technology industrial organization in the country, the State Economic Union (DSO) of Instrument Building (and eventually automation, too): a conglomerate of institutes and enterprises that were to unite research, development, and production of industrial control, automation devices, and medical apparatuses.[20] Yet these were complex machines that still had some years to mature and be implemented, while Popov was looking for a cheaper, smaller, and less demanding machine than a computer to capture markets before other socialist competitors did. This was the electronic calculator.

The calculator was the perfect machine for automating administrative work and an item in high demand in both the socialist and capitalist world. It was also simpler than a Vitosha. Iliev's Maths Institute, a proven ground of innovation, was given this new task with a deadline of another Moscow exhibit—Inforga—in May 1965. The field was still new, as the first electronic calculator had appeared in the UK in 1961, and only the Italians and Americans had produced any other such devices. Li͡ubomir Antonov was part of the team assigned to this task and remembers being unimpressed with existing models, such as the British "Anita," as they were not user friendly. Together with Stefan Angelov and Zhivko Paskalev (later Petŭr Popov), all trained in Dubna, they decided to make their machine capable of new functions, such as squaring and finding roots, that were lacking in previous world models. Based on germanium transistors built in Bulgaria with a French license, Angelov spent most of the time developing better algorithms for quick multiplication, as the Brits had solved the problem of multiplication through multiple additions.[21] By the end of 1964, the machine was complete; it was named ELKA—a diminituve of a woman's name as well as the first two syllables of the term "electronic calculator."[22]

The ELKA-6521 was unveiled to the Politburo on April 17, 1965, and was shipped off to the Inforga two weeks later. Popov presented it to the Soviet delegation, assuring Gosplan representatives that this was only the first of many models.[23] Bulgaria was ready for serial production, he stated, and the Soviets should sign orders for the whole upcoming five-year plan. These bluffs were typical of Popov, and the Soviets were sceptical. Surely this was a demonstration model like the Vitosha? Yet serial production started in Sofia in 1966 and then was transfered to the Danube town of Silistra in 1967, where the Orgtehnika factory would remain the home of the Bulgarian calculator and organizational technology for the rest of the regime. Despite Soviet reticence, the GDR and Czechoslovak delegations signed orders for the calculator immediately.[24]

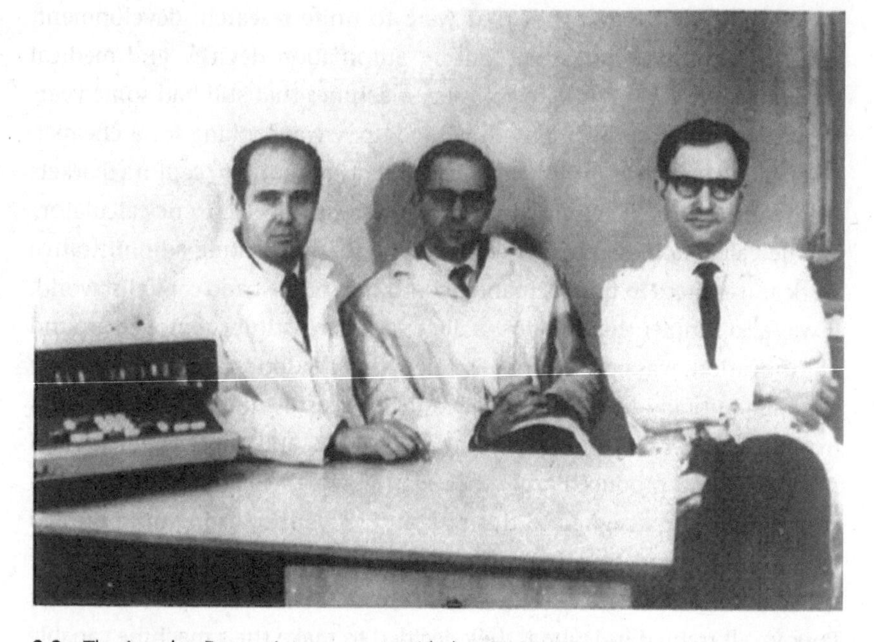

2.2 The Angelov-Popov-Antonov team with their ELKA-6521. (*Source*: Antonov memoir.)

Meanwhile, two new models—the ELKA-22 and ELKA-25—were developed, building on the first machine. These models were more streamlined and simpler to produce, with increased miniaturization, fewer cables, and a printing device—the first calculator in the world to have this novelty. Antonov had criss-crossed Europe to find a license for such a device, even getting caught up in a Vietnam War protest in Stockholm, before deciding it just wasn't to be found and creating one himself with the help of Ivan Stanchev.[25] Serial production commenced in 1967 and led to astounding success, including in the West.

The year the prototypes of the new ELKAs were complete, 1966, was a symbolic recognition by the state of the new industry. The Antonov team won the Dimitrov Prize, the highest medal for scientific and cultural achievement, the first in the sphere of electronics. At the same time, Popov ascended to the Politburo on the back of the success of putting the calculators into serial production and inaugurating a whole new industrial sector. While the Orgtehnika factory was still being completed at breakneck speed, Popov launched his assault on the Soviet market after the East German and Czech orders. He demanded high prices—1,200 roubles per calculator, starting a trend of high pricing in the Bulgarian electronics sector vis-à-vis the Soviets. Moscow countered that the GDR offered one for 730 (despite the device not yet being in serial production). Popov had started high in order to demonstrate power, but he was more than willing to undermine any competitor, and he agreed to a contract at the price of 700 roubles—access to tens of thousands of Soviet office desks would ensure profits at almost any price.

The year also saw the ELKA conquer the Hannover and Moscow fairs of informational technology, awing specialists with its unique printing device.[26] In Moscow, a thousand companies were represented, and Popov used all means possible to give the Bulgarians a 500-square meter space in the same pavilion as the US, France, and Sweden. The ELKA was the jewel of the exhibit, attracting interest from West Germans, who wanted production cooperation, as well as the Italian giant Olivetti. Importantly, Konstantin Rudnev, the Soviet Minister of Automatization, was convinced by the Bulgarians' promises and agreed to large imports in 1967. At the same time, he ensured that the Ministry of Electronics would deliver enough lamps to Silistra to offset the needs for Western imports.[27] The CSTP glowed:

> It must be said that our country passed the difficult test of comparison with some of the most advanced countries in the sphere of organisational technology and computer technics. The overall opinion of our Soviet comrades and of the representatives of numerous capitalist firms is that we have presented ourselves well.[28]

Silistra began serial production at the start of 1967, ensuring that the Bulgarians maintained their lead over Comecon allies, and the ELKAs impressed again that year at the Paris and Vienna exhibits.[29]

By 1968, serial production was achieving impressive results.[30] As the computer industry developed, the ELKA family continued to be the mainstay of electronics export in the early 1970s: The ELKA-22 alone accounted for 13 million out of the 22.7 million levs exported in 1970.[31] They were sold to Spain, Turkey, the UK, and Norway, one of the few Bulgarian machines that did well in the West.[32] In 1971, it was selling 20.5 million out of the nearly 56 million electronics exports, overtaken for the first time by another item.[33] In 1972, it was back to being the biggest exporter, with over 32,000 units sold for nearly 27 million levs (but now out of a much bigger overall export market).[34] As other electronic devices entered mass production in 1972, sales stabilized at around 20 million levs per

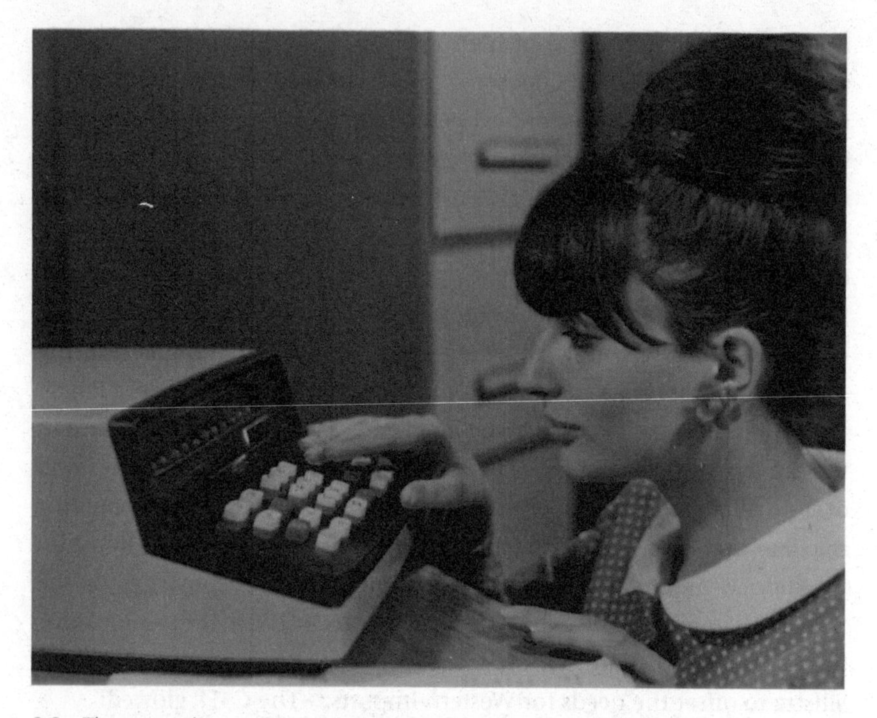

2.3 The game-changer ELKA-22. (*Source*: Central State Archive, Sofia.)

year,[35] but new models, including pocket models that became almost ubiquitous, continued being developed for production in the now fully functional "Orgtehnika" Silistra—from 1969 Li͡ubomir Antonov had been put in charge of a newly founded Institute for Electronic Calculators.[36]

The machine experienced the growing pains typical for an industry with no experience in the field. Popov's bluffs sometimes backfired as West German and Italian firms also placed orders of more than 10,000 units in 1966, learning of promises made to the Soviets—before serial production was capable of this level, and the USSR took priority.[37] The bigger problem was quality, especially for the more demanding Western customer. Snezhana Khristova recalls being sent to France almost immediately after the first exports there in 1967, spending two months fixing faulty ELKAs. She still had to send half of the devices back, and she blamed the poor conditions on the factory floor in Silistra, especially in the initial run.[38] This failure angered Popov, who arranged for a French engineer (who was also a member of the French Communist Party) to

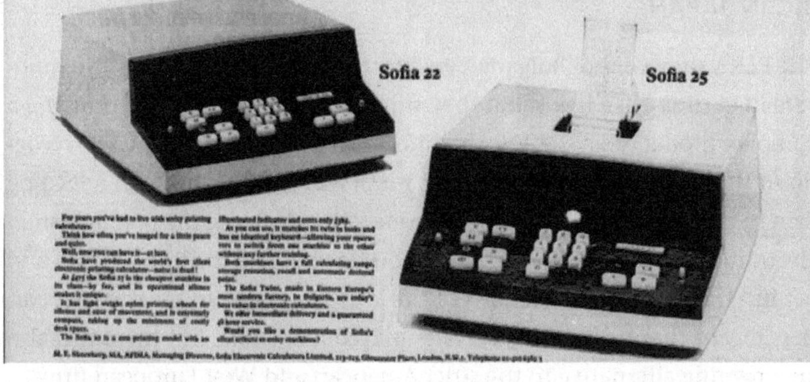

2.4 Advertisement for the ELKAs—styled as "Sofia"—in the *Financial Times*, 1968. (*Source: Financial Times* Archive.)

visit the factory and evaluate it. The engineer pinpointed problems that would vex the Bulgarian electronics industry throughout this period—poor worker training, including especially not wearing gloves. Sweaty palms resulted in perspiration and body oils getting into the solder joints, acting as a corrosive over time, no matter how airtight the machine was. Popov implemented the changes, yet faults remained high, and surprise inspections showed that two workers were still shirking the regulations. The subsequent dismissal of the two women earned Popov criticism for draconian measures, but he was adamant. There simply was no point in implementing high technology or Western licenses if Bulgarian working habits destroyed the good work from the outset.[39]

Regardless, the ELKAs were a huge success. They were the first mass-produced commodity of the electronics industry, the first to win Comecon and also Western market shares, and a sign of growing technological sophistication. Factories, institutes, and courses were created as part of its inception, and its successes were trumpeted and probably contributed to making electronics a sought-after specialization for young people. It solidified Popov's image as a strategist of the economy and science, earning him a place on the Politburo and thus immense power over the country's future development. But that was also due to his other "greatest hit," concurrent with the ELKA, which paved the way to true computer production.

BIG IN JAPAN

The ELKA showed that Bulgarian engineers could hold their own and innovate in cutting-edge fields, but they still had little experience of bringing a complex product like the Vitosha into serial production. The CSTP recognized this and knew that the fastest route was partnership with a leading firm in the sector. While the Soviets were key in training the first Bulgarian electronics experts, the know-how needed lay beyond the Iron Curtain and behind the embargo restrictions of CoCom. This need to break down barriers meant that the security services played a key role, but there was also an emerging alternative to the strict American and West European firms.

Japan's postwar economy is one of the miracle stories of the twentieth century, yet it was also an outlier in that its business stance toward the socialist bloc was not parallel to its geopolitical stance as a lynchpin of Cold War anticommunism in Asia. The alliance with the US masked certain anti-US sentiments, especially in the desire to escape the economic straitjacket of its ally's restrictions. While not recognizing Communist China, feuding with the USSR over the Kurile Islands, or being the unsinkable aircraft carrier of the Pacific, Japan developed a more nuanced trade policy. In the 1950s, for example, four private-sector agreements were signed between Japanese firms and the People's Republic of China.[40] Washington feared such moves were a precursor to political disloyalty and undermined Japanese desires for export markets in areas such as Southeast Asia. This, together with the access given to Japanese industry to the capitalist world, would lead to the deluge of exports to the West,[41] but the socialist world was also a lucrative, huge market. In 1956, Moscow and Tokyo officially ended the war, paving Japan's entrance to the UN and a different approach to Eastern Europe.

The new Japanese prime minister, Hayato Ikeda, turned away from foreign-policy issues toward bread-and-butter economic concerns after 1960, staking public support on an economic policy with the aim of doubling income within a decade.[42] This was part of his concept of dividing politics from economics, aimed at serving Japanese economic interests wherever they could be maximized.[43] Trade with the USSR boomed, reaching $281 million by 1962, and Japan would become the biggest Soviet trade partner among the capitalist states.[44]

The Japanese approach benefited Bulgaria, too. Meetings at the Plovdiv Trade Fair, and in China and India, sounded out the possibilities for reopening relations. And in 1959, Japan and Bulgaria officially renewed diplomatic contact.[45] A trade agreement was signed two years later, and Bulgarians now faced the problem of the quality of their exports. The embassy in Tokyo was also given the task of concentrating on what Japan could help with in terms of technical aid and whether they were willing to set up joint enterprises. In 1963, Bulgaria participated in the Tokyo International Trade Fair for the first time, showing off chemicals, tobacco, oils, pharmaceuticals, canned food products, and a few machines.[46] Some of these were sufficiently good to increase trade to $45 million over the decade, with Bulgarians importing ships and knowhow. Japan thus solidified as a potential capitalist partner in machine-building.

When the CSTP started searching for a partner in creating industrial-scale computer production, it logically turned to the embassy in Japan. The Bulgarian ambassador, Khristo Zdrachev, directed the efforts toward Fujitsu, as that company had proved willing to work with socialist states.[47] Its director, Tsenusuke Vada, had already visited Bulgaria in 1963 and met Popov, and he was interested in its efforts in the sphere. Thus in early 1965, Fujitsu representatives arrived in Sofia with catalogs and technical documentation, allowing the Bulgarian engineers to choose the machine best suited to them—the FACOM 230–30 digital computer, with its full range of peripherals and memory devices. In September, Popov returned Vada's visit, bringing with him Dimo Dimov, who was the future head of Izotimpex, the industry's export arm. The negotiations proceeded smoothly, and Bulgaria agreed to buy one or two machines, complete with programs, and peripherals for 20 more machines for $4 million. Fujitsu would deliver other parts, for them to be assembled in Bulgaria, and importantly, the Japanese company would train some specialists in its own factories.[48]

The first group of specialists included Stefan Angelov of ELKA fame and Blagovest Sendov, and they were in Japan by the end of 1965, studying both hardware and software. In 1966, the agreement was ratified by both governments, giving the CSTP the go-ahead to organize the industrial sites for FACOM's production. Originally tasked to the Committee of Machine Building, its lack of experience in such fine mechanics gave Popov the chance to transfer the production to the aegis of his own organization, creating the Central Institute for Computer Technology—CICT (the story will be picked up later in this chapter) and the ZIT factory (Factory for Computer Technology), under the watchful eye of engineer Ivan Marangozov.[49] Another group of specialists left for Fujitsu's factories in Kawasaki in autumn 1966 and stayed for a full year—the first CICT group trained in such advanced factories for so long, and this group managed to fully implement the technology in ZIT.[50] Their breakthrough was to work out how to create the machines using diodes and elements available in the Eastern Bloc—from the USSR, the GDR, and the new Botevgrad factory in Bulgaria itself. The ZIT-151 was born, the first mass-produced Bulgarian digital computer and a functional copy of the Japanese machine.

CICT specialists also trained in Fortran and Cobol, key programming languages, as well as in all aspects of computer architecture.[51] Dimov, who spent most of his time between 1966 and 1970 in Japan, recalls his impressions of Japanese organization. What struck him was a meritocracy based on education and results, rather than political contacts, as well as a slow and methodical decisionmaking process that resulted in no deviations once a plan was agreed on. He contrasted this to Bulgarian practices of quick decisions based on over-fulfilling quotas and the aim of "overtaking" a competitor, backed up by sloppy organization and habits.[52] Nevertheless, work progressed quickly, and in 1967, Dimov also started transferring know-how in magnetic discs. Four engineers from CICT spent months in the Nagano factory, mastering the arts of fine mechanics needed in this delicate production, the item that would catapult Bulgaria to its preeminent position in Comecon.[53] Eventually, by the 1970s, students regularly specialized in automation in Japanese universities, leading to curious moments, such as Toma Chuparov's stay between 1971 and 1976 while studying at Tokai University, when he lived with the family of Yasuyoshi Tokuma, the publishing magnate who owned the famous Studio Ghibli.[54]

The first ZIT-151 rolled out in 1969, by which time the Japanese connection had transformed the capabilities of Bulgarian electronics. Six years earlier, enthusiasm and fundamental work had carried the Mathematics

Institute to the completion of the Vitosha; now engineers had a dedicated institute and a purpose-built factory. But most importantly, they didn't have to work their way through the dark, scavenging ideas from disparate catalogs and textbooks. The Japanese giant was a mine of information on everything from general architecture to magnetic disc mechanics and the source code for the latest software. Tokyo was the gap in CoCom's armor that gave Bulgaria its chance to fight for its corner in the Comecon specializations that beckoned.

But Japan was more than just a site of technological innovation—it was the future. The country captured the imagination of Bulgarian elites. Popov and Dimov saw the potential for cooperation, but other key figures in Bulgarian economic policy got their start there, too. Nacho Papazov, ambassador to Japan between 1967 and 1971, succeeded Popov to head the CSTP until 1984, jumping into the position straight from Tokyo. He would go on to write a book on the Japanese economic miracle. Ogni͡an Doĭnov, the man who would displace Popov from the heights of Bulgarian economic power and management by the mid-1970s, also started his rise as a deputy trade attaché during Papazov's tenure in Tokyo. In his memoirs, he credits Japan as the place that gave him an in-depth knowledge of Western management techniques.[55]

The most important Japanophile convert, or "weeaboo" in layman terms, was Zhivkov himself. Becoming a "mini-Japan" is a slogan that the popular imagination associated with the period and with Zhivkov himself, conventionally dating it to his visit to the country during the EXPO'70 world fair.[56] As we have seen, this was not the start of Bulgaria's electronics mania, but it was the start of the country's Japan-mania. In practical terms, this was the beginning of Doĭnov's meteoric political rise. He was born in 1935, and after graduating as a heating systems engineer, he worked at the Ministry of Transport beginning in the late 1950s. His break came when the minsitry sent him as a deputy trade representative to Tokyo in 1965. Zhivkov plucked him out during his EXPO'70 visit and advanced him rapidly through the ranks: State Council advisor, head of BCP's Industrial Department, Central Committee secretary, Politburo member from 1977, and Minister of Machine-Building in 1981. Ironically, the man who would displace Popov started his political career the same way the father of Bulgarian electronics did when he was plucked out of the GDR.

The party boss himself was deeply impressed by the country. Despite being unofficial, his May 1970 visit was the first by a socialist state leader and was heavily covered by the media, keen to show off their own miracle to a high-ranking Eastern Bloc politician.

Zhivkov came with preconceived notions about capitalism, but he was quickly disabused by what he saw in Osaka. He admitted that what struck him was not meeting the emperor but the realization of "our deadly lag behind the advanced capitalist countries. And not only that. For the first time I felt oppressed by the shortcomings of the socialist system."[57] The visit "crushed" him, especially when he visited factories or met local captains of industry. The tempos were much higher that what he had seen

2.5 The Bulgarian pavilion at EXPO'70. (*Source*: https://socbg.com.)

elsewhere in the capitalist world. He recalls the "forest of chimneys" that to him spoke of a miracle, despite the pollution they caused.[58]

On his way back, he stopped in Khabarovsk and penned a letter to Brezhnev, waxing lyrical on Japan's achievement. In Moscow, Brezhnev harangued him over being blinded by their trickery, not seeing the workers' exploitation inherent there, but Zhivkov was unfazed—the industrial strength of Japan was obvious. "Up until Japan my impression that we were lagging was based on my political experience and of my knowledge of the socialist states," he recalls—and after further visits to West Germany and Italy, what had happened to him in Japan convinced him of the need to concentrate on high technology, exports, and contacts with the West.[59]

Found in a memoir, one could ask whether this was Zhivkov's true position at the time or a post-1989 justification of his own far-sightedness, but his report to the Politburo after his visit supports his claim. The Bulgarian economy was heavily criticized through the Japanese prism, contrasting the need for intensification and modernization with the achievements he saw in Japan. Bulgaria could not rely on its own powers to make up the lag and had to choose the best from the West, while enterprises must become better at actually implementing the technology, so they don't end up as "rags."[60] Production quality had to improve to compete in the world market, which required knowledge of the latest trends. Outdated documentation, sloppy work, and bad design all caused low productivity, but at the root of the problem lay the lack of information and contacts with the world.[61] Foreign innovations were more important than Bulgarian ones, and a further 50 million levs would be made available for licenses. The licenses were not to be "Bulgarianized" until full production was implemented—only then could tinkering be allowed.[62] This was a stark illustration of the regime's struggles to continue development once it had reached rough parity with the West—Bulgarian factories were now producing all the goods of the Second Industrial Revolution, having been able to build on decades of achievement and choosing the proven technologies. They were now in the middle of an unfolding information revolution: Choosing the best available technology was now riskier, as many of the innovations were still being implemented in the West, unlike previously, when technologies tried and tested for decades could safely be chosen. It was clear that Bulgaria was a small state and had to rely on foreign achievements to get ahead.[63] The urgency of the Japanese example gave new impetus to the intensification debates but was grafted onto an industry that was already growing before the trip.

FROM PIECEMEAL TO POLICY

Both the ELKA and the Fujitsu know-how took off around 1965, when there was no umbrella organization to put into practice Popov's vision for a high-technology sector. The problem of organization was just as pressing as the intellectual obstacles. The first step was to create sites that could carry out the R&D of hardware and software that a computer industry needed. The Mathematics Institute was simply not suited to the task, despite its early importance for the sector. Other BAS institutes that were created to further the ideas of cybernetic governance, such as the Institute of Technical Cybernetics, created in 1964 on the basis of an earlier section on tele-mechanics, were aimed at industrial automation and eventually robotics. The ITCR would become a key intellectual center for future developments, including the personal computer, but it was not solely focused on computers, and it was also housed in the academy.[64] Popov's successes convinced the Politburo that a dedicated institute was needed, and in 1965, the Central Institute for Computer Technology was created (CICT; *T͡Sentralen Institut po Izchislitelna Tehnika*). The Mathematics Institute group became the core of this new institute,[65] and it was inaugurated on March 1, 1966, starting with 233 scientific workers under director Boris Borovski, with I͡Ulzari and Dimitŭr Atanasov as deputies.[66]

CICT had purview over both hardware and software, with its first annual plan focused on algorithmic languages, economic information databases, and programming automation, as well as the practical tasks of implementing ELKAs into wide production. A team also worked on a prognosis of worldwide computer development up to 1980 to serve as a guideline for future research.[67] Personnel grew quickly, with 250 more staff added in 1969 and 280 in 1970, as Comecon specializations were won.[68] CICT grew to become the spine of the industry, and by 1972, it would become the biggest scientific institute in the country. By the 1980s, it employed over 3,000 people, most of them highly qualified specialists and an unprecedented concentration of engineering talent for Bulgaria's size.[69] Angel Angelov's directorship between 1968 and 1975 saw most of the expansion, before he moved on to ITCR, where he would oversee that institute's growth. His tenure at CICT coincided with the creation of the ES series of computers in Comecon, which called for the organization of multiple sections in the institute: magnetic tapes, magnetic discs, controllers, interfaces, programming, terminals, design work automation, and other areas of expertise. Boîanov recalls the key intervention that Angelov made in the development focus of CICT, which would ensure the maximum profits this industry would bring the country:

for long nights, engineer Angelov used to gather us and five or six of us would discuss it over a vodka (it was cheap back then—1.73 levs) or over coffee, what was most productive to specialise in. After those discussions we took a decision that turned out to be strategic. In a computer system at that time, the biggest number of devices were discs and tapes, and as all central processors, from the slowest to the fastest, use the same peripherals, it was decided that if Bulgaria specialises in discs and tapes[,] we would export a lot and to all countries.[70]

Angelov himself recalls the story a bit differently, giving Ivan Popov his due as well as another expert who helped him calculate the future success:

There was actually fierce competition [over] who could make what in that sphere. Bulgaria was successful because we had made a calculation of what costs how much. We, from Bulgaria, specifically calculated the cost of the components of the entire enterprise of a "calculation center": the central processor and the peripheral devices. According to our calculations, the central processor, in its smallest version, would cost about $200,000 (it could do what a single PC today could do for less than a $1,000). We said that up to 2 million dollars were the costs of the peripheral devices, which was exactly the case. From these peripheral devices, there were two kinds: 1) memory devices on tape or 2) memory devices on magnetic discs. We showed these two kinds comprised over 80% of the cost of a "calculation centre." Three people were responsible, a triumvirate: 1) Prof. Ivan Popov 2) Mikhail Krinkov (who was an independent, not a party member, very intelligent and competent specialist on matters of automation), and 3) myself.[71]

The ultimately lucrative decision to focus on memory devices was the key to Bulgaria's huge market share throughout the coming years, and it seems that just three men made the country billions—but of course, they only showed the way forward, over vodka or not.

Popov also organized the information infrastructure to help CICT, as 1966 also saw the creation of Central Institute for Scientific-Technical

Information and the Central Scientific-Technical Library, giving engineers access to the latest periodicals and publications in all technical fields. Thus by 1966, the country had the research framework for its industry, but this was not yet matched by the production sites: old barracks were housing the first part of the Silistra factory, while ZIT was still not fully functional. A purpose-created organization to develop the sector was still lacking.

The new Politburo member made it clear to his colleagues that they were in a whole new world and that this was a strategic, economy-defining sector. A dedicated organization would carry out two tasks—creating and designing the means of cybernetic governance and implementing them into production. If both were under the same umbrella—one that could see the computer through from cradle to grave—the tempo of growth would be quicker, as there would be no friction between factories under the auspices of CSTP, the machine-building ministry, BAS institutes with production bases, and so forth.[72] Based on this report, a DSO was created in "Computational and Organizational Technology" (IZOT in the Bulgarian abbreviation, from *Izchislitelna I Orgaziit͡sonna Tekhnika*), which was responsible for the research, design, production, implementation, servicing, and sale of this technology—both in Bulgaria and abroad.[73] On February 17, 1967, Bulgaria had established its huge industrial electronics conglomeration, the heart of the industry until the regime's end.

IZOT united scientific and production sites that operated at the time—the CICT, the Central Institute for Elements, Central Design Institute "Orgproekt," Central Experimental Base at Gabrovo, the Base for Technical Development of Organizational Technology at the Silistra factory, the Orgtehnika factory for ELKAs, the Training Centre for Qualification in Computing Technology, the ZIT factory in Sofia, the typewriter factory in Plovdiv, and the industrial enterprise "Office Equipment" in the capital. This was not enough. Popov made it clear that the country had to create at least three new factories in memory devices by 1975 to ensure exports worth 469 million levs by that year.[74] These factories would have to be created without Soviet aid, and quickly, if Bulgaria was to become a Comecon leader. A fund for computer technology development was created at IZOT, financed by the enterprises's own profits. This fund financed new production and the modernization of existing factories, which had to upgrade quickly due to new world developments. Money would also be used to train new cadres, including sending many individuals on extended trips to the West.[75] The capital investments up to 1975 were set at 192 million levs.[76]

Significantly, IZOT was to be self-financed—its profits were to be invested back into the sector, to develop it and keep it up to date. Its anticipated profits were not, at least in theory, to prop up less-productive sectors of the economy. This took a lot of effort on Popov's part, fighting against a conservative Politburo. Most members were in power due to participation in the wartime resistance rather than to any particular expertise, and many were not as willing to listen to experts as Zhivkov was. To overcome this problem, Popov put them through the wringer, organizing a two-day seminar in February 1968, a year after IZOT was founded but before it had full financial backing. Attendance was mandotaroy for all members and candidates of the Politburo, leaders of the Central Committee departments, the Secretariat, high-ranking members of the Bulgarian Agrarian National Union,[77] and the first secretaries of all district committees and people's councils. The highest echelons of the BCP were thus put back in the role of students. The lecture plan was arduous: three hours on the latest Western methods of organizing scientific research, presented by Popov himself, and two hours on the latest computer developments; this assault was to ensure they understood how the "rest" of the sectors financed profitable sectors. The second day familiarized the BCP with the implementation of computers into society and economis, including a two-hour lecture on creating a national system for economic data processing. All was capped by Popov, who compared the developments of technology abroad and at home.[78] The content and structure of these two days made the aim clear: The BCP was to understand that if it didn't back Popov's idea, it would be left in the dust of a whole new industrial revolution.

The offensive continued in 1969, as Comecon specializations kicked in. Modern computing, Popov reported, needed a modern base for integrated microelectronics[79] that lowered costs, lead to miniaturization, more applicability, and the future automation of all aspects of life.[80] He enlisted the head of the State Planning Commission (SPC) to underline the need for microchips, which were replacing transistors as the basis for all electronics in the future. Fatigued from all the technical language, the Politburo nevertheless acted quickly and in October created another DSO—Electronic Elements—based in Botevgrad, where a semiconductor factory had been operating since 1963 and had been producing silicon transistors since 1967.[81] It now expanded to duplicate IZOT's role, but in the elements base, it obtained its own research institutes and long-term development plans. The production of electronics had already grown by a staggering 48 times since 1965, when the first ELKAs were built. And by 1975, it was hoped the DSOs could produce a fifth of all Bulgarian machine exports.[82] In fact, by the end of the regime in 1988–1989, electronics and related long-range communication equipment made up 25 percent of all industrial production in the country.[83]

A week later, the Politburo voted on the massive expansion of IZOT to fulfil Comecon needs as well as to stay in tune with the September 1969 theses on the implementation of scientific-technological progress. The effect was a veritable take-off. By 1970, seven new factories were created under its umbrella, and very quickly, more were built for the elements base. The factories fell into three distinct echelons. The first echelon consisted of factories that created the elements that went into the finished products, such as the Shumen Factory for Instrumental and Non-Standard Equipment (ZIENO), the Ruse Factory for Printed Boards (ZPP), the Blagoevgrad Factory for Mechanical Constructions (ZMK), the Factory for Magnetic Heads at Razlog (ZMG), the Gorna Oryahovitsa Factory for Magnetic Powder (ZMP), the Harmanli Factory for Power Devices (ZTU), and the Mekhatronika factory in Gabrovo. A second echelon created more complex machines, such as peripherals that could be sold in their own right—the Stara Zagora Factory for Peripheral Devices (ZPU); the Plovdiv Factory for Memory Devices (ZMD-Plovdiv); the Pazardzik Factory for Magnetic Packets (ZMD-Pazardzik); the Registration Technology Factory at Samokov (ZRT); and the Plovdiv Typewriter Factory (ZPM). The third echelon consisted of the computers themselves—the ZIT factory in Sofia, tasked with manufacturing the ES large mainframes; the Elektronika factory, also in the capital, which concentrated on SM; the Orgtehnika in Silistra, which created the ELKAs; the ZZU Factory in Veliko Tŭrnovo, which took care of modems, network devices, and teleprocessing systems; and eventually, in the 1980s, the Microprocessor Combine in Pravetz, Todor Zhivkov's home village-turned-town.[84] To this list must be added key factories outside IZOT but closely tied to its functions, especially the semiconductor plant in

Botevgrad that created transistors and then integrated circuits, the flagman of DSO Electronic Elements. Later, as microcomputers entered production, factories such as Analitik in Mihailovgrad (which built monitors) or the Instrument Building Plant in Petrich (which created matrix printers) also became key auxiliaries to the system. IZOT not only created a whole host of new factories but also became the center of a galaxy of enterprises in related sectors that supplied it and were part of the electronics industry.

The territorial distribution, such as building factories in backwaters like the towns of Razlog or Harmanli, was a deliberate decision to integrate high technology into a wider territorial development plan that would in theory lessen the urban-rural divide and move jobs out of Sofia, thus easing the capital's housing shortage.[85] It also necessitated better logistics and production coordination, as a hold-up in one factory, especially in the elements base, could hold up all the others. This reasoning was at the heart of IZOT's philosophy, which aimed to reduce the need for outside equipment and elements to less than a third of the total requirements, a huge difference from the early 1960s, when the prototype machines required elements that were not produced in the country.[86] As an aside, it is notable that when in the 1980s, the party undertook its last and most ambitious development plan—to economically uplift the mountainous region of Strandzha-Sakar in the southeast of the country—it integrated electronics so as to attract young specialists to the area, setting up enterprises producing electric cables for the industry as well as a whole branch of the ITCR and the Sofia Electronics Technical College!

In May 1970, IZOT got its definitive normative documents and structure, reflecting its increased importance. It contained new directorates—Systemizot and Izotimpex—to handle software and export of all devices, respectively. It also created a bureau to provide services and support for its products: Izotserviz.[87] Its new mission was not just to create computers but also integrated automated systems of governance, the cybernetic twist that reflected the Politburo's obsession with economic automation. This was the obverse of the obvious need for hard cash that Izotimpex and Izotserviz reflected, a tension between huge exports and relatively little implementation inside the country that would remain until the end of the regime. IZOT was a homegrown beast but with an outward-facing purpose.

Comecon and its heart—the USSR—were the ultimate raison d'être of the conglomerate. Popov justified investments with the aim not of the electronification of the Bulgarian economy, but with the possibilities of a huge and captive market. IZOT's growth was huge after 1969, but its creation in 1967 was the primary reason that Bulgaria managed to ensure the conditions for that growth in later years.

BESTING YOUR TEACHER

The socialist world offered dual possibilities: a place to learn and train, and a place to capture. Many learned the trade in places such as Dubna and Moscow, or in the GDR (as in Popov's case). The group of students who studied at Moscow State University and formed the first generation of the Mathematics Institute's computer expertise were the forerunners of a Bulgarian influx into Soviet universities and institutes in the 1960s. As head of the CSTP, Popov turned to its Soviet counterpart—the GKNT (*Gosudarstvenyi Komitet Soveta Ministrov SSSR po Nauke I Tehnike*)—to secure technical assistance, aid, and bilateral agreements. The GKNT had been an invaluable source for Bulgarian industrialization through the 1950s, but now it could be used to focus on the electronics sector. Annual agreements between the two bodies identified areas of primary importance for both economies, which determined training cooperation, too. Reflecting the industrial power of each, nearly all of the expertise was provided by the Soviets. As a snapshot, the 1966 plan called for cooperation in 83 areas of expertise, of which the Soviets were to take the lead in 71. They accepted 219 Bulgarian specialists in another 95 areas, while the Bulgarians could only host six Soviets in two areas.[88] But importantly, the areas were moving away from staples, such as metallurgy, and toward electronics. The 1966 plan contained "Automation of Production" as a priority theme for the first time, and specialists learned about Soviet computerization of metallurgical production.[89]

From these early stages, the GKNT noted that the Bulgarian economy was recording steady growth, including in priority sectors. A report noted that

> the Bulgarians are excellently informed about the conditions of and state of research, design and construction organisations in the Soviet Union. At the same

time, as a rule, Soviet organisations do not know of the technical innovations that Bulgaria carries out.[90]

This knowledge allowed the Bulgarians to identify with pinpoint accuracy which Soviet technologies or enterprises to request as hosts or aid; this accuracy was not reciprocated by Soviet industry, which didn't really know what Bulgarians could help them with. The GKNT noted that this asymmetry was a problem, as the Bulgarian "tempo of development has not overtaken just countries equal to Bulgaria in economic potential but also highly-developed capitalist countries." The country was now heavily investing in avant-garde areas.[91]

By 1967, the Bulgarians, riding the waves of ELKAs and the creation of IZOT, were demanding much more computing assistance. This was not so much in technology but in its application to modeling, programming used for automating accounting, city waterworks, industrial control, the cement industry, and the measurement and testing of mechanical construction.[92] The Soviets had experience in these areas, agreeing to help IZOT, but they had their own problems. Soviet computer research was still largely tied to scientific institutes and had limited industrial application. When Victor Mikhailovich Glushkov, Soviet cybernetics' luminary, lobbied for the creation of a nationwide network to automate information processing and allow a rationalized, streamlined, and incorruptible command economy, his Politburo opponents countered that computers could turn the lights on and off in chicken coops and play Mozart to the poultry to increase yields, and that was enough.[93] Like Popov's CSTP, the GKNT was a place where avant-garde ideas were discussed and led to criticism of the Politburo's conservativism. Konstantin Rudnev, who led the committee until 1965 before taking over the Ministry for Instrument Building (and providing Bulgaria with the elements for ELKAs), was succeeded by Dzhermen Gvishiani, who became an even bigger champion of Soviet science and its internationalization. Seizing on the CPSU's 1962 decisions, he lobbied for real coordination of Soviet computing within a framework of international socialist cooperation. A 1969 report, addressing questions raised by the Council of Ministers, noted multiple shortcomings of Soviet industry in this sector. Even leading Moscow enterprises were not using computers effectively for most of the year. Often the machines came without programs, rendering them useless. A further 60,000 people needed immediate training to get automation off the ground.[94]

> Because of the lack of a central governing organ, the slower development of Minpribor (Ministry of Instrument Building), the insufficient number of computers and their inability to be used in ASU, the process for creation of government systems has to a large extent been ungoverned, with irrational usage of the already deficit cadres. There is also a lack of unity in the ASU created.[95]

The Ministry was attacked for being clearly unable to supply all that the economy needed. Nine out of 12 factories in instrument building and nine out of 15 in the radio electronics industry had not even started constructing their buildings, let alone any machines. All this while the USSR needed, by the GKNT's calculation, at least 20,000 machines of MINSK-32 caliber by 1975. Importantly, the Cold War loomed over these discussions, as a full 40 percent of these investments were to be made in the defence sectors—electronics was tied to the competition with the US.[96] This gargantuan appetite for computers by the Soviet defence sector would be a boon for the Bulgarian industry and a crippling blow to Soviet civilian economic efforts. The report was the natural conclusion to GKNT complaints over years, finishing with a strong recommendation to train up to 160,000 automation specialists by 1975, speed up construction of factories, and concentrate on "materials needed for using automated systems of governance (special papers, perforated cards, magnetic tapes and so on)."[97] It recognized that in the contest to win an armed race with the US at the same time as automating everyday life, the USSR could not go it alone—Soviet science had the know-how but not the production capability to both reach communism through automation *and* defeat the US.

The first steps toward computing cooperation with its allies were taken in 1964, when Comecon created its first commission on computer design.[98] Only after the plans for socialist division of labor were torpedoed in the late 1960s did intergovernmental cooperation take place, however. In 1965, a subcommittee on radio and electronics was created in the Permanent Comecon Commission on Machine-Building, trying to coordinate states' plans and preparing specializations.[99] It noted that over 30 different computers were being produced in the Eastern Bloc, most of them incompatible with one another. Polish software could not be implemented on a Czech machine, and there were simply no economies of scale possible. There was also no real international socialist computing market—no nation could set up true production, as it had no incentive to do so.

It was to combat such divisions that the Intergovernmental Commission on Computer Technology (ICCT) was created in 1968. Popov played a key role in this true start to socialist integration in the sector, having lobbied for the body in Moscow numerous times while preparing Bulgarian industry for this moment. The ICCT's head was the deputy director of the Soviet Gosplan, Mikhail Rakovski, and the allies provided a representative at the rank of deputy minister or above. A coordination center was created in Moscow, with a rotating presidency and a permanent representative from each state.[100] Popov represented Bulgaria into the 1970s, becoming a towering figure by virtue of being the only socialist minister to be present. Of course, he had also ensured that Bulgaria entered the ICCT possessing IZOT, a trump card that others did not expect.

The first and most important decision was on technological unity. Soviet scientists had kicked around the idea of a unified series of computers since 1966, but only now could they get their allies on board. Some wanted to base the series on the BESM or MINSK, others on the British ICL—a move the Poles supported, as their Odra computer was compatible. The British had early successes in placing computers in the region, but by the late 1960s, they were in competition with IBM.[101] The American company had changed the game with the 1964 introduction of its System 360, a mainframe series suited to various tasks and company sizes. It became the workhorse of many businesses due to its speed and compatibility—you could always upgrade to a new machine if your needs changed and not lose any of your old data. Popov was one among many who were fans of the system, and he vocally supported the like-minded Soviet colleagues. The IBM 360 won the day, becoming the basis for the new ES system (*Edinnaya Sistema,* or "Unified System") of socialist computers. The ES would be based on a proven world standard, able to use the know-how of the most successful American machine, and it would be able to use programs created by and for IBM. Together with these larger mainframes, the ES was complemented by a parallel series of SM machines (*Sistema Maliyh EVM,* "System of Minicomputers"). While the ES equipped the large computer centers for economic, social, industrial data, the SM was used inside enterprises, in smaller centers or research labs. They, too, were based on Western standards for the same reasons as the ES, in this case, the PDP series by Digital Equipment Corporation. In the following decades, the series would move to compatibility with VAX, the PDP's successor.

These choices had immense consequences for the socialist computing industry. Many will unfold over the following chapters, but they should be highlighted here, too. Using one standard created a large, transnational computer market, as a machine produced in Bulgaria could be applied everywhere. American know-how meant that it also plugged computer developments into wider international networks of knowledge. Due to Cold War realities, this choice—to bring socialist computing in line with an American standard—meant that espionage also became a key part of the industry, always needing to purchase and find those components or information that IBM would not sell. Compatibility with IBM saved costs, because the hard work of creating new architectures or programs could simply be copied, and you didn't have to start from scratch in hardware or software. Ultimately, this would devastate the impetus for domestic fundamental research in the area; it increased costs due to espionage; it tied the whole socialist bloc to a single system and deincentivized upgrading due perceptions of sunk costs, as the 360 series was superseded over the years. The machines were also not designed with the countries' elements base in mind, so the machines could not be designed on the basis of existing realities, leading to increased costs (often resorting to the purchase of Western chips) and problems. Relying on foreign standards ultimately doomed the industry to perpetual catch-up.

Nevertheless, the immediate benefits were obvious, and the work on creating ES and SM machines started immediately. The first meetings on determining the unified elements base and standardizing documentation were held in October 1968. In these meetings, each nation held its own, and ICCT minutes demonstrate the clash of national interests that hid beneath the veneer of a unified socialist world system. Both the USSR and the GDR proposed different element systems based on their own productions. Moscow quickly found out that it was behind in areas such as printed circuit boards, its own production inferior to that of the GDR—and surprisingly to everyone, behind the Bulgarians too, who had created a specialized factory within IZOT.[102] Document transfer was also key to ensure that everyone had all the IBM specifications, with the aim that by 1972, absolutely all specifications had to be unified across the bloc. The Bulgarians suggested that documentation coordination was the first hurdle to overcome, proposing a centralized library to help each industry.[103] By the following year, this coordination was happening in earnest, each nation passing on what it had on the IBM machine—the Poles had documents on the input/output interface, the East Germans on chip sets, and the Soviets glossaries of terms and algorithms. By this point, even the Poles had abandoned the idea of trying to change the ICCT's mind, and only the GDR tried to raise the question of a different approach based on their own systems.[104]

A Council of Main Constructors (SGK, from the Soviet abbreviation) was also created, becoming the highest decisionmaking body. In 1973, it was split into three sections, one each for the ES, the SM, and for their application in the economy. Multiple temporary working groups functioned under these umbrellas, solving hardware and software questions. The first Bulgarian representative was the CICT director Boris Borovski, but he was quickly replaced by Angel Angelov, who attended up to three meetings per year in Moscow. It was up to him and his deputy—Stefan Angelov—to fight in the Bulgarian corner when divvying up the ES and SM pies.[105] Angelov held this position until 1980, the longest tenure by any SGK member, helping him become a senior figure in the key decade for cooperation in socialist computing.

Bulgaria entered the fray with a powerful industrial and scientific conglomerate, IZOT, but its capabilities were still being built. Angelov had already been a key figure in working out that peripherals and especially memory devices were the most profitable part of the industry (as they would be compatible with any of the new series of processors) and that the planned economies and their vast militaries would require countless megabytes of storage. Alexander T͡Svetkov, the ex-director of ZMG Razlog, recalls the Soviet defence industry buying huge amounts of discs that they did not immediately need, as a redundancy. "They were locked away in underground bunkers . . . the back-up in case of a nuclear war," he recalls, because electromagnetic pulses could damage many of the data centers of the Soviet military and strategic rocket forces. These backup tapes and discs would then be needed, to support the military effort. Thus, memory would be bought in quantities previously unknown by the USSR, even more than was needed by its central planning tasks.[106] Memory would thus be the key aim of Bulgarian interests in the SGK meetings.

Popov had another set of clear instructions. It was imperative that the country get at least one of the processor projects, as this was key for both market and internal needs. The ES mirrored IBM and had a range of processors, small to large, suited to different needs. Each country was to make its case for why they were well suited to create a processor. Popov demanded having a processor in production, as it was a high-value good, was prestigious, and would create domestic capacities to equip computer centers for the BCP's automated dreams. Popov instructed Angelov to promise the world in Moscow, in a daring case of bravado:

> You will emphasize that we have Japanese know-how; that many people have specialised throughout the world. You will multiply everything by ten, just so we can take this specialisation. If other countries promise three years, you will promise a year and a half. If they say—two years, you will say one year. . . . It doesn't matter that it might not happen, let us start and after two years, even if we haven't finished it, we will be so far ahead that we will have already secured the positions.[107]

The gambit worked. Much like the earlier ELKA bluff, Popov's case was helped by that early success as well as by the existence of IZOT and the possession of Fujitsu training. Little Bulgaria had come out of the dark in a few short years and had vaulted over its vaunted allies. It was saying that it could do anything the rest could do better, faster, and more reliably. And there was reason to believe it, as the country showed off new factories and licenses.

To everyone's surprise, Bulgaria received four specializations from the get-go—one full computer system and three types of memory devices. The computer was the ES-1020, a middle-sized system, to be created in cooperation with a Soviet institute in Minsk. The ES system included several machines—the smallest going to Hungary and the largest to the USSR—but the ES-1020 was of sufficient size to equip national calculation centers, so it could fulfil domestic needs in that sphere, too. Most importantly, however, were the three memory specializations—the ES-5012 magnetic tape device, ES-5052 magnetic discs, and ES-5053 removable magnetic disc packs.[108] Ratified in December 1969, this decision marked the end to debate about specializations and the start of real socialist integration in the area, probably the most successful of such Comecon efforts.

Integration was a weapon in a Cold War that had increasingly moved to the sphere of technology and economics. The capitalists had to be overtaken in living standards but importantly also in terms of production per worker. The parties recognized that only a huge integrative effort could best allocate the resources for this struggle, in the search of what Lenin called a "unified world co-operative."[109] The BCP noted that the bloc had immense scientific-technical potential and industrial capacity. Its most important sectors had attained "overtaking" speeds, especially in electronics.[110] The poorest areas had been accelerated out of a timeless rural sleep into the automata age:

> As a result of the socialist industrialisation Bulgaria, Poland and Romania turned from agrarian into industrial-agrarian countries. . . . Machine-building, which is the material basis of mechanisation and automatisation in the national economy, is now one of the main sectors of the national economies.[111]

The Politburo noted that while its specialists were negotiating ES deals, Comecon produced 33 percent of world industrial output (as opposed to 17 percent in 1950); Bulgaria itself had an annual growth of national income per head that was double the European average (an increase of 8.7 against 4.3 times over the period; higher even than that for Japan).[112] The USSR was an immense resource reservoir, providing 90 percent of needed oil, 85 percent of iron, and 60 percent of cotton. It possessed masses of specialists who could maximize a client's capabilities, while offsetting shortages in primary products.[113] The leadership continued noting other important aspects of socialist integration: 41 percent of Bulgaria's machine exports were in the "specialized goods" category, items they were solely responsible for, and this was the highest share in Comecon. Its trade with other member states constituted 72 percent of all trade, above the average, and this amounted to 330 roubles per head, placing it third behind the GDR and Czechoslovakia and well ahead of countries like Poland. Links with Moscow were even stronger, as the country took 58 percent of all Bulgarian trade, and Sofia in turn imported twice the average amount of goods, being a veritable poster child of Soviet modernity. At the same time, countries like the GDR were stymying cooperation, wanting to protect technical leads, and leading to overall bloc indebtedness vis-à-vis the West, along with irrational duplications of purchased licenses.[114] The CSTP noted with satisfaction that such irrationality was recognized by Comecon's higher bodies, and so were the mistakes of the 1950s, when specialization deepened the industrial-agricultural divide in the bloc, with the GDR taking a full 70 percent of machine specializations, leaving Bulgaria with just 7 percent.[115] The creation of new sectors such as electronics had diversified the field and led to the removal of such discrepancies.

These assessments of Comecon in 1969 are important, as they highlight the importance of this framework to Bulgaria's economy, which operated within the logic of an accessible, huge market. The BCP was aware that the Soviets had an interest in helping Bulgarian progress, as it was a Balkan "display window" of their model. The central theme for the party was, however, sharing and integration, to avoid approaching the West in a piecemeal fashion in trade talks. The agreement in electronics was a first step, demonstrating the ability of all countries to divvy up tasks in the hope of making up for lost years against Western computing. Post-1968, Comecon was increasingly integrated, with specialist commissions popping up in other areas of the economy. The culmination was the 1971 Comprehensive Program for Socialist Economic Integration, enshrining integration into the organization's guidelines. Pricing was fixed for five years ahead, corresponding to the states' own five-year plans and administrative pricing mechanisms, a step that moved away from world pricing and created a more closed trade bloc, with its own trajectory. There was a mechanism to keep pricing in line with world prices by determining moving price averages on the world commodity market over the preceding five years—thus creating a lag. In the coordination of plans, autonomy was retained (after Romanian obstinacy on the question in the early 1960s), but if multilateral or bilateral agreements were signed, these were to be taken into account in internal planning. There was, however, no superior joint body that could enforce these actions. Joint projects and investments were encouraged and provided for, easing financial burdens on the poorer states. The clearest and strongest emphasis was placed on science and technology, with specialization to be facilitated by easier and larger technology transfers within Comecon.[116]

This was a clear attempt to converge economic development in the bloc through bilateral and multilateral cooperation and investment. It introduced joint participation in more projects as well as forecasting and prognosis, yet this effort found it difficult to square the circles of different levels of autonomy at the enterprise level, where more contacts were desired: a Hungarian enterprise had more than a Soviet one, for example. Yet it fostered circulation of capital, technology, and even labor as Bulgarian timber workers went to Komi in the USSR or Vietnamese guest workers arrived (on whom the bloc relied in later years). Bulgarian elites, like others in the region, saw increasing socialist integration as a counterweight to the growth of the European Economic Community (EEC). The BCP was worried that its wish to benefit from Western trade and technology would become increasingly difficult with Western integration, because its customs union and other policies were seen as devised to hurt the socialist camp.[117] The country was acutely aware that in the 1960s, over half of its exports to Western Europe remained in agricultural goods, which had to change, especially in light of the EEC's agricultural protectionism.[118] Overall, this convergence and integration within the bloc should be seen as constituting a functioning and real Second World, an alternative modernity that faced the Western challenge in a coordinated manner.

In practical terms, the level of assistance and cooperation skyrocketed as the ES agreement and 1971 Program came into effect. Immediately in 1969, Bulgarian and Soviet teams met to discuss the design of the ES-1020 and storage devices,[119] while Bulgarian specialists were accepted in topics such as automation in research and design work or automated control systems in cable factories.[120] It wasn't a one-way street, with the Bulgarians passing documentation on 640 technologies and training over 1,500 Soviet specialists in areas such as the well-received Maritsa typewriters (key for office automation and built in Plovdiv, Bulgaria), a massive change to just a few years earlier when almost no Soviet specialists went the other way.[121] Ivan Popov still found the Soviets underestimating Bulgarian capabilities in science, as the Soviets were ignorant of the country's high-density wiring that equipped the rest of the bloc.[122] Unlike them, he and the CSTP were very well aware of Soviet scientific achievement and packed the cooperation plans with cybernetic and computing topics. Automation of construction, correlation analysis, and enterprises; computer usage in media programming; fundamental research on the use of electronics such as for magnetic field manipulation; the usage of MINSK-32 computers for production optimization, automation of information services and training of cadres for computer centers; and computer-controlled mine ventilation—these kinds of questions dominated technical cooperation plans from this period onward.[123]

The GKNT was starting to notice that Bulgaria was investing in cybernetic devices and its applications, and the country had opened up to the West to do this.[124] However, the committee also noted that the smaller state was gaining advantages in cooperation with the USSR. Popov's CSTP strongly opposed any efforts to base work on contracts rather than on the bigger cooperation agreements, or any attempt to place things on a market rather than a "fraternal" basis. In May 1971 at a joint meeting on Soviet-Bulgarian cooperation, Popov had outright denied any cooperation based on financial balancing. He stated that high-level decisions meant that the Soviets had agreed on free transfers of knowledge to Bulgaria, and just because the latter had agreements based on financing with the GDR, this didn't hold for the Sofia-Moscow relations. In essence, multilateralism was fine if it created markets and the framework for specializations, but bilateralism ruled the day when it came to technical cooperation.[125]

The GKNT also noted that the Bulgarians initiated talks and always directed them to areas that "interests only the Bulgarian side."[126] Between 1968 and 1972, over 40 percent of all information packages and items passed to Bulgaria were free, many of them concerning electronics.[127] When in the mid-1970s the GKNT managed to convince the Politburo to push through contracts for technology transfers, Bulgarian documentation sent to Moscow declined rapidly, even if it remained active with countries such as Cuba![128] Popov always noted that the basis of free transfer and Soviet help was backed up by decisions made at the highest levels, where Zhivkov continued to be a master of sweet-talking Soviet leaders. As with Khrushchev, he wooed Brezhnev, once again suggesting that Bulgaria join as a sixteenth republic. He exploited Brezhnev's love of hunting and the finer things, entertaining him at Bulgarian lodges. It was in one of those lodges, in Voden in 1973, that he laid out Bulgaria's needs over the coming years.[129] He accentuated Bulgarian loyalty in a region with two NATO states and in which the other socialists were mavericks. He noted that historically, Bulgaria was the most backward of the Eastern Bloc, and that the BCP's policy had lifted it into modernity. Its future depended on Soviet help, and Brezhnev's agreement secured Bulgaria's subsequent development. But Zhivkov's request that the talks be kept secret from other countries reveals that although Comecon was a lucrative market, it was not one where decisions were always made by planning bodies. Zhivkov's personal politics were key in securing huge subsidies, and by 1980, Bulgaria was annually receiving 5–10 million cubic meters of oil, 6 million tons of coal, 5 billion cubic meters of natural gas, and 1.5 million tons of steel.[130] While the Soviet oil was often resold at world prices, Bulgaria was also guaranteed Soviet markets for its industrial and agricultural goods.

The superpower's GKNT was thus, paradoxically, in a weaker position than its client's CSTP, allowing Popov to push through cooperation on Bulgarian terms. Soviet scientific capabilities continued to be harnessed to its ally's needs, while its markets remained captive. Electronics and cybernetic governance were the order of the day in terms of cooperation, with 30 enterprises to be given ASUs in the 1971–1975 cooperation plans, together with a transfer of all algorithms for MINSK-32 machines.[131] Glushkov's huge Cybernetic Institute in Kiev helped the ITCR with mathematical modeling programs and city planning automation systems.[132] But Bulgaria was also helping the Soviets in these areas, reflecting a marked change in capability since the 1960s—by 1970, it was training Soviets in computer-aided design of road networks and city waterworks.[133] Two thousand Bulgarians were to be trained in the USSR, but a full 900 hundred Soviets in Bulgaria.[134] The growth in Bulgarian ability to help the Soviets and the latter's interest in the Balkan state's computer capabilities is as much a testimony to the importance of the industry as are the export numbers.

By 1972, a permanent group on ASU design was created between the two countries, as it was the area of most intense cooperation over the past years.[135] By 1974, there were talks of interlinked, mutually compatible, and fully automated systems of exchange for scientific-technical and social information by 1980. This would make coordination in all important sectors easier, and would fulfil the aims of Comecon:

> The main task of the consultations is the international division of labour in scientific and technical research with the aim of maximising the acceleration of scientific-technical progress of countries, the choice of complex multi-sector scientific-technical problems, a preliminary look at the state plans of economic development of the USSR and Bulgaria and the important sector problems that

can be solved through a higher technological level of production and thus its efficacy.[136]

Electronics and automation were the technological fields that increasingly intertwined Soviet and Bulgarian science and fostered integration on a wide level. If the lens is on technical policy, the Second World was a well-functioning space of shared knowledge and material culture. These strengthening channels led to the supremacy of this united scientific front in 1977 when the GKNT suggested a step that mirrored Zhivkov's earlier ingratiating moves toward Moscow:

> The widening of the exchange of information of materials between the USSR and People's Republic of Bulgaria should be organised on the principles that are used in the exchange of information between the different republics of the USSR.[137]

If not politically, then scientifically, Bulgaria was functioning as the sixteenth republic. Having created outsized capabilities in the electronics research sector, it also could draw on the almost unlimited capacities of the USSR. This was not subservience materialized but profit extracted, unlike Zhivkov's moves. Soviet research institutes were key partners in creating the numerous ASUs that the party desired for the economy. Comecon's integration was moribund in some areas, but not so in Soviet-Bulgarian technical cooperation, which grew closer and closer. It played a decisive role in helping the Bulgarians automate production of as well as improve the ES devices they specialized in.

The CSTP always bet on the Soviets, who showed "an unlimited interest" in cooperation (unlike the Hungarians or Romanians, who had to be pushed).[138] The committee was more than happy to adapt the country's science information systems to the Soviet model, unifying standards and programming languages, in order to freely access on a reciprocal basis the Soviet data universe: "in essence what we are talking about is for the strategy of our scientific policy to become a part of the strategy of the scientific front and policies of the Soviet Union."[139] This can be read as a surrendering of independence, when it was anything but; it was a road consciously taken by Popov, the CSTP, and the party itself. Soviet capacities were endless—in terms of providing natural resources, technical assistance, and markets. Scientific integration and standardizing production meant that Bulgaria could economically be treated like a Lithuania or Armenia, with all the benefits this brought. The specialization within Comecon created the dual fields in which Bulgarian capabilities could grow—a nice boost for its industry and an ever-growing infusion of information from Moscow. Bulgarian science was entangled in Comecon. Bulgarian industry captured Comecon's markets. Loyalty and political orthodoxy in the Bulgarian case are often seen as sycophantic or obsequious. If interpreted through the economic and scientific policies of the country, however, they were rational beyond compare. Bulgaria was the small tail that wagged the giant dog.

THE PAYOFF

IZOT ramped up production in the early 1970s, starting a trajectory that lasted until the end of the regime. An enumeration of all the devices would not tell much of the story, but the narrative arc fleshed out in Appendix B (where the reader can find more about the capacities of the devices) highlights the importance of these devices to both Bulgaria's export policy and its internal sociopolitical goals.

The Bulgarian ES-1020 computer was unveiled at the Plovdiv Fair in 1971, winning a gold medal and shown to the world before the Soviet computer was—another Popov stunt to make sure he got one over the superpower.[140] The Plovdiv factory started producing the ES-5012 tape devices that same year, and the Stara Zagora plant churned out its first serial run of ES-5052 discs in 1972. That same year, the ZMD-Pazardjik started mass production of ES-5053 disc packages, which could be reinstalled into the ES-5052. The ES-1020, demonstrated in its null series in Plovdiv, started mass production at the ZIT-Sofia plant in 1973, together with an ES-1020B modification that included add-ons and peripherals.

This computer, which would become the heart of many Bulgarian and worldwide computer centers, was equivalent to an IBM 360/40 series. This configuration had already been superseded in the West by the IBM 370, but it was still the most modern within Comecon.[141] The processor and some of the discs were co-developed with the institute in Minsk—the Soviets co-developed many of the processors—but the implementation and serial production was left to each country, with IZOT pulling ahead of the Soviets due to an existing base and licenses.[142] There were Bulgarian innovations in production, such as the use of the novel Balevski gas counterpressure casting system to create the aluminium discs needed for the drives. The team that constructed the tapes and discs received the Dimitrov Prize in science.[143]

The sixth five-year plan, spanning 1970 to 1975, was the time when the industry made the leap to mass production. Capacities increased to reach 1975 production volumes that satisfied all predictions. ZZU-Plovdiv produced 150 million levs worth of goods, ZIT-Sofia—93 million. Certain comparable factories in Japan were producing around $30 million (the official exchange rate being around 1.17 levs to the dollar) in the same year. Measured per worker, a laborer in Sofia's ZIT produced 71,000 levs worth of goods, while in Japan, the figure was $19,000.[144] We must allow for inflation in these internal numbers, as is common in socialist accounting at the highest level, but there can be no question that this was by far the most productive branch of Bulgarian industry and in those years, under Popov's leadership, it operated at a world level of volume and productivity. "Factories paid back for themselves within a year and often less," Markov and Vachkov proudly proclaim.[145] By 1973, Bulgaria had implemented into production all its specializations, placing thousands of items on the socialist market and increasingly on the world one, too. The BCP praised it with its recognition as the undisputed champion of all economic sectors in terms of efficacy.[146] Popov was not as dizzy with success, however, warning that the country could not rest on its laurels, as the next generation of machines was beckoning, and other countries were also securing Western licenses. Bulgaria had won primacy, but it could lose it if it didn't make sure that, as in the late 1960s, it entered the next negotiations better prepared than its partners.[147] Bulgaria was at this stage lagging behind the West but had managed to enter production at an almost contemporaneous level; the next generation could leave it many years behind the country wasn't willing to think as progressively as it did in the 1965–1969 period.

This early success catapulted Popov ever higher. He became the minister of all machine-building between 1971 and 1973, then the deputy president of the Council of Ministers up to 1973, and eventually the vice president of the State Council. By 1973–1974, as the first processors and discs were being sold, the career of the worldly technocrat had reached its peak, a testimony to his organizing the most effective and profitable industry in the country. His reward was paramountcy over all economic and scientific policy, Zhivkov's favor, and—arguably—the position of second most powerful man in the apparatus. In those years, the electronics industry was separated from machine-building, having its own ministry. Iordan Mladenov, until then deputy head of the CSTP, was its first leader.[148] It became responsible for DSOs such as IZOT and Electronic Elements. In 1978, Vasil Hubchev succeeded Mladenov before the ministry was rolled back into a united Ministry of Machine-Building and Electronics in 1981, reflecting the weight given to the sector—equal to the whole of the rest of the industry that had obsessed socialists for decades: steel, chemicals, and power. In 1984, the "Electronics" was dropped when Ogni͡an Doĭnov was head, but this did not reflect a change in policy.

Doĭnov had not only caught Zhivkov's eye on his Japan visit, but also would be the man who deposed Popov.[149] The professor's success was his undoing—precisely because of his "fatherhood" of the electronics industry and his high standing in the socialist technocratic community. Zhivkov was a wily politician whose longevity was predicated on preventing any Politburo member amassing enough power to challenge him. There are no indicators that Popov ever contemplated that, even at his zenith in 1973, yet Zhivkov was already planning his removal from the peaks of policy: positions such as the one Popov held in the State Council up to 1976 were a demotion. In July 1973, at a Politburo meeting, the young and unknown Doĭnov was the conduit of criticism as he read a scathing report, which laid out the need for change in the electronics and scientific policy sectors. He argued there was no clear long-run strategy for the machine-building industry, where heavy industry was underinvested in. Popov was aghast, but Zhivkov had thrown in his lot with Doĭnov in the meeting, a preview of what was to come.[150] In 1974, Doĭnov replaced the professor as deputy-head of the Council of Ministers, assuming greater power over the economy, and at the Eleventh Congress of the party in 1976, Popov was kicked out of the Politburo to the relative obscurity of chiefdom over the Scientific-Technical Unions.[151] By 1977, Doĭnov was a full member of the highest body of the party, a position he held until 1988.

The young technocrat had his own pet obsessions, such as shipbuilding and heavy industry. The latter would lead to the disastrous fiasco of the Radomir industrial plant in the 1980s, but Doĭnov was not blind and would not give up a good thing. Despite his criticism of Popov, he in fact held up the latter's policy and championed IZOT, which he recognized as being on an upward trajectory. As the party focused on cybernetics and computerization in its political programs from the mid-1970s onward, electronics production grew under Doĭnov's suzerainty. Zhivkov did not want the industry to suffer; he only wanted to prevent its manager from becoming a political rival.

The seventh five-year plan invested in new types of peripherals, processors, higher-density discs, MOS integrated circuits, mini-computers, computer numerical control for industrial machines, electronic tills and office equipment, and automated telephone exchanges. This was the full gamut of goods that were needed to keep up exports but also automate society.[152] By 1977, electronics made up 15 percent of all exports and was rising. Doĭnov called for 50 percent of industrial machines to be computer controlled by 1990, and for 25 percent of all metalworks machines to be robotic by 1985.[153] Between 1971 and 1977, the industry grew three and a half times, exports—five times.[154] Despite Popov's absence, the sector had a life of its own.

A new generation of machines and improved models entered production in those years. The ES-1035 processors were 14 times faster than the ES-1020 series, suited for more intense economic and social governance calculations; they entered production in 1977.[155] It could be connected to ES-2335 matrix processors, boosting analytical power and calculation speeds by up to a factor of 100.[156] Higher capacity discs and changeable disc packages were another jump in the sphere and started production in the same year. The IZOT-0310 minicomputer, a clone of the PDP-11, was another entry of that year, suited to office administration and research labs. This was part of the SM series, and other machines of that type were in production by 1980—the SM-3 and SM-4, also for workplace automation, and exported throughout Comecon. These machines could now be equipped with the ESTEL system of teleprocessing, implemented in ZZU-Veliko Tĭrnovo in 1976. This series was upgraded until the end of the regime, reaching the ESTEL-4 level, allowing users to connect computers through the telephone or telegraph network, creating the possibilities for remote data processing, and importantly—the dream of local and national information networks.[157]

The next five-year plan—the eighth, in 1980—saw the creation of ambitious and costly Elektronika-8 and Avtomatika-8 programs. These Doĭnov-led proposals were presented as a way to move to the next level of production, especially aimed at a qualitative jump in memory devices. The reports warned of continuing problems in producing discs with more than 100 MB capacities, leaving the space for other states to overtake Bulgaria. The CSTP highlighted that difficulties producing ES-1035 upgrades cost over 20 million levs per year, jeopardizing another specialization.[158]

2.6 Crated ES-5052 discs ready for export to Kyiv, USSR. (*Source*: Central State Archive.)

IZOT demanded a lot more capitalist currency to make up all these lags, an important part of the increasing indebtedness of the country thoughout the decade.[159] The need to secure another decade of Comecon dominance made Avtomatika and Elektronika priority programs, with magnetic disc production slated to grow by a factor of two. The new discs were to be made according to the Winchester technology standard, which would allow for capacities in the 625 MB range and beyond—absolutely unprecedented in the Eastern market.[160]

Over 465 million levs were to be invested in the program, with the expectation that it would be paid back within three years or less through exports.[161] The focus was not just on money but also on better utilization of resources, as the CSTP noted that 32 percent of all modernization projects were not implemented on schedule.[162] Another highlighted problem was the preference given to building or expanding factories versus investment in new machines, with the Bulgarian ratio being 60:40 compared to the world ratio of 20:80. This of course reflected the expansion of production that the small state was undertaking to meet Comecon needs, yet the plan for the decade was to reach at least a 30:70 ratio.[163] Investments in both programs thus grew rapidly and by 1982 involved 58 large-scale automation projects in industry for that year alone, aiming to create discs of up to 50 GB capacity by the 1990s. The investment was ramped up, reaching 1.8 billion levs over the five-year plan, which was expected to create the basis for a much longer and sustainable development of technology over decades.[164] In 1980, at the start of the program, 358 million levs were earmarked for R&D, recognizing the need to focus on domestic developments as well.[165]

The money spent did lead to tangible results. The Stara Zagora factory implemented the production of larger discs, up to 625 MB capacity, which became the largest produced anywhere in the Eastern Bloc and catapulted the factory to its premier status not just within IZOT but also in the whole of Comecon. New text-processing IZOT-1037 machines were introduced for the growing service sector. By the end of the 1980s, the IZOT-1014E (ES-2709) machine could lay claim to being a supercomputer, achieving over 120 million op/sec, with the ability to be boosted further by a network of processors—faster than the Elbrus machines that the Soviets used to equip their space and nuclear missile control centers. The machine would equip parts of the Soviet space program, the nerve center of the terminal stages of the Venus and Halley Comet-bound Vega missions. Others appeared in China, Vietnam, and India.[166] A member of the development team, Krasimir Markov, recalls how this project was led by Stoi͡an Markov (no relation), the last rising star of the Bulgarian electronics community, deputy-head of machine building and the CSTP in 1984–1985, head of the CSTP after its renaming as the State Committee for Research and Technology in 1987–1988, and candidate member of the Politburo (the youngest ever) between 1986 and 1988.[167] A Doĭnov protégé, he was to forge his own post-socialist path through his role in technical policy at the end of the regime.

Listing all devices produced during these years would not help the narrative, but it should be noted they ranged from full computer centers to floppy discs and tiny elements, such as magnetic reader heads.[168] The volumes were impressive and most importantly, profitable. In 1980, for example, 14 large ES-1035 computers, 40 ES-2635 matrix processors, 1,000 large-capacity discs, and 80,000 removable disc packages were produced.[169] The upward trend can be seen in the table later in the chapter, but thousands of discs were churned out each year—the true star of the industry. Alongside this was the huge growth in automation system production, computer numerical control devices, and robots—and eventually the creation of the PC industry, with thousands of machines produced and huge repercussions for the dream to automate workspaces, home life, and labor.

This chapter has situated this success in the export-driven logic of the Popov and the party, and the existence of huge markets. Thus export figures can best tell the story of the industry's importance. Calculators were the first cash cow of the industry, especially as other factories were being built into the early 1970s. Even when reduced as a share of total electronic sales, between 1971 and 1985, over 487 million levs worth of electronic calculators were exported to the socialist and capitalist worlds.[170]

This was but a drop in the ocean of IZOT sales. Table 2.1 illustrates the total production and profit levels for select years. Table 2.2, based on Izotimpex accounts, shows the continuing upward trajectory of two decades of fast growth. What becomes clear from these numbers is the weight of Comecon in exchanges, and especially the USSR as the single-biggest captive market and trading partner. The capitalist market remained a perennial dream, an object always out of reach, a source of debt rather than a

Table 2.1 IZOT production and profit (in millions of levs)

Year	Production	Profit
1970[a]	55	−1.5
1973[a]	335	139
1976[b]	688	250
1979[c]	No info	435
1984[d]	1518	706
1986[e]	2100	1008

[a]Dimitrova, *Zlatnite Desitiletiia*, p. 48.
[b]TsDA f. 1003 op. 1 a.e. 16 l. 1 (IZOT Economic Data 1976).
[c]TsDA f. 1003 op. 1 a.e. 19 l. 8 (IZOT Cost Reports 1979).
[d]TsDA f. 1003 op. 1 a.e. 27 l. 1, 26 (IZOT Annual Report 1984).
[e]Dimitrova, *Zlatnite Desitiletiia*, p. 49.

field for sales. The developing world was the replacement, of sorts, becoming a market for electronics after the mid-1970s. Accounts for these sales are sometimes separate, coming through the Ministry of Foreign Trade, so the table gives an imperfect picture of the scale, but the point is clear: sales would never reach the level of the socialist market for which this industry was built and designed. The developing world market remained important in other ways than purely cash, as a field of learning how to market and "do business," for the lack of a better term.

The capitalist world remained a dream due to embargoes, prices, and quality. Comecon was simply more profitable as prices were favorable to the seller. Originally, they were fixed for the duration of a full five-year plan, but even after renegotiations, they were updated every two years rather than annually. This was immensely favorable to Bulgaria, as the electronics industry between 1960 and 1990 was especially fast moving. Things created one year were obsolete the next. Freezing prices for years in advance was an advantage IZOT had in the logic of the Second World that no Western firm had in the First. IBM could not sell its 7.25 MB drives at the same price five years later, when they were out of date; IZOT could—and did. Even two years of frozen prices were enough to rack up enormous profits. The Comecon pricing system played right into the party's hands, and the CSTP always resisted Soviet attempts to remedy this

Table 2.2 Izotimpex exports (in millions of levs)[a]

Year	Socialist Export (of which USSR)	Capitalist Export	Developing World	Total Export
1968	2.6 (2.57)	3.7	—	6.3
1969	9.3	0.6	—	9.9
1970	22.1 (15.5)	0.55	—	22.6
1971	—	0.78	—	56
1972	113 (95)	0.9	—	114
1973	238	1.4	—	239.5
1974	312	4.2	1.8	318
1975	353 (294)	2.1	2	358
1976	476 (364)	2.9	2	481
1977	(312)	8.4	—	544
1979	706	15	—	721
1981	—	11.5	—	953
1982	—	—	—	1,040
1984	1,290 (1,000)	10.8	—	1,301
1985	1,554 (1,195)	10	—	1,564
1986	(1,446)	12	—	1,831
1987	(1,636)	13.7	—	2,078
1989	2,212 (1,869)	11.9	—	2,240
1990	1,329 (1,236)	3.7	—	1,332

Note: A dash indicates no information available on the exact number, rather than "no export."

[a]Statistics are from f. 830: for years 1968–1976 they are from op. 1 a.e. 89–96; for 1977–1987 they are from op. 2 a.e. 20–28; for 1988–1989 it is from op. 2 a.e. 36–37 (Statistical Reports of Izotimpex).

system. Its internal reports are very clear about: a 1981 memo to Papazov on the next stage of memory device development stated that:

> In relation to the leading capitalist firms in these areas we will have a lag of around 4–5 years. Cutting down this interval is impossible, and a possible faster fulfilment relative to the developments of ES series of the socialist countries wouldn't make economic sense.[171]

The logic of profit was also the logic that worked against innovation and meant that IZOT's East-facing nature pushed it further and further behind the Western firms. The Iron Curtain created a closed world, which made IZOT so powerful, profitable, and ultimately backward. The pricing of the ELKAs was a clear example of this—one ELKA-22 cost the Soviet user 969 levs, but the French customer paid 593 levs for it.[172] The price could drop down to 200 levs, as it did in Turkey.[173] Even different Comecon members were charged differently, depending on negotiations and volumes—an ELKA-25 cost the Romanians 1,345 levs but the Soviets 910.[174] These pricings meant that into the 1980s, IZOT's returns were much higher in the socialist than in the capitalist world—an average of 111.51 percent profit in the socialist market versus 41.4 percent in the capitalist. Certain factories, such as ZMD-Pazardjik, could reach profits of 256 percent for some of their goods; while certain floppy drives were down to 15 percent profitability in some markets, due to falling world prices.[175] Some devices, such as the IZOT-1036S computer, were very expensive to produce, due to the need for imported elements, sometimes

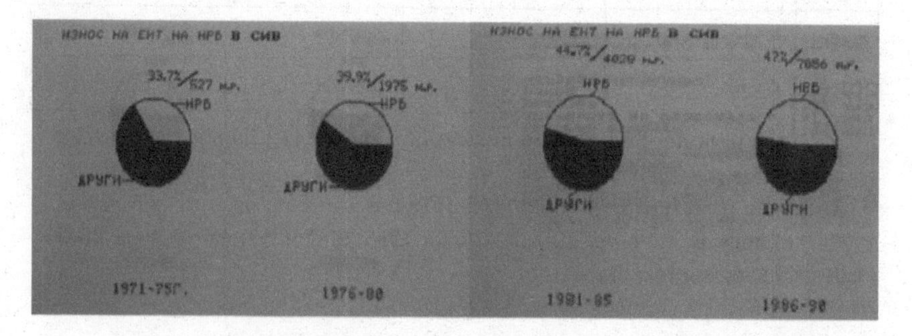

2.7 The growth of the Bulgarian share of Comecon electronics exports, 1971–1990. The white area of each pie chart is the Bulgarian sector. (*Source*: Central State Archive, Sofia. TsDA f. 1B op. 68 a.e. 1836 l. 200–1 [Politburo Electronic Development Report 1986].)

up to four times more than planned, meaning that none were produced in certain years, such as 1985.[176] Other devices, however, fetched enormous prices: a single 635 MB disc cost the Soviets 580,000 levs; a 317 MB disc—604,000.[177]

This industry was strong in terms of the logic of the Second World's planned economies and fraternal division of labor responsibilities. Despite lags, as long as it was the first to introduce a mass-produced Winchester drive, Bulgaria was ahead of its allies. As long as this logic operated, the industry boomed. Sites such as ZZU-Stara Zagora became behemoths of industry, sometimes producing around half of all the sector's exports in the late 1980s.[178] Most importantly, Bulgaria was the undisputed bloc leader from the very start, responsible for 33.7 percent of all exports within Comecon during the sixth five-year plan (1971–1975), rising to 44.7 percent in 1981–1985 and projected to reach 47 percent in the 1986–1990 period, as a 1986 Politburo report noted. In 1979, the ICCT had planned that IZOT would produce 45 percent of the bloc's electronics by 1985,[179] and this had been achieved. The Western media was also taking notice, even when it was prone to exaggerate the Bulgarian share—*Der Spiegel* reported that Bulgaria was responsible for 70 percent of all Eastern Bloc advanced electronics![180]

British foreign secretary Geoffrey Howe recalls Zhivkov's boasts that "Bulgaria is [doing] well, because it has colonies and the biggest one is the USSR."[181] Computer exports are testimony to the fact that this was not a joke. Bulgarian electronics' victorious march across the East was brought about by a far-sighted strategist who convinced the Politburo that this approach was the golden ticket out of agricultural backwardness, a political leader willing to listen, and an international market closed off to world trends and operating on its own terms. Popov had created the golden goose, and for his efforts, he was removed from his post. But his successor added rather than subtracted from IZOT. From the Vitosha to the age of robots and microcomputers, Bulgaria managed to catch up to the world in principle if not always in quality. The disc drives were less reliable, the computers slower, the monitors fuzzier, but IZOT produced the full gamut of devices needed for any socialist state wanting to enter the information age. The numbers tell the story of this industry's importance for Bulgarian finances, as well as the need to see the Second World as an integrated market that allowed for capture and profitability. The seeds of its end have been sown in these pages and will be picked up in chapter 7 and the conclusion, but first, another question must be answered—that of the mechanisms through which this Western technology made its way into Bulgarian institutes. The technological Iron Curtain was increasingly penetrated—often illicitly—with wide ramifications for the scientific and technological community of the electronics sector and beyond.

3

ACCESS DENIED: SPIES, TECHNOLOGIES, AND CIRCULATION ACROSS THE IRON CURTAIN

Aficionados of the James Bond franchise prize the elusive book *Avakoum Zakhov versus 07*. The 1960s paperback pits a Bulgarian superspy against a barely changed Bond figure after the author, Andrei Gulyashki, failed to secure the relevant rights to use the iconic name from the Fleming estate. The Bulgarian spy wins, becoming probably the only rival spy who has defeated Bond. Zakhov was one of the Bulgarian superspy heroes, together with author Bogomil Raĭnov's legendary but tragic Emil Boev, and the communist partisan Major Deĩanov from the television series "Every Kilometer." These spies lived in novels and on TV and cinema screens, and they were the country's manifestation of the quintessential Cold War cultural artefact—the spy story. Bulgarian spies on screen stole Nazi secrets or crisscrossed European resorts in search of diamond treasures. Their real counterparts in the Bulgarian intelligence services were much more active and prosaic, being more interested in IBM business plans and memory devices made of aluminum.

The computer was born in the US and the UK during the 1930s and 1940s, not in Bulgaria. The West remained the space of the leading innovations in the sector. Von Neumann architecture, the microchip revolution, Winchester hard-drive technology, the personal computer, programming languages, source codes, peripherals, and computerized controls for industrial machines were just some of the technologies that Western firms produced. Even Japan, a country more open to Bulgaria than most Western countries, was not the source of these innovations most of the time. Sofia was dependent on these technologies in its quest to catch up. Domestic innovation and development were always part of the industry, but if a state with the human and capital resources of Japan went down the route of reverse engineering, smaller and poorer Bulgaria would have to bet even more on that horse.

The Iron Curtain was a real impediment to this quest. The embargo of so-called dual-use technology such as electronics, which could be used for military as well as civilian applications, was one of the barriers that caused the two worlds to develop differently and to even *look* different. As tools of calculation and planning, computers had obvious applications to the space, nuclear, and arms races, especially once miniaturization made them integral components of a new generation of precision munitions, fly-by-wire avionics, fire control systems, missile seeker heads, and beyond. This is not to say that socialist regimes did not pursue all legal means to gain technology from the capitalist world. The story up to now has already included the use of licenses, purchases, and joint work through schemes such as those run by the UN technical aid agencies. Times of détente also made such joint efforts easier after the late 1950s, with technological, scientific, and cultural exchanges seen as precursors to diplomatic recognition. However, whatever the case, much of the know-how remained behind the embargo walls. Moreover, Bulgaria was always seen as more suspect than others, as it was the most loyal Soviet ally. Miroslaw Sikora has talked about how Poland benefited from a "differential" in terms of US policy—Bulgaria didn't.[1] Even then, by the time of the 1980s and the second Cold War, the whole of the Eastern Bloc suffered from extra restrictions.

Bulgaria, like other countries in the bloc, thus turned to its intelligence services as an integral tool to gain access to the latest resources. The Committee for State Security, *Dŭrzhavna Sigurnost* (DS), and its Scientific-Technical Intelligence (STI) section were an increasingly entangled part of the story of the electronics sector. Again, this is not a uniquely Bulgarian story, as Sikora has shown how Polish intelligence—as others in the region—became focused on civilian technology, including electronics, and quickly decided that only superpowers like the US and the USSR could depend on their own strengths to develop such industries.[2] In some ways, the ES series and the KGB's own focus on electronics proved that the USSR was also not capable to a large extent. However, the prism of the Bulgarian archives show how STI should be seen as a real, important, and underrated conduit of knowledge transfer during the Cold War. This is because the story of spying is not just one of industrial espionage but of colonization: of both trade and exchange in all spheres of life, and surprisingly, of the intelligence services working in the civilian economy. By the 1970s, it is hard to disentangle what is a licit and illicit channel of communication and trade with the West, as the DS set up foreign firms and used all available channels to gain information, including work exchanges or specializations. To an unprecedented level, it was also under the purview of the CSTP and reliant on civilian knowledge, planning, and evaluation. The question arises: Why is it important to look at such action as "legal" or "covert," when the logic of the Cold War drove a particular evolution of knowledge acquisition that interrelated spies and civilians to such a large extent? The Iron Curtain created the closed worlds that Paul Edwards talks about through consciously crafted mechanisms of isolation and control.[3] However, these closed worlds interpenetrated to a much larger extent than expected, and through the world of spies, the boundaries of what is "inside" and what "outside" the socialist world began to blur.

As this chapter demonstrates, espionage also blurred the lines between socialist and capitalist thinking, knowledge, and material worlds. The channels created to the West by the STI were key for the managerial class that controlled its goals and practices, allowing STI to connect to the emerging world of international business networks. That this was done through the intelligence services would have wide repercussions for its post-socialist transformation. However, as thousands of civilian specialists became entangled in STI work through bureaucratic mechanisms, intelligence work diffused new ideas into much wider circulation than just the walls of DS. This chapter thus argues that while a tool of oppression, the security services were also a tool of knowledge production and became indispensable for the Bulgarian computer industry to become a truly internationally connected and embedded part of the economy. Facing East for its markets, it had to face West for its technologies and practice.

THE BARRIER: COCOM AND THE RATIONALE OF STI WORK

The creation of the Coordinating Committee for Export Controls (CoCom) is tightly linked to the origins of the Cold War, Western antagonism toward the USSR, and increased US links with its Western European allies. The Marshall Plan and promised military assistance to Europe drove a number of acts in the late 1940s, among them the NSC Decision of December 1947, the Cabinet decisions of March 1948, and Section 117 (D) of the Economic Cooperation Act of 1948.[4] An unprecedented step in US economic history, it was accepted after Soviet opposition to the Marshall Plan had proven the USSR to be a "threat to world peace and US security." As the coup in Czechoslovakia sharpened enmities, and dock workers in New York refused to load ships to the USSR, the Marshall Plan negotiations included requests that all Western European allies stop trading in embargoed goods (list "R") with the East. By late 1948, the British had fallen in line, and while opposition by the Dutch or Luxembourg persisted, the Korean War made everyone fall in line. It was estimated that European acceptance of the trade restrictions would lead to a loss of one-third of exports and one-third of imports in vital areas of trade with Eastern Europe (around $1 billion per year, or 1/3 of the Marshall Aid value).[5] The CoCom restrictions would ebb and flow over the decades, depending on thaws and wars, but it would always remain a powerful obstacle to Eastern technological development.

It was not insurmountable, however. As Chris Leslie points out, there were multiple failures, and there was always the important question: Is the technology worth protecting, or is it the innovation culture that made it flourish? This question hangs over the whole discussion of socialist computing failure.[6] CoCom, however, made the endeavor expensive. Some Bulgarians frankly see it as a tool to bankrupt the socialist world. Alexander T͡Svetkov, head of the ZMG Razlog factory, often had to deal with acquiring embargoed goods in the manufacture of magnetic heads. For him, CoCom was "a way to make us pay $1 million for a technology that cost $200,000 on the open market. They knew someone would always sell it to us, but it just ensured we paid a higher price!"[7] Krasimir Markov holds a similar opinion: "you could always find a businessman ready to put his head in the bag for triple the price, as greed is a universal human quality, and control was such that you can never catch everyone—neither there [the West] nor here [Bulgaria].[8] CoCom could not stop every firm or businessperson from selling to the East, and there were high-profile controversies, such as Siemens' sales to the GDR.[9] Some technologies would never be available on the open market, others were too prohibitively expensive, so increasingly, STI became the only way to get the industry what it needed.

The DS's First Directorate (foreign intelligence) created a scientific-technical department in 1959, but it recorded little activity in its first years.[10] The early 1960s were a watershed moment in this sphere as well as others for the DS, as the organization moved from overt repression and being purely a tool of party control to one that set up wide networks of agents abroad, and became increasingly professionalized.[11] The KGB's own section on scientific-technical intelligence was raised to the status of Directorate in 1963, under the command of the Military Industrial Commission, which set its sights firmly on the US defense establishment.[12] The Bulgarians followed suit as they often did, because Soviet intelligence had become a veritable tutor for the service.[13] In 1964, the section was thus transformed into the Seventh Department of the First Directorate, raising its profile significantly and signaling a new chapter in its existence. Its first head was Raĭcho Asenov, who was replaced by Ivan Ivanov in 1965, and in turn by Angel Dimitrov in 1972. Its last head, Georgi Manchev, took over in 1979 and commanded it until 1990, the years of its ascendency to the level of a full directorate, even if under the auspices of the First Directorate.[14] Its work was also put on a new footing with the 1963 servicewide reorganizations, when the Minister of the Interior set out the goals for STI work as "gaining scientific-technical informations[15] of practical usage for the economy of Bulgaria and socialist members of Comecon" as well as fundamental research questions and above all, military innovations of the West. While defence issues were to take precedence, on paper, at the inception of the department, the head of the CSTP was also required to send annual plans informing STI of what the economic priorities of the country would be.[16] Very quickly, this civilian side would overwhelm the military direction of intelligence work. As Khristov notes in his articles on the issue, the growth of STI and its priorities were tied from the start to whoever led the CSTP and set its agenda. In the early 1960s that was Popov, and STI's growth is inextricably tied to the evolution of the Bulgarian electronics industry, which would become the main client and user of "informations" from the directorate.

THE RISE OF THE SPIES

It all started modestly, with just ten operative agents and some technical support staff in 1964.[17] The purview grew quickly, and so did the staff, with another 30 officer and two sergeant positions added in 1966.[18] This growth was helped because the deputy head of DS pointed out that even with so little staff, STI had achieved a lot in just two years.[19] It steadily grew to nearly 100 officers and 10 technical personnel by 1980,[20] the year it was raised to the level of a directorate in the First Chief Directorate through a secret ministerial decree that acknowledged its importance to the economy.[21] The same decision expanded its staff to 187, with a full 160 officers working abroad, and over the next five years, the plan was to create 105 extra staff to work in the STI center full-time, another 50 to be attached to various organizations in technology and the economy in the country, a further 95 to work outside the country, and 27 extra technical staff to help with the analysis of information.[22] Overall, by the early 1980s, the STI directorate thus expanded to over 300 full-time staff, with hundreds of operatives abroad and deep links to most trade and scientific organizations in the country, including the CSTP.

The military aspects quickly fell by the wayside, even as the documents throughout the 1960s paid lip service to that goal. Almost immediately, more and more of the pages of the annual plans were dedicated to questions and tasks related to the national economy. By 1967, the main sectors of concentration were radio-electronics, atomic energy, some avionics, chemistry, biotechnology, and the newest methods of "organisation of production and the application of mathematical methods and computer technology to the governance of labour."[23] The staff was increasingly well trained, with about a third having undergone more rigorous training in Soviet academies on scientific intelligence work abroad.[24] As the STI and its staff expanded, the language of national economic interests became enshrined in every annual plan, to the detriment of any Warsaw Pact goals. A 1976 instruction for STI work during the year is a clear indicator of this focus and is representative of all plans after the mid-1960s. Its first bullet points are worth citing in full to show the complete absence of military preoccupations for the Bulgarian STI service. In 1976, the priorities would be:

[The] Acquisition of secret scientific-technical information fundamental for scientific investigations, having a key role for the development of science, technology and economics.

The acquisition of constructive and technological information for the solution of concrete problems linked to the development of our industry and the introduction of new effective productions.

Acquisition of scientific-technical information and models in the area of agriculture.

The acquisition of scientific economical information, needed for the long-term and perspective planning of the development of the national economy.

The acquisition of concrete economic information for the economic position, foreign-trade and price politics, financial and credit relations; financial, trade and industrial integration of the major countries under investigation, as well as on questions of industrial co-operation and scientific-technical co-operation.[25]

The informations gathered were not just material objects but also business plans, market prognoses, insider trader information—everything that could help the isolated state get ahead in international trade. In effect, STI became the market research arm also of foreign trade organizations, passing on the best information to position them better in negotiations with capitalist companies and governments. This economic focus can of course also be seen as a reflection of the whole change in competition after Stalin's death and in the 1960s and 1970s in particular, as production rather than military confrontation became paramount:

The development of international relations, the transfer of the main struggle of the two systems into the sphere of economics, science and technology, the decisions of the 10th Congress of the BCP and the following party and state decisions created new, extremely important and responsible tasks for our intelligence.[26]

As science was to be integrated into the economy as a productive force, electronics and cybernetics became an all-encompassing slogan for the BCP. STI was thus increasingly responsible for all economic intelligence rather than purely scientific-technical information throughout the first decade of its existence, and it became explicitly responsible in 1974:

Almost without exception all operative workers in the STI line can actively gain economic information too and vice versa—workers in the economic department have the possibility of gaining scientific-technical information. It is objectively wrong to divide one type of information from the other. Scientific-technical information cannot be complete if it also doesn't include production-economic information. The division of the tasks and leaderships in carrying them out, the division of the information itself, the difficulties in coordination of this activity that exist in the current structure and so on objectively hinder the raising of the effectiveness of intelligence work.[27]

This change of focus was reflected in the change from the more traditional geographic division of work to an economic sectorial focus. In 1976, four sections, grouped around specific technologies, were separated to organize agents' work. The first was on military technology, machine-building, metallurgy, transport, energy, and construction; the second on chemistry, microbiology, agriculture, and light industry; the third on economics; and the fourth on electronics.[28] It is worth pointing out that only electronics got its own branch of STI, while numerous industries were grouped into the same section at times. The geographic span of STI was also different from the traditional work of foreign intelligence, which often focused on Greece and Turkey. Bulgarian STI focused disproportionally on Japan and the Third World as areas of high technology or on countries where Western technology could be sourced outside CoCom restrictions. Thus, countries like Japan and Sweden, who played looser with the embargoes, were grouped together with India as a huge market for First World firms.[29] Countries such as Nigeria, Libya, and Iran were also areas of active interest due to the logic mentioned above: Bulgarian enterprises needed to know how to position themselves in markets that the country was very interested in.[30]

The service was also increasingly professional, with regular department-wide evaluations on completed work every three months, and weekly and monthly agent reports on their own work. The most useful "finds" were inventoried every three months, and results of cooperation with other intelligence services were evaluated every six months, helping streamline further cooperation.[31] The growth in paperwork also meant that STI and the DS as a whole became one of the first computerized organizations in Bulgaria, putting in practice their electronics focus. By the early 1980s, STI was using the ISKRA ("Spark") system, the catchier name for "Automated Information System of Counter Intelligence," the conglomerate of computer means and programs for gathering, storing, processing, and retrieving information for intelligence agency users. ISKRA collected and stored information from the full variety of activities of the intelligence community, ranging from agents to cover firms and information gathered within the country, aiming to create a "unity of information" in conditions of an increasing avalanche of such.[32] It drew on the KGB's "Photon" system, which had been in operation since 1966, which, however, seems to have been much more problematic and ran into an inherent institutional conservativism.[33] The ISKRA underwent constant modernization, allowing for the distribution of information to the right users, cross-referencing of data, and a classification system that allowed agents to see lists of sources suitable to particular tasks (such as existing information pertaining to electronics plans for the year). The Bulgarians aimed at creating complex algorithms rather than linguistic-based databases, to streamline the searching process. The system was based on a control computer in the main center of the Ministry of the Interior, linked to STI offices through satellite SM-4 terminals that ran through encrypted lines. A test volume of "false information" was created to test the system's capabilities, and special in-house software was also created to facilitate the network connections, something the KGB was impressed by and wished to try out.[34] One of the only surviving normative documents on ISKRA is from 1990 as the service tried to reorganize during the transition to democracy, when rules were added that it could not be used to harm the rights of Bulgarian citizens, but a glimpse remains of its original purpose. The document states that an important task is to prevent the overlap of operative activity or information gathering that has already been carried out, thus increasing the economic efficiency through this rational system. Moreover, by the end of the regime, the data were no longer only quantitative but analytical, providing reports to agents, too.[35] ISKRA was thus part of maximizing STI's returns, which helped make ISKRA an organization that itself was at the country's cutting edge in electronics.

As STI organization grew, so did its reach. Operations were expanded in Africa, East Asia, Latin America, and the Middle East, and its scope and successes became magnets for other socialist services, growing in stature relative to the state's size. By 1983, it was helping Mongolia set up its own STI section, as the state wanted to follow the Bulgarians in its focus on civilian rather than military matters. The Bulgarians passed information on structure and organization, but they also offered three-month training courses for cadres as preliminary steps to the fuller Soviet 10-month course.[36] Other countries, such as the Cubans, availed themselves of other STI expertise, such as the organization of work *within* Bulgaria, aimed at extracting information from foreigners and technical experts who visited it.[37] Vietnam's STI, created in 1980, also sought help with structural issues as well as specific help with reconnecting with Vietnamese agents they had lost contact with in various countries.[38] These contacts testify to the fact that by the 1980s, the state's STI was modern, active, and in a position of mentorship to newer services, a far cry from its own subservient tutelage under the NKVD and KGB in the late 1940s and 1950s. Once it focused on civilian issues, its cooperation with fraternal agencies, especially those of the USSR, also illustrate a much more independent path than that assumed by the DS as a whole. There is no doubt that Bulgarian secret service remained closely tied to the KGB and its Cold War missions, but its focus and annual cooperation plans also reflect its role as a facilitator of the computer industry's positions within Comecon.

COOPERATION IN AN ELECTRONICS KEY

Much like Comecon divided economic tasks, socialist intelligence agencies sought to amplify the amount of technology gained through cooperation. These annual plans also reveal the national interests that often ran counter to Warsaw Pact commitments, actual cooperation, and (most often) Soviet wishes. Bulgarian dealings with the Soviets in particular show the clear importance of the Soviet nuclear and military umbrella, which created the space for the smaller states to focus on profitable areas, such as computers. As Charles Tilly memorably put it:

> As the twentieth century wore on, however, it became increasingly common for one state to lend, give, or sell war-making means to another; in those cases, the recipient state could put a disproportionate effort into extraction, protection, and/or state making and yet survive. In our own time, clients of the United States and the Soviet Union provide numerous examples.[39]

The Bulgarian military was armed largely by the USSR, so there was little need to develop cutting-edge equipment to counter NATO, as the Soviets provided the advanced fighters, precision artillery systems, submarines, and other hardware needed. The domestic military-industrial complex grew immensely during the period, becoming a comparably golden sector to computers, but it too was export oriented—its assault rifles, shells, missiles, and armored vehicles sold to national liberation movements or friendly regimes, especially in the Arab world. By the early 1980s, the positive trade balance with the Arab states was $1 billion per year, mostly because of the weaponry exports.[40] While many of these weapons also equipped the domestic armed forces, the Bulgarian military industry did not have to produce the full gamut of arms needed by a late-twentieth-century military—it could simply rely on the Soviets. Thus, STI had no incentive to concentrate as much of its focus on military matters as the USSR did, carrying the burden of Warsaw Pact military research and production. The Minister of the Interior in 1972, Angel T͡Sanev, summarized the directorate's thinking and its goals most clearly by expanding on this strategy:

> What political processes there are in a capitalist country—I have no affinity for that, maybe I am mistaken. And for our intelligence workers to circle around these questions only, that is playing at intelligence. That is a waste of power and means. There are new times in intelligence work in the world. Gone is the time of separate intellegences, where each country aimed at knowing the secrets of the other. Now other powers solve big questions. That is the USSR, the socialist camp, economic power. In questions of war—that is the rockets of the Soviet Union. They solve the questions.[41]

As some officers objected, stating the country should have its own military intelligence focus, he shot it down: "Let's look at things realistically. How will we gather information for military production? How do you enter into a military factory, how is it guarded?" (a passage which the STI reader underlined and marked as "Correct! Realistic!" in his handwriting in the copy of the minutes).[42] Under the Warsaw Pact umbrella, Bulgarian STI was free to "create a mature socialist society and develop Bulgaria as the leading socialist country in the Balkans, so it can be an example to other Balkan countries. Our task is also the fuller satisfaction of the material needs of our people."[43] This logic also drove STI's search for electronics information in its cooperation agreements with its socialist partners.

The KGB was unsurprisingly the biggest and most important partner. Between 1966 and 1971, they passed on 816 "informations," totaling over

155,000 pages of documentation, in exchange for 687 Bulgarian reports (57,000 pages) and 44 chemical and microbiological strains.[44] The Bulgarian STI establishment also possessed a key asset for the Eastern Bloc, agent "Delon" (named after the French heartthrob), who was a great source of American and French military secrets, and at least for a time worked through the Swiss residency.[45] Together with agents "Hans" and "Frederick," he was moved to exclusively military tasks "of interest to the Soviet comrades" in the early 1970s, who praised him highly.[46] In 1974–1975, he alone sent over 1,000 reports to the Soviets.[47] By 1976, after another 505 key packages gained by him, he was passed over to the Soviets, allowing him to send his findings straight to the KGB[48]—an extremely productive asset, which earned STI valuable points with the Soviets. "Delon" gained important information straight from the US Department of Defence, such as "instructions to the US Army and Navy, materials on the developments of the aviation industry, as well as fundamental and military developments in electronics, chemistry and nuclear energy." In addition, agent "Frederick," another key asset, passed on information on polymers applicable to aviation technology, data on NATO avionic instruments, artillery and antiaircraft specifications, heavy tank armor designs, and other valuable material.[49] This intel was still not enough, as in 1976, internal STI evaluations stressed that their achievements "are still not enough and don't fully cover the tasks of the information plan" vis-à-vis Soviet expectations.[50] This was a common thread in Soviet-Bulgarian intelligence relations. The year of Delon's greatest success, 1976, was also the height of military information gathered by the DS—35 percent of all such information gathered by the agency[51]—an outlier compared to most years, when it could be as low as 6 percent.[52] The Bulgarians freely admitted in their internal reports that any focus on military secrets was Soviet driven rather than serving any domestic military requests.[53]

Instead, the KGB became a source to squeeze as much electronics information out of as possible. Reflecting economic specializations, STI called for coordination in areas of common interest, specifically mentioning computers, as the "USSR, GDR and Bulgaria are responsible for the creation of [computers that are] different in their parameters."[54] Fraternal countries were noted as having "valuable scientific-technical information related to those machine models that are being developed by other countries," so the division of tasks in espionage, as well as the free exchange of information already gained, had to reflect the economic task divisions.[55]

The Bulgarians followed this guiding principle with great success. Already in 1968, the Soviets passed on 1,700 pages of electronic documentation, and 16 manuals separately, alongside 1,000 photos and one working prototype of a computer device; all received the highest evaluations by the Bulgarians.[56] By 1970, electronics dominated the exchange, accounting for 147 out of 212 informations received, far more than the second-place sector, which received only 39 reports.[57] In the later 1970s, up to 39 percent of all informations were on electronics, a huge boost to STI's achievements.[58] Between 1980 and 1985, it received 306 electronics (out of a total of 1,094), second only to chemistry and microbiology combined. Most of these received extremely high evaluations, as the KGB managed to supply documentation on Western CPUs used in the upgraded ES-1037 computer; operation systems for 32-bit machines; and database processing packages that were integrated into Bulgarian software sold to Comecon users.[59] This was always a two-way street, with the Bulgarians passing 469 electronics informations (out of 2,011 total) to Moscow, reflecting the vast cache of knowledge that the Bulgarians had amassed through their own activities.[60] In 1986, the KGB was even more forthcoming, providing 136 informations that were *all* evaluated as answering key needs of the economy—new integrated circuits, Winchester prototypes, mathematical models of MOS schematics, and 1.25-micron chip technology. But this was not all—a staggering 828 informations were passed in a separate "computer" category, which usually meant full documentation packages together with a prototype or working model, a massive boon for development. Important items included M-80 processors, which were used for the IZOT-1014 supercomputer launched that year, and programs for VAX-compatible machines that were key to the development of the next stage of the SM minicomputers in the Elektronika factory in Sofia.[61] In 1987, electronics virtually monopolized exchanges with the KGB, accounting for 1,044 out of 1,233 packages of materials passed.[62] In many ways, this covert exchange can be seen as the parallel and just-as-important sibling of Soviet technical assistance discussed in chapter 2—it is not an exaggeration to say that by the end of the regime, the KGB was another research arm used by the Bulgarian industry.

The East Germans were another close ally, perhaps building on the generally close DS-Stasi relationship, which stretched to the cooperation of capturing—or worse—GDR citizens trying to escape the bloc through Bulgaria's southern borders. In scientific matters, the Germans also heavily focused on electronics and high technology at the expense of military secrets, eventually harnessing such information to their own Robotron's attempts at creating an advanced chip industry.[63] The military aspect was so uninteresting to either side that in their first cooperation agreements from 1970, they agreed to not even bother sharing military intel with each other, but just with Moscow, further evidence of the satellites' interests running quite counter to those of their superpower backer. The closeness with the Germans rivaled that with the KGB, as the Stasi trained some STI experts in the usage of their own in-house computer systems.[64] Berlin also suggested joint financing of expensive purchases from foreign traders willing to overlook the embargo, something never suggested by the Soviets.[65] Both sides were very open with each other in terms of sharing their successes, and both services evaluated the others' achievements highly—all work was evaluated as mutually beneficial by both the GDR and Bulgaria.[66] By the early 1980s, Bulgarian spies had bolstered German efforts by passing on discs in the range of 317 to 800 MB, as well as prognoses of West German electronics development.[67] They got much back, with 216 out of 258 informations in 1982 being about electronics, reflecting how the Bulgarians too had successfully steered the German cooperation into their desired sphere, and the payback for their openness with a competitor.[68] A 1985 STI internal report informed higher-ups that "the information exchange with STI of the GDR is developing mainly in the direction of "Electronics and Computing Technology," as agreed by the leaders of the two STIs," with future developments targeting microbiology and chemistry, as well as certain sectors of machine-building—all areas that the Bulgarians were trying to focus on in the 1980s as part of a drive to accelerate the economy.[69]

Other services were also helpful, bolstering Bulgarian efforts and amplifying the information that entered the industry far beyond the state's own efforts. The Czechoslovaks were seen as helpful, despite concerns over the quality of their cadres after the 1968 purges of the service.[70] Yet they were responsive to requests, with around a quarter of informations delivered in 1982 being in electronics, including valuable complete terminals, processors, and operating systems, a veritable coup.[71] By the mid-1980s, the help was evaluated even more highly, especially in the areas of robotics and microchips, where the Czechoslovaks delivered documents on the latest devices in the US, Japan, and Sweden.[72] In Latin America, where the STI's own network was less developed, the Bulgarians grew close to the Cuban services. Their allies had wide networks among Cuban émigrés in the US, including some who worked in electronics research, as well as well-developed operations in Mexico, which became a safe haven for meeting American sources. This helped make computing information the bulk of exchange between Havana and Sofia, too.[73]

Yet national rivalries did play into this sort of cooperation, too. While the East Germans or Czechoslovaks calculated that what they got back from the Bulgarians was enough to offset their helping an economic competitor, troubles with the Poles and Hungarians show that there was no seamless socialist intelligence cooperation. The Bulgarians themselves sent back less information than they received, and reluctance was the operative feeling among all those sharing. In dealings with the KGB, this can be explained by smaller staffs and thus reduced capabilities, but when dealing with the Czechoslovaks, for example (especially in the immediate post-1968 period, when their service was gutted), there was a concerted policy of extracting as much as possible while giving as little in return: 1983, the Czechs got barely 50 percent of what they sent the Bulgarians (77 versus 143 materials),[74] in 1985—even less (78 versus 162).[75]

The Hungarians and Poles, however, played the same game, and were more problematic. Into the early 1980s, the Hungarian services were not interested in exchanging much with Bulgaria, sending a mere 11 informations in 1976 and 33 in 1980, in return for at least triple that amount arriving from Sofia.[76] Bulgarian persistence eventually worked, with bilateral meetings highlighting "fraternal shortcomings" by Hungary and increasing the volume of exchanges, with 210 sent by Budapest in 1981 including important data on computer monitors.[77] The Hungarians were noted opponents of Bulgarian proposals in the ICCT throughout the 1970s, trying to bolster their own industry, and thus their own intelligence was withheld. By the 1980s, the Hungarians had switched to a tactic similar to that of the Bulgarians, encouraging allies to help out with exchanges, and from 1983 on, a third of all information was in electronics, including important software source codes.[78] The Poles, however, remained obstinate and didn't switch tactics. Since the 1960s, their own services had made electronics a priority area.[79] Between 1969 and 1973, only five informations—in total—were passed to Sofia, as both countries were vying for Comecon specializations.[80] At these early stages, much like the Hungarians, the Poles saw themselves as rivals, with each information being able to help Bulgarian industry in its relative infancy. The Bulgarians tried to spur reciprocity as they did with the Hungarians, passing on more information than what they received, but even when the Poles started responding after 1974, the computer information was often not that requested and was practically useless.[81] Analysts bitterly noted in 1984 that "the information exchange doesn't correspond to the real capabilities of both countries," and both sides had supplied each other with "too little of those [informations] that can solve a specific task fully."[82]

STI work in the Eastern Bloc was thus both a reflection of economic specialization and the philosophy of pooling resources to catch up and overtake capitalism. Rivalries and the chronology of specializations often were reflected in cooperative work or the lack thereof, while political closeness also paid dividends in intelligence sharing, especially with Moscow and Berlin. Bulgarian STI could thus benefit from a much larger pool of knowledge and agent networks than its own admittedly formidable efforts could achieve. Their own efforts and intelligence sharing combined to make these technology transfers often the more important source of Western knowledge, exceeding the usual technical assistance channels, as hundreds of items and hundreds of thousands of pages of documentation circulated between Sofia and other agencies each year. Yet, all this was being done in service of a civilian economy that was increasingly, both deliberately and unwittingly, controlling STI tasks.

IN THE SERVICE OF THE NATION

Ivan Popov's name is cited when talking about the birth of both the computer industry and scientific-technical intelligence in the Bulgaria. From the start of its existence, STI passed its informations to a special section of the CSTP, which evaluated it and implemented it into the country's plans. This section didn't work very well at the start, as it had a single worker, but Popov's reorganization of CSTP work in 1965 expanded it. In a secret report to Zhivkov, Popov highlighted the importance of STI work in overcoming the embargo and using "the experience of leading capitalist firms and research institutes . . . for the needs of socialist construction."[83] The CSTP thus expanded its analysis section that year, needing more people to both process the information and "efface" it before it reached Bulgarian users, obscuring the illicit channels through which it was obtained.[84] It is worth noting that Popov's stature even in the intelligence community in these years was so great that the Stasi asked that some of its documents on computing to be passed directly to him rather than go through the usual STI channels.[85]

Popov's vision is what harnessed STI work to electronics, under the purview of his CSTP policies. The annual work report from 1968 highlighted the fast pace of contemporary technological revolutions, leading to a constant process of economic reconstruction in industrial states. Every economy sought to protect its secrets even from allies in the search for advantage. To achieve this growth, there was a hunger for cadres in much of Europe, who sought short-term contract workers, often at low wages, allowing Bulgarian scientists and technicians to be placed abroad for short periods of time. The CSTP also had a specific understanding of the nature of the scientific professions:

> The wide scientific-technical exchange and cooperation on an international level to a large extent removes the nationalist character of many scientific-technical achievements. This leads to a lessened feeling of moral and patriotic responsibility among the people who have to share or pass on these innovations to the representatives of another country (even when it is a socialist one). They don't feel that this action harms the interests and security of their own country.[86]

Science was an international language, and STI could exploit the eagerness of intellectual workers to share their insights. Bulgarian technical intellectuals are of a similar opinion when asked. Petŭr Petrov, the deputy head of ITCR, recalls that a great deal of sharing was not STI-led but freely exploited by him and other Bulgarians who were operating in their own version of the scarcity economy, isolated from the latest Western developments.[87] He remembers, first of all, a much freer ability to travel and work in the early 1960s than most compatriots enjoyed—both due to the state's encouragement of its technical cadres to gain experience on the world stage, and due to the need for qualified workers in the West. He spent the whole of 1963 in Vienna, as after two botched operations, his wife needed specialized treatment there. He accompanied her and found work as a technician for the city communication network—"I wanted to find a job as an engineer, not just a technician, but as you know things were more complicated then . . . to give you an engineering job they wanted you to stay for five years which would have meant I'd need to throw away my [Bulgarian] passport."[88] His stay was a valuable source of connections too, as he made friends in the Austrian and West German technical communities. Afterward he specialized in East Berlin and used his connections to visit an IBM computing center in West Berlin, where a German friend took him around the premises, showed him the mainframes, and introduced him to colleagues. He recalls the visit as based on his curiosity and his friend's enthusiasm to show him where he worked, a shared passion between two professionals. With amusement, he notes that it took some time before he was asked where he was from—and when Bulgaria was mentioned, he was ushered out of the building gently but quickly.[89] His friendship had gotten him inside a building that would have been much harder to access for an agent. These connections also helped him in the late 1960s, at the Hannover Fair, where he visited the Texas Instruments stand but was denied a catalog of modules due to CoCom restrictions. He simply sought out a friend who was manning the Siemens stall, who delivered the catalog to him 10 minutes later, circumventing the embargo without a second thought. "I often received two Siemens magazines, one was on new developments, also an IBM one . . . I had friends there, over time we had become close friends, so they sent me the magazines regularly."[90] Professional friendships and networks were, as the CSTP suspected, channels open to exchange where personal relations could trump political restrictions. In Petrov's case, such considerations don't seem to enter the mind of his Western friends until another engineer points out that the person he is taking around IBM is from the enemy bloc. STI activities were thus paralleled by an undocumented but wide-ranging personal exchange system, a network of friendships complicated but not defined by the ideological conflict. Petrov even states that sometimes these personal connections were more infomative than STI intelligence:

often they [STI] collected things which weren't useful at all . . . one day, someone from that service brought me some work in German, because they knew I spoke it . . . and I look at it, and I recognise it . . . and I remember pulling it off my shelf and turning around, he tells me it is secret, and I say "oh I didn't know that, look, here it is" . . . so yes, they were somewhat stupid.[91]

Under Popov and the CSTP stewardship of intelligence work, it was no surprise that electronics became the main focus. The service's actions were integral to each annual plan's success, starting with 1969 and the national specialization, when STI delivered magnetic memories. From the very start, the industry would thus rely on STI, and this is reflected in that year's CICT report, stating that "the stance of CSTP is to not carry out [its] own scientific research work."[92] Throughout the next two decades, the institute would carry out its own research, but the fundamental fact was that it remained heavily dependent on the state's spy operations. In 1970, for example, 288 informations were gathered in computing, electronics, and related fields, while chemistry was the next most commonly scrutinized sector—and a very distant one at that, with only 65 acquisitions. At the same time, only two military informations were gathered.[93] In 1973–1974, as Popov's presidency of the CSTP ended, the whole of State Security was reorganized, and its new normative documents enshrined its efforts to help "with its means, forms and methods for the development of the national economy" alongside the political goals of defence of the socialist state and its laws.[94] Doĭnov, who inherited this conglomerate of science and intelligence work, expanded its efforts greatly, seeing them as the core of his ambitious Elektronika and Avtomatika programs. He reaffirmed the focus on computing, and he expanded STI efforts in acquisitions of robots and automation systems.[95] Popov's vision found its natural conclusion under Doĭnov, as the service became a veritable purchasing arm of the industry.

There were numerous successes that contributed to the growth of the computing industry. One of the first was in the field of the elements base, when a French engineer—codenamed "Bor"—was recruited while in Bulgaria as part of the French contract to help with the construction of the Botevgrad semiconductor factory. In 1965, Bor supplied the service with transistors as well as matrices for mass production, which were not part of the Bulgarian deal with the CSF company.[96] Another French engineer, codenamed "Tŭrkovski," showed the latest semiconductor schematics to a Bulgarian agent, who copied them.[97] The very base of the industry was thus inextricably tied to STI work alongside the agreed-on contract with the French—both the legal help and the ability to get the latest Western developments that were not part of the deal were important to the setting up of the factory.

The search for technology took agents beyond Western Europe, where much general intelligence work was done. By 1967, three agents were dedicated to scientific intelligence matters in Japan—again the illicit channel ran parallel to the aboveboard partnerships.[98] In Europe, the UK, and Austria were other areas of early success in electronic spying by 1970.[99] Agents gained experience and widened networks in these countries, targeting not just engineers but also secretaries, librarians, and those working in technical archives and specialized collections, as new sources for blueprints and plans.[100] Professionalization was also increasing, and by 1972, most agents in leading countries of interests were graduates of university-level physics or electronics course.[101] Over 60 percent had also undergone specialized training in the KGB school in Moscow, while the rest underwent a three-month accelerated course in Bankya, Bulgaria, based on Soviet methods. Significantly, 93 percent of these agents could work in at least one foreign language.[102] The difference between the 1960s and 1970s was marked due to these improvements: In 1971 STI acquired 16 times more informations than it did in 1966.[103] Twelve items of this long list, including magnetic memories, were expected to create an economic effect of 60 million levs when implemented—in terms of new production as well as costs saved on research.[104] STI increasingly turned to noncapitalist countries, recognizing them as spaces where embargoed knowledge was more easily obtained. By 1980, the Indian residence[105] had become a key channel, supplying more information than did countries such as Spain.[106] Agents operated extensively in Iran, Mexico, Tunisia, and beyond.

IZOT's memory device business was particularly dependent on STI's achievements. In 1974 the services managed to deliver every list in the annual plan that was designed to upgrade Bulgarian production for Comecon's next five-year period. This included 29 MB discs, 2x100 MB discs, and 4000 magnetic reader heads to cover domestic elements' shortfalls and ensure the production of 400 magnetic discs for Eastern Bloc users which would otherwise have not been complete.[107] But sometimes the operations went far beyond mere devices and manuals, as was the case for a spectacular coup in 1977. An unnamed American company producing discs and magnetic heads in the range of 29 MB was leaving Portugal in the wake of the Carnation Revolution, because of the fear that the country might go communist. The new government offered the whole factory for auction, and it was purchased and inventoried by an STI agent. IZOT deemed the factory to be a gold mine: two cargo planes had to be chartered from Spain, as well as a 15-ton truck, to ferry all this American equipment back to Bulgaria. Some of it was even smuggled out in a diplomatic car by Bulgarian officers. The haul included 286 objects of interest, the most audacious and large-scale electronics operation that survives in the existing archives.[108] Portugal seemed to be a gold mine for Bulgaria at this postrevolutionary time, it seems. Robert Pastorino, the American commercial attaché in Lisbon at the time, recalls how a factory for the US company NCR was taken over by a workers' collective. NCR produced electronic tills and registers, as well as computers, but not discs—thus it is unlikely to be the company in the operation above. Yet it drew the attention of Bulgaria as the workers, when meeting Pastorino, threatened that their "friends, the Bulgarians," were willing to provide the parts that the Americans now refused them. The Bulgarians also demanded money for those parts, however, and not in the worthless escudo.[109] In the mid- to late 1970s, Portugal thus seems to have been a potential gold mine of American technology for STI—as well as a source of money.

The Iron Curtain recedes further once you take stock of the annual STI reports, which read like shopping lists of computing goods for IZOT and other factories. Full documentation for 200 MB discs; the latest integrated circuits; microcomputer manuals; testing equipment; software; printed boards—this is just a 1978 snapshot, with each item earmarked for the Elektronika-S program.[110] In 1979, IZOT received 1,300 intelligence reports through STI work.[111] By the end of this program, STI's own efforts—not counting intelligence passed on by fraternal agencies—had acquired 38,000 pages of technical documents and 102 working models of embargoed goods.[112] Overall, by 1980, STI had acquired a massive 4,416 reports on electronics to serve the ambitious program, or 68 percent of all informations gathered: Electronics was the undisputed king.[113] Successes continued: In 1982, STI acquired a matrix processor that was stringently embargoed, with only 16 existing prototypes in existence at that time, which became a key input for the ES 2335 processor. For that year, it calculated that this acquisition saved the electronics industry 20 million levs in research.[114] Aims for the mid-1980s centered on creating the largest discs in the Eastern Bloc, in the 635 MB range, and IZOT managed to achieve up to 25 percent higher profits through intelligence efforts in updating production.[115] The decade's fledgling microcomputer industry was also heavily dependent on STI supplies of microchips, processors, and operating systems.[116]

There were, of course, multiple failures, too. Canada had been an important site of intelligence work, as it was

> the most favoured partner of the USA and there are practically no limitations in the trade and scientific-technical exchange, as well as in border or visa regimes between the two countries. The level of research activity is extremely high and doesn't lag behind the USA at all.[117]

Its market offered all the advantages of the sought-after American one, with much less of the risk. This changed in 1980, however, when a successful police operation hit the residency, and it never recovered.[118] In 1983, a Bulgarian trade representative in New York was arrested while trying to obtain nuclear secrets—one of many such periodic arrests of spies throughout the world.[119] Despite such shortcomings, it is not an exaggeration to call STI a lifeline for Bulgarian industry. A great indicator is how it took over control of such major residences as Milan, Frankfurt, Japan, and Canada, whose political work fell by the wayside.[120] STI also had overwhelming participation in the Norwegian, Austrian, Swedish, Indian, Mexican, and Singaporean residences.[121]

The focus on technology is of course part of the Cold War struggle for better production and thus a sort of socioeconomic victory of one's political/economic model, but it was more obliquely related to the geopolitical struggle than the usual intelligence work that socialist intelligence services are associated with. Subordinated to the strategy of economic modernization, the service became enmeshed with the civilian economy in complex ways. Increasingly colonizing ministries and foreign trade organizations, due to the need to provide covers for its agents, the CSTP nevertheless found that it needed the civilians as much as the other way around. STI work was intricate, and agents rarely had the scientific skills to properly evaluate items or analyze world trends when making their plans. Without overtly recruiting thousands of technology experts, the service had to come up with a way to make use of their brains.

THE SYMBIOSIS

The agents and scientists increasingly became symbiotically mutualist organisms, benefiting both sides. The CSTP always recognized that the STI in-house analysis section was too small to deal with the huge tasks that an advanced economy required. The intelligence service also knew that it didn't have enough specialists to coordinate plans and implementation of acquired technologies. The electronic sector of STI was well above the average for Bulgarian intelligence work, with most agents speaking at least one Western language by the 1970s and having some technical education. In fact, the DS as a whole reported in 1974 that STI work was very heavily slanted toward English-speaking electronic specialists, to the detriment of other scientific areas or languages.[122]

The overall level of DS training was often woeful. A study of the professional biographies of 47 leading officers of the service, done by Metodiev and Dermendzhieva, reveals a continuity at many levels between the supposedly more professional services of late socialism and the politically chosen one of the 1940s and 1950s.[123] Political loyalty and class background remained key to a swift rise through the ranks. Professionalization was evident, and by the late 1970s, the officers with higher education outnumbered those with primary school education only. Yet 475 of those officers were still trained only in the higher police academy, rather than possessing other university degrees. The most glaring problem remained languages—by 1978, only 2,028 officers spoke a foreign language versus 5,263 who didn't. Even worse, the number included Russian speakers, by far the most prevalent foreign language.[124] Grotesquely funny mishaps occurred, such as an agent not understanding the meaning of "doctor honoris causa" and reporting that a target was traveling to the US to meet a "Dr. Homoris Cauza."[125]

The situation for STI work was definitely better, however. From 1968 on, agents who were to be sent to key countries, such as the US, West

Germany, or Japan, were to undergo specialized KGB training. They also had to know a language, with reports of the time showing that all of those working in countries such as Italy or West Germany knew the language of the country at a good or excellent level. Strikingly, even Japanese proficiency was to be found, a scarce talent in those years—Petŭr Bashikarov was a student of Japanese, and an STI agent. He would go on to be a deputy minister of foreign trade and an ambassador to the country by the 1980s.[126] STI agents abroad thus did draw on the best that the DS had to offer, often at the expense of other departments. Another indicator is the professional biography of Georgi Manchev, the longest-serving and last head of STI, during its peak in the 1980s.

Born in 1941, Manchev's family had taken an active part in the wartime communist resistance. In school, he continued the political tradition, being active in the Komsomol youth and then graduating from the school for reserve officers in Shumen, Bulgaria, completing his national service as a senior sergeant in a reconnaissance platoon. He also had the technical training often lacked by his colleagues in DS, studying precision engineering in the prestigious Kiev Polytechnic Institute between 1960 and 1965. On graduation, he actually spent a year as a scientific worker in the BAS Institute of Electronics. The following year, he became an officer in the DS First Directorate and was immediately put in the STI line, heading a residency in West Germany. After completing the KGB school in Moscow, he was embedded in various trade organizations, and by 1971, he was in the most coveted placement—the New York residency. After 1976, he was the STI representative to the Ministry of Electronics, being a key part in liaison between the two during the Elektronika-S program. Around this time, US counterintelligence had blown his cover, so he was moved to Sofia headquarters, where his career continued its upward trajectory. Fluent in English and Russian, possessing technical training and having even been a scientific worker, having experience in intelligence work in the leading capitalist countries as well as knowledge of Bulgarian ministries, he became head of STI in 1979. During his tenure, the service expanded its electronic program, and his professionalism and importance was attested to when he rose to deputy head of the whole Intelligence Directorate in 1986. Briefly, after the fall of communism, he was the deputy head of the whole service.[127]

3.1 The future Major General Georgi Manchev as a trade representative in New York, 1972. (*Source*: arz.bg.

STI was thus led by a competent individual who understood both its intelligence and scientific missions, and the service had more well-trained agents than did the rest of the DS. Yet it never had enough officers to deal with the huge swathes of information it received, even after its 1980 expansion. The volume was overwhelming, but so was its nature—STI analysts were acquiring everything from computers to metallurgical information to chemicals and pharmaceuticals, nuclear secrets and even new breeds of livestock for agriculture. It became clear that only the country's civilian research capabilities taken as a whole could analyze the results of all these acquisitions. The solution to the problem of harnessing this capability without making tens of thousands of scientists agents of the DS was found in 1973 through the creation of the Centre for Applied Information (*T͡Sentŭr za Prilozhna Informat͡sii͡a*) or CAI, described as "a holistic system for scientific-technical and economic data," to organize the work of the First Directorate, orient its future plans, and ensure the rational usage of the information gathered.[128] CAI was also given the task of facilitating the work of foreign trade organizations by gathering and evaluating data on the "enemy's" financial-credit situation, to strengthen Bulgarian positions in negotiations over licensing or trade.[129]

CAI would go on to enlist the help of leading lights in Bulgarian science as evaluators of incoming information to determine what was valuable. It also created STI positions for each ministry, to liaise with it on the annual plan's drafting and fulfilment. CAI thus facilitated a permanent communication channel between STI and Bulgarian research institutes, planning bodies in ministries and enterprises, the universities, and BAS. As a cover, its institutional home was the CSTP's own Central Institute for Scientific-Technical Information, the state's paramount scientific information service. Peter Petrov states that many at the higher levels of institutes such as his knew of CAI's real function, but it never struck them as an interesting fact—just another Cold War reality: "As you know, copying was just how it was done."[130]

It started operation with 25 staff and had ensnared 666 specialists in its network by 1974, but it boomed immediately thereafter, and by 1976, there were 4,256 specialists on the books, with over half cleared to work on the most sensitive and secret information.[131] The last archival audit is from 1984, when CAI had over 7,000 specialists involved in its work each year, and given its growth rate, it is safe to assume that the number was even higher by 1989.[132] As a comparison, broadly speaking, 75,000 people were working in "science and scientific services" at the time, which as a broad category included lower-level technicians, library staff, doctoral candidates, and others who would not be likely to be under the CAI purview, so well over one-tenth of scientific workers in middle and higher levels were part of this network.[133] Engineers, researchers, professors, and higher-level specialists in research departments of enterprises were all involved. Their task was to receive the information through CAI and evaluate it, as well as participate in setting next year's targets. They had to grapple with a persistent problem of the socialist economy—how to shorten the road from acquiring a technology to implementing it in production.[134] The Central Institute was a valuable asset, helping in planning for the Elektronika and Avtomatika programs, often crossing out obsolete or superfluous technologies from STI lists. Its actions had positive effects—in 1979 they improved the acquisition of goods related to the state plan goals by 46 percent.[135] After the ISKRA system was completed, CAI's own computerized "subsystem of information for limited distribution" was evaluated as the most wide-ranging of any system available to intelligence services in the bloc, reaching the vast majority of users needed by the service.[136]

The relationship between STI and CAI became increasingly at odds, as the service saw the center as acquiring too much power and often evaluating things too differently from their in-house analysis section. A 1986 report betrays the feeling that the spies were in the service of the civilians, with agents having no independent power over CAI-floated directives. This was even more galling for them, as CAI was often late meeting its deadlines. Some agents even joked that "if you wanted a certain material to fall through, run it through CAI."[137]

The center did indeed have the superior position in evaluating goods and plans, backed up by CSTP and party directives, but it was ultimately also beholden to the institutes and people it relied on for expertise. CAI was always envisioned, from its inception, as a cybernetic feedback mechanism for its clients in national science.[138] Documents from CICT reveal the relationship that is often obscured in the intelligence archives. The leading computer institute treated CAI as its in-house information service and was often openly critical of its failures. While thanking them for their 1981 deliveries, CICT in the same breath requested an even longer list of IBM products for 1982, including items unrelated to its ES obligations. Others, such as 635 MB discs, were explicitly portrayed as indispensable to retaining primacy in Comecon and presenting a new model at the next meeting of the Council of Head Constructors.[139] In 1983, even larger volumes were requested, using past CAI successes, such as the acquisition of data that allowed them to circumvent Siemens patents, to demand similar help in new microcomputer patents.[140]

Civilian science, in the guise of CICT, was also not afraid to strongly criticize the intelligence-service center. In 1986, CICT reprimanded the center for sending it lists with generic names, such as "Graphics" or "Pulsar," which told specialists nothing about their usage or provenance.[141] There were also financial interlinkages that are apparent in the archives. Requests for certain VAX systems by Digital Equipment Corporation came with notices of about how much Western currency it would cost and what percentage of it CICT would supply.[142] CAI, and through it ultimately STI, were to foot the rest of the bill and in effect to act also as the procurement and financial officers for the institute. CAI was also often subject to extremely urgent requests, to be done within a mere week or two, such as 1987 requests for microcode loaders and diagnostic equipment. In the same breath, CAI was criticized for listing inflated prices for the goods it could get, as CICT willfully ignored the extra costs that intelligence agents had to pay due to the embargoes or the cost of its own operations.[143]

The trend was, however, one of general improvement and responsiveness on the part of CAI. CICT often thanked the center for its efforts, as in 1988, when they evaluated CAI's help as invaluable in keeping Bulgarian industry competitive and wining new markets in the USSR and Czechoslovakia, as well as helping "intellectualize" the economy. Teleprocessing system ESTEL's newest upgrade was also thanks at least in part to CAI efforts, ensuring good exports to the USSR, GDR, and Hungary.[144] The center always took care of its image among users, understanding its symbiotic relationship with them, and thus often turned its own criticism on the intelligence services as in 1978, when it harangued provincial police departments (who also participated in spying) for providing users with information that was openly available, thus undermining the STI system's reputation as an information supplier.[145] The very system it employed also made it an object of study by other agencies, who often raised questions in annual cooperation plans about its experiences. The Hungarian services launched a full-scale study of CAI organization and effects in 1981, in preparation for create their own center.[146] Even the Poles, often at odds with the Bulgarians, imported practices of evaluation and implementation, due to unsatisfactory organization in their own services.[147]

There was another even more direct way that intelligence meshed with the intellectual community. STI "colonized" institutes and ministries, using their postings as covers. The CSTP itself was the first such organization to be used, with its international relations and scientific exchange departments being used from 1967 on. Positions also existed in the ministries of machine-building, energy, chemistry, and foreign trade; in the radio-electronics and automation-related institutes; and in every foreign trade organization.[148] "Economic advisor" or "deputy trade representative" positions often were STI positions.[149] By 1971, 30 agents worked in Bulgaria as well, dealing with foreign experts who came as part of license agreements or to attend scientific conferences. To help with this, agents were embedded in BAS, CICT, and Izotimpex (IZOT's export arm).[150] These embedded agents fed back to CSTP decisions on who to send abroad, as agents could report on who had a loose tongue, who was in contact with

Western agencies, who andwas a potential leak—after all, STI was also a part of the massive DS repression apparatus, having the power to block travel and break the careers (and ultimately, the lives) of those intellectuals deemed unsafe.[151] Dimitŭr Stoĭanov, the Minister of the Interior between 1976 and 1990, saw STI work as potentially fraught, because it relied on technical intellectuals who often had no party experience or "blooding" in disciplined political work and demanded liberal applications of the "Soviet Chekist" experience.[152] The service was thus an important part of evaluating who would get the coveted specializations abroad, for both security reasons and the need to best fulfil its annual plans. CSTP covers allowed six STI agents to go abroad in 1980, but also another nine cooperating civilian intellectuals. All of them were selected due to their specialized skills and acceptance at important sites, often in electronics—one went to specialize in Columbia's computer faculty, another to the same department in Aachen Technical University, a third to the Illinois Institute of Technology, and a fourth to Japan on a computer specialization; others went to SUNY, CUNY, and Britain on a Leverhulme stipend in other high-technology areas.[153]

Manchev's biography also hints at the entanglement with ministries, where agents were placed as "advisors" to liaise between the civilian and intelligence plans. The Ministry of Electronics resisted in 1974–1975, feeling that CAI and CSTP were sufficient for coordination, and rejected the original agent, Major Milcho Galchev, who was qualified to be a "senior specialist" but not an advisor. DS called this bluff and appointed Galchev at that level, as the Ministry scrambled to find a way to block the agent's entry into the higher ranks of the institution. The Soviets had similar problems with their ministries, depending on a Ministerial Council decision to sway them, which was applied to the Bulgarian case, too.[154] Secret decision 141 of June 1975 thus created the positions of "advisors" in the ministries of electronics (which Manchev took), machine-building, chemical industry, and agriculture. They were to help with "questions concerning foreign expertise in the areas of science and technology."[155] Doĭnov himself assigned an agent as an advisor to the Council of Ministers, ensuring STI access to the highest echelons of economic decisionmaking.[156] The service was thus part and parcel of all levels of the scientific and economic communities, in service to these communities but also actively participating in their structure.

The cross-colonization of the economy and the intelligence apparatus was such that by the late 1970s, it is almost impossible to disentangle what is a legal and what is an illegal source of knowledge transfer between the West and Bulgaria. STI was instrumental in the creation of what Khristov calls "the empire of foreign firms" from the 1960s on, which numbered 450 by 1990.[157] Joint enterprises, projects, and companies were a way to circumvent the embargo, both in acquisitions but also in sales for hard currency. This story was also part of the rise of a managerial class that increasingly operated in the West both physically and mentally, in search of the manna that would boost the economy or their own interests. At the same time, CAI ensured that a parallel electronic professional class emerged and was kept abreast of world developments. We will come back to these stories in chapter 7, but they are worth keeping in mind when evaluating the role of spies in the national economy and above all in computing.

The 1980s were a decade of increasing trouble for communism and intelligence work, with the two being closely related. As détente broke down, and then Presidents Carter and especially Reagan ramped up the "second Cold War,"[158] work in the West became both more risky and more expensive. A 1981 defection by Colonel Vetrov of the KGB's Directorate T revealed the scale of Warsaw Pact espionage in the technology sector. The resulting "Farewell Dossier" revealed the identities of over 200 Soviet residents in the West and led to the CIA assessing that US science was supporting the enemy's industry. A deception operation of immense proportions was started, feeding the KGB faulty designs and false information, as well as arresting many key players.[159] Unverified legends abound, such as a 1982 gas-line explosion in Siberia, supposedly caused by a computer virus in a component sold through Canada.[160] What is certain is that CoCom restrictions were tightened, especially for dual-use technology. Bulgaria was not much affected by the Vetrov defection, but the country was swept up in the aftereffects and blacklisted as the closest Soviet ally. Its agency was already scandalous in the West, after the 1978 Georgi Markov assassination and the suspicions of its involvement in the 1981 attempted assassination of John Paul II. After Vetrov, too, Bulgarian ministries found it much harder to trade legally. The US Customs Service launched Operation Exodus in late 1982, intercepting high technology exports to the East, operating both reactively (reviewing export documents, searching cargo) and preemptively (investigating trade deals). By 1985, US Customs had seized 4,400 cargos worth over $300 million.[161] The effect was clear—STI work became much more expensive and prone to failure, and the agency was unable to carry out all its tasks. The effects were clear in terms of economic and scientific retardation. In June 1989, the US Office of Scientific and Weapons Research prepared a report for the CIA Directorate titled "Soviet Bloc Computers: Direct Descendants of Western Technology." The report concluded that the region was likely to remain dependent on acquiring Western know-how into the 1990s, compensating for domestic shortcomings, and that in some areas (such as disc peripherals and super mini-computers), they were up to 15 years behind the West.[162] It noted inflexible command economy planning; compartmentalization of knowledge, which restricted access to it; poor research coordination; and bureaucratic disputes, all contributing to the slow implementation of innovations.[163] The report's 56 pages also listed a comprehensive tally of Western technology copied by the East, such as "many critical parts illegally imported from the West" for Bulgarian PCs that were IBM compatible.[164] No matter how much they succeeded, STI simply had too much to obtain and was finding it increasingly hard to do as the 1980s progressed.

Analyses of the ultimate failure of STI to improve the Bulgarian economy hinge on the poor implementation of technologies by the enterprises and sectors. Spies' acquisitions simply got stuck between the gears of a grinding system. Metodiev, for example, holds that the intelligence services were far ahead of the civilian economy, and thus it was the latter that failed the former in its inability to put information into production.[165] To that should be added the costs of the materials being acquired, which became prohibitively expensive by the 1980s—the electronic sector's needs were part of the worsening financial situation of the country. The more expensive the goods acquired, the less profitable the goods created were, which was definitively noted by the end of the regime.

Understandably, many surviving participants in the industry are reticent to openly talk about any links with the security apparatus, a fear born out of being implicated as an informer on friends or even an indirect participant in repression. However, if talk is directed toward the general role of STI in their work, responses become more candid. Alexander T͡Svetkov, the director of the ZMG Razlog, the factory that produced write/read heads for magnetic discs, admits the huge importance of STI work for his own production: "After all, they [the West] were ahead of us while we were ahead of the USSR," and to maintain those positions, intelligence work was crucial. He recalls how his factory acquired Japanese heads that could read on a micron level, something unheard of in the USSR, and "they [Soviets] were amazed—how could we have it while they had no such prism? They still wanted to take each disc apart."[166] Stoi͡an Markov, the last director of CSTP and thus the last strategist of Bulgarian socialist science, dismisses the question of STI work being sensitive with the comment that "we all did it," referring to all participants in the Cold War.[167] This view is also shared by specialists further down the managerial hierarchy, such as Nedko Botev and Boyan T͡Svonev. The two were the lead developers of the CICT's sections on magnetic discs and tapes and winners of the Dimitorv Prize in 1971 and 1974, respectively. They shared the view that STI work was natural, not shameful, and was necessary in the conditions of ideological and geopolitical confrontation. "There was the embargo and CoCom . . . I could read CoCom lists, but we could only buy the lowest capacity discs and devices . . . and this, in fact, in this isolated system, created the conditions for this market. Because, after all, it exists objectively as progress carries on regardless," T͡Svonev remarked.[168] CoCom was what enabled the closed world of socialist market exchanges, where Bulgaria could carve out its paramount position among other late starters in the game—something that would have been impossible if it went toe-to-toe with the US or Japan in the 1960s. At the same time, no embargo could stop the march of history, as computers were the core of the Third Industrial Revolution, or the Information Age, whichever title the specialist chooses as a label for this time. STI work was natural for any country operating in a Cold War world, which at the same time wished to create a modern economy and society.

By the late 1980s, Zhivkov had found that Gorbachev was unwilling to play the Brezhnev game and reduced the flow of assistance to Bulgaria. At the same time, an internal report reveals how STI analysts saw the potential for future work in these changed circumstances. In a 1987 report in which the KGB thanks the Bulgarians for invaluable help in many fields, the Bulgarian writer evaluates this as recognition of the country's achievement and suggests further investment in this field, in line with the decisions of the Thirteenth Congress of the BCP and Forty-first Session of the Comecon.[169] By assigning more agents to this work, Bulgaria would continue increasing the professionalization of its cadres and gain better control of the flow of technology between West and East. Surprisingly, he suggests that such a lead allows Bulgaria to keep an eye on developments not just in the USA but also in Soviet industries. Bulgarian intelligence was thus invaluable for predicting the future developments of Soviet techno-politics and preemptively changing the state's course to better take advantage of these developments. This report could superficially be read as a child seeking approval from a parent, yet the writer's interpretation of Soviet needs was also a recognition of STI's close ties to the most important Eastern computer industry and its ability to allow Bulgaria the space for its own endeavors.

Ultimately, despite many failures, STI work was indispensable to the computer industry, and was increasingly dependent on civilian expertise, too. In a world with obstacles to the free flow of information, technology and access to it were highly politicized. STI was a channel of know-how transfer on a huge scale, and its choices of focus or cooperation with allies made it very different to the usual socialist security services' political goals. Under Popov and Doĭnov, this branch of the DS became a window to the West and an extension of the national economy. It was highly responsive to scientists and technicians, putting manuals, software, and whole embargoed computers on their desks. Its obstinate refusal to focus on military technology demonstrates how even the most loyal ally can make independent choices and follow its own objectives. By the 1980s, Moscow was in some ways just another market to be monitored.

Legal and illegal are, in the end, useless categories for the socialist scientific community. The Iron Curtain was a geopolitical fact that had to be circumvented or punctured. This was done to an unprecedented degree at the service of civilian economic interests, further calling into question the distinction between security service STI work and "the civilians." The transfer ensured Bulgarian domination in Comecon, its own world of rules and restrictions at least in part created by the CoCom realities. All that was created with this acquired know-how was to be sold, but not just in the

East. Other rules operated when selling to the West, or the Global South, and different things were learned through these interactions, especially in the sphere of business dealings. These trade dealings outside Comecon's peculiar world were in some way as important as STI work in learning about the world. It was in the emerging markets of the newly independent, developing world that the Bulgarian computer industry would open yet another window to the global economy. And to that we turn next.

4

ROSES AND LOTUSES: BULGARIA'S ELECTRONIC ENTANGLEMENT WITH INDIA

"I come to your land of roses from my land of the lotus," Indira Gandhi effused in her speech at the official gala dinner during her November 1981 visit to Sofia. Addressing the elite of Bulgarian society, not least Todor Zhivkov himself, she dedicated a Chair of Slavic Studies to the recently deceased daughter of the leader—Liudmila—before turning her attention to the heart of what her visit was about:

> your country and mine are engaged in economic development. We have pursued different paths but the goal is the betterment of our peoples' lives. For us, with our burden of size, the climb is steeper. Also there have been many obstacles, from inside and outside.[1]

While Zhivkov may have been in little mood to talk business in those personally difficult days, the rest of the Bulgarian elite definitely was not. Economics and technology were at the heart of the joint Bulgarian–Indian declaration after the visit, celebrating 14 years of intensive contact. This was Gandhi's second visit to the country, after her inaugural one in 1967, when she made Bulgaria one of her first ports of call after her election. By her death in 1984, India had become Bulgaria's biggest diplomatic target in the developing world, with the countries meeting 14 times at the highest level (foreign minister or above) and with two visits each by the heads of state. Helped by the cultural politics of Liudmila Zhivkova,[2] India also was economically important—the biggest potential market of the nonsocialist world, a potential gold mine for a regime that was always in search of convertible currency.

The socialist market was the most important for the Bulgarian economy, but increasingly, decolonization and the growth of world trade created another space for profit. The regime of course had an ideological reason to entangle itself in the Global South, driven by a commitment to helping national liberation and anti-imperialist movements.[3] Participation in this new world was the sign of a mature, advanced society, and while Zhivkov never harbored the grandiose dreams of his neighbor Ceausescu and his fantasies of Romania as a world power, Zhivkov still courted and bankrolled many anticolonial or newly independent leaders. It was all proof that Bulgaria was now on the world stage as an industrialized and urbanized country that could be copied, rather than just copy other countries.

Bulgarian technicians and engineers increasingly became a common sight in construction and technological projects throughout North and parts of sub-Saharan Africa, the Middle East, and Asia. Often they were engaged in urban and infrastructure development, agricultural modernization, and industrial and military enterprises.[4] However, as the Bulgarian computer industry grew, it also turned to the developing world as a space in which to prove itself. The regime was always looking to improve its nonsocialist trade, and Bulgarian computers found it hard to crack the Western market, where competition was stiff, technological demands were great, and the embargo already put them in a disadvantageous position from the start. The developing world, however, was a different matter. Regimes and users there still sought the best technology, but for political and financial reasons, the inferior Bulgarian products could make their impact in places with less experience in the electronics sector, which was true for many of the newly independent countries. This was even more marked in explicitly socialist states outside the immediate Eastern Bloc, which favored Bulgarian computers as the most readily available in the socialist market.

These openings to the Global South have often been explored through the prism of development.[5] Discussions have centered on the US contribution and its modernization themes, focusing on efforts to tackle overpopulation or increase agricultural yields.[6] Often these efforts focus on the competition aspect. As Nick Cullather summarizes it:

Projects were designed for 'display' to produce statistical victories or as carefully staged spectacles dramatizing the fruits of modernity. They were also composed, usually from inception, as 'models', formulas to be replicated at later times and other settings. Finally, narratives defined lines of conflict in development politics. Models were pitted against each other as tests of allegiance and modernizing prowess.[7]

Socialist modernization was a competitor, something to be proven correct and beneficial for all humanity, and thus a source of prestige. Fruitfully, Unger and Engerman have called for ways to move beyond the focus on the US and modernization theory. First is the decentering of the nation through a focus on international organizations, such as the IMF. Second is to go beyond the 1950s and 1960s, or the so-called "American decades," and look at the 1970s and beyond, when the efforts of Europeans and other actors took off. Finally, the eye-catching heavy industrial projects are not the only faces of development—there were a multitude of ways to modernize.[8] This chapter takes the last two approaches seriously, by looking at both the electronics sector and a small state's efforts to enter this world. However, socialist states did not participate in some of the international organizations that have a bearing on development, such as the IMF. Such a focus again privileges the First World's effort. This chapter argues that there are other ways to decenter the nation, namely, by looking at substate rather than suprastate actors, such as companies and enterprises.

Socialist development literature has expanded in recent years, adding support for such an argument. Hilger, in his overview of East German policies toward India, emphasizes the enhanced economic contact between the two states as a priority for the socialists.[9] Despite the anti-imperialist rhetoric, trade and bilateralism were key—an approach paralleled by their West German counterparts.[10] It is this line of argument that this chapter picks up—India as a space to do business and learn how to do business, rather than as a laboratory for development and modernization.[11] The various conceptions of development, seen as a solution to the problems of backwardness through economic progress and social engineering, or development as a clearly ideological battlefield between the two Cold War sides, are not the focus. They do, of course, play a role in the argument, as India itself modernized and sought to hew its own nonaligned way, while Bulgarians made certain grandiose claims about the merits of their own technology and its applications. Bulgarian development experts, however, were not the ones who worked in the electronics industry—they were instead participating in creating new agricultural or heavy industrial enterprises. Yet not everything that foreigners did in India has to be seen through the explicit rubric of development. Instead, this chapter shows how participation in the developing world markets had an impact on late socialist economic thinking by pushing it to adopt business practices that were often seen as capitalistic. Socialist enterprises had to adapt due to competition, and they took back what was learned in India to the way they marketed and traded in the socialist world. If we are to answer the questions of how we make capitalism without capitalists, it is this Indian experience of the developing world as a market rather than a laboratory that is illuminating.

The electronics industry, created for profit by a socialist state that was facing the dual problems of uncompetitive exports and stagnating social productivity, was international by its very nature: The computer was not created in Bulgaria, so those tasked with its creation and selling had to be mobile and competitive. In India, they faced a protectionist state that was hewing its own path to national modernization.[12] The Indian market was also more open to Western technology. While Comecon specialization agreements allowed Bulgarians access to huge markets whatever the qualities of their goods and services, in India, they had to compete with the giants of the electronics industry, be they American, Japanese, British or even Soviet. In India, the Second World met the First on the grounds of the Third. These encounters changed the way that Bulgarian socialists did business throughout the rest of the world. The Global South was not just a place to sell, or to develop, but also somewhere to learn how to be modern businesspeople and capitalists, an important space through which the modern ideas of the emerging Information Age could feed back into socialist enterprises and thinking.

This chapter thus follows the Bulgarian computers sold throughout Asia, Africa, and Latin America, as the regime sought hard cash. In its encounters with India, however, it is not development efforts that tell the whole story of East-South exchange. The computer had to overcome Indian protectionism, Western competition, and Bulgarian incompetence. The chapter thus follows these encounters with the markets in the Global South, which pushed the industry to modernize in terms of its business practices, and it shows how these dynamics impacted Bulgarian conduct elsewhere, too.

FACING SOUTH

India was gargantuan, important, and fascinating for the Bulgarian regime, but it was not the only newly independent state that it was involved in—the Arab world was another area of concentration, a space for Bulgarian arms and construction specialists. Throughout the 1950s and 1960s, the prolonged end of European imperialism created fertile ground for a rapidly modernizing and increasingly self-confident state. Bulgaria was finding itself able to furnish the machines needed for modernization but importantly also credits, expertise, and education that newly independent states sought. In that the country was no different to its other socialist ally, even if started late, again, compared to states such as Czechoslovakia, which were delivering arms worldwide since the late 1940s (most famously, to Israel). With that logic, a 1960 Politburo protocol stated that the first job of an independent state is to form an army, and the protocol defined a goal of exporting arms to Iraq, Indonesia, Tunisia, Guinea, Ghana, Morocco, North Korea, Vietnam, China, Mongolia, and the United Arab Republic. This would alleviate hard currency problems, and the acquisition of Bulgarian arms would capture a market that could be flooded by other goods and specialists.[13] This goal helped pave the way for the creation of Texim in 1961, a foreign trade firm founded in Liechtenstein, becoming the behemoth of Bulgarian trade during that decade on the back of arms sales.[14] Bulgaria became the first country to deliver arms to the Algerian national liberation movement, a 10,000 ton ship arriving in the country after a Council of Ministers decision in April.[15] Weaponry opened the way for wider Bulgarian trade with the South.

Involvement became deeper and more wide ranging than just guns, with a permanent exhibit of Bulgarian produce in Port Said planned, as Egypt was a major market.[16] Engineers and medical professionals also became a much more common sight in North Africa, and by the mid-1960s, they numbered 230 in Tunisia and more than 700 in Algeria.[17] Urban planning and engineering bureaus won contracts over competing Polish and Czech projects in Afghanistan, with Kabul demonstrating a preference for

Bulgarian technicians, citing professionalism, punctuality, and competitive pricing.[18] Throughout the 1970s, the state's focus expanded beyond the Middle East and North Africa, becoming a bigger presence in developing countries. Specialists helped Angola with agricultural reforms of their Portuguese-inherited fazenda coffee plantations, as well as setting up medical and education services, canning plants and food industry sites, and metalwork enterprises.[19] Tanzania received expertise on metal-cutting factories, irrigation pump production, and technical schools.[20] In Ethiopia, specialists worked in civil planning, public health, higher education policy, and agricultural and industrial development plans.[21] Even a cursory glance at the archives reveals the names of most newly independent Asian and African states. It was telling that in the 1970s, Bulgaria's airline started new direct services to places like Lagos, Salisbury (today's Harare), and Colombo. The party's thinking in these openings to the world intermingled ideological and economic concerns, as the states "freed from colonial yokes" were both allies in the struggle against capitalism and "beneficial economic partners and in practice should occupy second place [after the socialist world] in our foreign economic and trade activity."[22] The Global South was, sometimes literally, a gold mine. Bulgaria sought economic benefit from places that were more open to its goods than the capitalist market, while at the same time building prestige as a country that was now demonstrably modern. Building schools or factories in new states was a self-defining exercise, too.

It also became a provider of credits and education. Despite its own debt problems, Bulgaria became a creditor to many developing countries. Examples include a loan of 5 million levs to Mongolia in 1968[23] and 3 million levs to Tanzania in 1972 to finance the building of factories.[24] India received a $15 million credit in 1967 as part of Bulgaria's presentation of itself as a prosperous partner to the huge South Asian state.[25] Altogether, between the sort of "inauguration year" of 1961 (which marked a more active opening to the developing world) and mid-1976 (the cusp of increased involvement in Africa, especially Angola), Bulgaria extended $600 million in credits to the Global South, $390 million of which went to the Arab world. Only 27 percent had been utilized by 1977, however, so the Ministry of Foreign Trade pushed for a stricter framework on what could be purchased by such arrangements, as well as prioritizing machine exports over services.[26] This was an indicator of the continual problem of placing finished goods elsewhere, and of Bulgaria's international credentials often based on credit more than on goods.

Out flowed the credits, in flowed the students. Bilateral agreements with Sofia encouraged such exchanges with many developing states. Bulgarian higher education was one of the regime's biggest success stories, and by the 1970s, it led Comecon in admissions of foreign students per capita. In 1972 there were, in absolute numbers, as many foreign students in Bulgaria as in much larger Poland, and twice as many as in Romania.[27] A special accelerated language school was set up, the Nasser Institute, and students were accepted into the priority areas set by their home countries: Algeria, Guinea, and Peru wanted engineers; Yemen and Bangladesh medics; Iraq and Afghanistan agronomists.[28] Education was not just to create specialists but also to inculcate the right worldview:

> The main goal for us with regards to the foreign students is to prepare good specialists, creatively thinking personalities, ready to actively fight against imperialism and neo-colonialism in the name of the national independence and social progress of their countries and nations.[29]

Sometimes the opposite effect was achieved, as was the case with some Mongolian students. In 1973, a report stated that much of the state's intelligentsia was trained in European socialist allies, where living conditions were better, and once back in Mongolia, they indulged in drunkenness and developed anti-party feelings: "the difference between those who sit on the camel all day, drink kumis and live like how their ancestors did, and those who educated themselves in the capital of an European socialist country, is huge."[30] Such prejudiced conclusions, however, did not stop Bulgaria from educating increasing numbers of students, who became another channel for increased Bulgarian interaction with the world.

The party, however, always saw increased trade as its priority. Trade exceeded 1 billion levs in 1978, of which 760 million were Bulgarian exports.[31] Trade was particularly strong with countries that exported commodities the regime needed, such as rare food items, metals, and above all oil: Algeria, Iraq, and Iran each took in over 100 million levs of Bulgarian exports in 1980, and Libya lead the way with over 350 million.[32] The developing world never managed to become that second biggest market, but by the 1980s, it did mature into a significant trading partner, accounting for around 2 billion levs of Bulgarian exports and just over 1 billion of imports in 1985 (out of around 14 billion levs in each category).[33] Importantly, while trailing the capitalist market, the balances were positive for Bulgaria, which was always untrue for trade with the West. Such trade also increased Bulgaria's presence in world trade and brought the country success and recognition that was often harder to achieve in the West due to embargoes and quality.

PROFIT AND PRESTIGE: ELECTRONIC PATHS IN THE GLOBAL SOUTH

Computers, the crowning glory of Bulgarian advancement, increasingly became a part of the state's strategy of profit and prestige in the Global South. As the domestic industry took off in the 1970s, electronics followed the path trod by rifles, civil engineers, and medics. By the 1980s, the IZOT files bear witness to a wide-ranging presence of Bulgaria's goods, placed in countries both socialist and not: Algeria, Egypt, Iraq, Iran, Syria, Libya, Mozambique, Angola, Zimbabwe, Nicaragua, China, Vietnam, North Korea, Cuba, and many others.[34] All this brought money, but also luster for the party, as Bulgaria presented itself as an advanced, developed society at technology's cutting edge. A few deals out of the many that the archives can trace illustrate the general aims and effects of such trade.

Prosaic as it is, profit was again a leading reason for entering many markets. The Ministry of Foreign Trade made electronics a priority to expand exports to Syria and Libya in 1977, after Bulgarian technical experts had been in the country for some years, assisting or leading many infrastructure projects.[35] The reports stated that machines must become the leading share of exports there, with computers being the leading edge of this campaign. "Our products must be realized against convertible currency" the report went on to say, as currency was one thing that the regime lacked the most while needing it to purchase the latest technology from the West.[36] Egypt was another target, with trade projected to reach 220 million levs by 1980. Bulgarian computers were already present, so pure export was not needed; instead the ministry was trying to focus on joint bureaus to service and upgrade ES-1020 machines. This would foster confidence in

Egyptian users and improve the chances of further sales and Bulgarian-equipped computer centers, such as with the new IZOT-310 machines.[37]

New markets were opened beyond the usual Middle Eastern and North African ones, such as those in Sub-Saharan Africa. In 1981, an ES-1035B computer center was sold to Mozambique.[38] This was the newest processor that Bulgaria produced, and it would not only bring in money but also create a whole new market where there had not been one—the sale was also justified on the grounds that it would bring a continuous need for spare parts, peripherals, and discs.[39] The trade representatives noted that the state had just created a Centre for Data Processing, and equipping it ensured a captive future market. Mozambicans were flown out to Sofia to meet representatives of KESSI, the Bulgarian committee for creating a nationwide information network. It worked, as the ES-1035B sale went through. A similar tactic was employed with the Nicaraguans, who also bought an ES-1035B computer center in 1981.[40] Elsewhere, the Bulgarians were more devious: In 1983, they gifted Zimbabwe an IZOT-1007S computer center, together with free training for local specialists, who were flown to Bulgaria.[41] These actions were calculated partly as an internationalist move, but also as again setting up a market that would be expanded later.[42] The tactic succeeded, as by 1987, Izotimpex was the best-performing foreign trade organization in the country, selling computers and copiers worth 3 million deutschmarks, often at three times the price they would fetch on the capitalist market. This success was used as a springboard to expand into neighboring Tanzania.[43] Angola, Nigeria, and Afghanistan were also seen as potential cases for the same type of expansion through offering SM-4 minicomputers; the ES-1035B system was deemed as the right way to enter Ethiopia and Yemen.[44]

Prestige and profit mingled in each of these sales, as every success in equipping a developing state with computers was proof of Bulgaria's own success in the space of a generation. The scientific-technical revolution had become by the 1970s the paradigmatic ideological content of both Soviet and Bulgarian socialism, as we will see in chapter 5. If Mozambique, Zimbabwe, or Nicaragua wanted to enter the age in which computers were key to national development, Sofia was the place to look to for assistance. This entanglement had been evident in previous Bulgarian efforts in Africa, when telecommunications experts were sent to Angola in addition to the irrigation and medical experts who had been requested by the government to set up the basics of its economy and social provision. Bulgaria insisted on being well placed to provide the country with the most modern radio and TV stations, too, which would be key to consolidating the regime.[45]

The last decade of the regime brought increased sophistication and more range, as robots and personal computers also entered production and more and more Bulgarian workplaces. The foreign trade ministry thus also put such electronics on display abroad. When the Bulgarian Trade and Commercial Chamber didn't include any electronics in its plan for the 1980 National Exhibit in Libya, IZOT and the Ministry of Foreign Trade insisted on revising the project. Swinging the pendulum the other way, the exhibit was flooded with ES-1035B machines, the newest ESTEL teleprocessing systems, SM-4 minicomputers, and CAD systems.[46] This effort reaped benefits throughout the decade, as by 1987, IZOT was preparing the export and installation of complete computer labs in Libyan colleges and technical schools, as well as training the nation's specialists in computing and automation.[47] The two sides signed an agreement for the joint development of programs and other "means of information technology," with Bulgaria taking a leading role and training Libyan programmers.[48] By the end of the regime, Bulgarian informatics specialists were sought after in the socialist world, with Vietnam requesting help from Sofia rather than Moscow in the application of microcomputer systems to national economic tasks.[49]

In two significant ways the developing world market was linked to the domestic one—in technological and business terms. Bulgarian socialism was much more open to experiencing realities in this world rather than the West, which was closed to it thanks to CoCom. But developing technology for these local clients also overlapped with developing the same technology back home. While birthed as an export industry, Bulgarian computing spawned an intellectual boom in the country, opening up vistas of automation for the regime and its experts. These were of course part of worldwide trends, including in the Global South. The first Bulgarian computerized supermarket opened in Sofia in 1977, with a central computer linked to both inventory and electronic tills.[50] Its fate will be discussed in the following pages, but it is important to note that it was not an isolated story domestically: In 1975, Bulgaria had signed a deal to build 54 supermarkets "with a high degree of automation" in Libya. The Ministry of Electronics was to participate and seek ways to create the most modern shop for the Libyans, complete with automated inventory tracking and accounting.[51] The planning and building of this massive project in North Africa was thus connected to domestic computerization—Libya was a field to try out the new technology and a test of what could be done at home too. Computerization of warehousing and inventories was a priority area from the start of the Bulgarian industry and made the country a preferred partner in such endeavors. The IZOT-0310 system (a copy of the PDP-8/11) was an area of concentration from 1971, precisely to automate such work, as were the program packages for it.[52]

Similarly, in the 1980s, Bulgarian education began its computerization with classes in programming, while the DKMS (Dimitrov Komsomol Youth Union) opened a network of computer clubs in towns and schools. In this case, these domestic programs were also exported abroad as a showpiece of Bulgarian personal computer development and its accompanying educational possibilities, and as a model for education in the new age more generally. DKMS club programmers and pedagogues were thus sent out to set up a computer center with 20 Bulgarian PCs for the Hanoi branch of the Vietnamese Communist Youth Union's Central Committee. With Bulgarian help, it grew to be the biggest computer center in the area, and the only one that the Vietnamese party deemed professional enough to train programmers, concentrating its software education at the center. At their request, Bulgarian experts' contracts were renewed to train more local programmers. The teachers delivered 52 lessons per course through a local translator, lecturing on BASIC DOS, and Bulgarian software. While the computers themselves worked well, only one of 20 floppy drives were delivered in working condition, so often they had to improvise while awaiting the technical documentation that would allow local repairmen to service them. Still, it was a success, testified to by being given the task to automate certain internal administrative duties for the party's Central Committee.[53] A similar club was set up in Pyongyang.[54] The state was also contracted to equip up to 4,800 modern classrooms for the study of electrical technology, automation, and mechanics in Nigeria, a reflection of IZOT's growing prominence as a trusted supplier of advanced education equipment too.[55]

Non-European socialist states were also willing partners. Kim Il Sung had stated to Zhivkov in 1973 that small states, such as North Korea and Bulgaria, should concentrate on their own scientific strengths rather than following multifaceted developments like the Soviets did, so small partners needed to supplement one another by cooperating.[56] By the end of the decade, North Koreans were thus importing whole computer centers from their Balkan socialist brothers.[57] In 1985, when Pyongyang turned toward creating a domestic electronics base, it was specialists from IZOT and CICT who helped plan the regime's new factories.[58] The huge Chinese market was cultivated, too, with IZOT producing specialized advertising brochures in Mandarin.[59] From 1985, Bulgaria participated in developing computer-controlled telephone exchanges, printed boards, automated warehousing operations, and computer keyboard production.[60] Work was often hampered by poor planning and a Bulgarian feeling that the PRC wanted to get their hands on documentation and licenses without giving much in return—of course, the Bulgarians conveniently forgot many of these were copies of Western work anyway.[61] Still, as 1989 and the regime was drawing to a close, the state was preparing to cooperate on the manufacture of magnetic memory ferrites as well as microprocessor systems for industrial tasks.[62]

The final, key, element to this engagement with the Global South was the possibility of procuring high technology and cheaper, Western-licensed elements to supplement the poor quality ones of the socialist bloc. No innovations or expertise seemed to be able to offset the poor reliability of red microchips, and no amount of engineering effort could stop machines grinding to a halt. The rise of the "Asian Tigers" in the 1980s, however, meant Bulgarians could circumvent CoCom more easily, as the new powerhouses were more willing to trade than even Japan had been in the 1960s. In 1986, Izotimpex was looking to source $13.5 million worth of FANUC robotic elements from Singapore. The island tiger looked like a great partner due to prices, ability and willingness to deliver embargoed goods, short delivery times, and a wide range of products.[63] These features were often key in solving production bottlenecks in the last months of annual plans, when domestic elements factories did not make their deliveries. The Global South was thus not just a market to buy from, but also a highly specialized industrial supplier of electronics too, sometimes saving Comecon's own plans.

These various examples show the constellation of possibilities for the computer industry. The prism of the biggest development market, however, allows the issues and possibilities to crystallize. India was identified by the Bulgarians as the highest prize in trade terms, inaugurating close diplomatic, cultural, and economic ties. Trade with India sometimes eclipsed that with Middle Eastern countries (where oil revenues often inflated the trade figures heavily), but importantly, the country retained a deep draw for Bulgarian experts of all hues. Here they encountered a self-consciously protectionist state that aimed at fostering domestic production in a labor-saturated economy but also a wide range of firms to compete with—Indian, Western, and even their socialist allies.

THE ROSE MEETS THE LOTUS

The first treaties concerning economics and trade between the two countries were signed in 1956, but the links became more active following 1967, after Gandhi's accession and her openness to the East, as well as Bulgarian industrial growth, which allowed it to offer more industrial goods after the 1950s.[64] On May 2, 1967, both the Bulgarian CSTP and the Indian Council for Scientific and Industrial Research (CSIR) signed an agreement on scientific-technical cooperation, which established a joint committee to coordinate research and exchange in the area, during the visit to India by Tano Tsolov, the executive director of the CSTP.[65] Provisions were made for personnel exchanges between institutes and universities, as well as a commitment to general scientific cooperation, including electronics. A $15 million Bulgarian credit was also provided, aiming to increase trade, especially in machines, over the next few years.[66] The same year an exhibition in New Delhi advertised Bulgarian technology to Indian firms for the first time, and another one followed in Madras in 1968. Telecommunications devices and computers were the centerpiece of both.[67] Slowly but surely, Bulgarian electronics were making their appearance in the lucrative market.

These activities were part of the increasing links between the two countries at the highest level, as the newly elected Indira Gandhi made Bulgaria one of her first ports of call on her first international trips abroad, visiting Sofia in the middle of October in 1967. India became a priority area for Bulgarian propaganda in Asia from the following year, being put in first place among the countries on the continent,[68] reflecting the leadership's desire for closer links with the huge state and access to a growing and potentially gargantuan market. The economic obstacles outlined in her speech in Sofia created a general line held by both states during her tenure, pushing for a greater division of labor. Bulgaria sought raw resources and scarce luxury goods, while hoping to place more and more machines on the Indian market, including computers; the Bulgarians managed to get the Indians to agree to a 1974 "Program for Long-Term Co-Operation in the Area of Computing Technology."[69]

At the start of such cooperation, Bulgaria faced a major challenge—protectionism. Here it was dealing with a country with its own history of developmentalist politics. The Indian National Congress's decision to follow socialist-style national development dated to its third congress in 1938, while the British Colonial Development Act of 1940 had also set the newly independent state on a commitment to state involvement in the economy, which it would reverse fully only in 1991.[70] The Indian government created a "List No. 1," which listed imports that were to be rejected outright under its 1951 Act of Industrial Development and Regulation, with the paramount item on this list being any electronic and computational device.[71] Thus, at the start, Bulgarian technical exports were mainly in agricultural machines, such as tractors. From the start, however, they searched for loopholes to get their golden export into India. In its 1967 long-term program for foreign trade with India, Bulgaria noted the lack of involvement by Elektroimpex, the state union charged with exporting electrical goods, and stipulated it should form its own long-term plan to break onto the market.[72] Elektroimpex exploited the slow progress of India's own electronics and electrical industry from 1969, managing to export first televisions and then signing an agreement with Telefunken-India to export radios, too.[73] The Ministry of Foreign Trade noted this favorably and suggested that the company should seek to create further links by asking for payment not in cash but in Indian component goods or ferrite materials needed for the construction of these devices, thus creating a return avenue for more complex Bulgarian goods. Bulgarian technical diplomacy in the late 1960s thus aimed at subverting India's protection of its domestic producers to secure larger profits.

From the cooperation agreement onward, the two countries achieved concrete results in the sought-after "division of labor." The Indian representatives accepted a program that set out plans for mutually complementary industrial specializations on both sides. Foreign Trade Ministry documents list electronic elements that India produced that could be imported to ease the work of the fledgling computer factories. In 1970, Kintex[74] was pushed to find areas in computing and telecommunications where each country would agree on importing specific items from the other for a period of up to five years, and thus stop specializing in that particular area. Bulgaria sought radio tech imports from Siemens-India but carved out a bigger niche for its own goods, especially transistors, television, and analog computer elements, attempting to make Bulgarian factories indispensable to Indian industry.[75] India was part of a wider Politburo strategy, also aimed at the Arab states, to "create the basis for long-term binding of some of our sectors of economic activity on the basis of partial division of labour between our country and the interested developing states."[76] Little by little, Bulgaria was thus cracking the Indian market and creating the space for its own exports.

Sofia's clout with India was also helped by its diplomatic stances during the 1971 crisis in South Asia, which pushed New Delhi much closer to the socialist bloc. Bulgaria, like its allies, recognized Bangladesh immediately, and as the US State Department put it immediately after the conflict, the USSR (and the rest) had backed the preeminent power in the region and were now to reap increased clout.[77] Gandhi noted in a later meeting with a Bulgarian politician that "friendship is felt in such heavy moments, and we were assured that in this difficult for us moment, Bulgaria remained true to its principles."[78] India was thus placed even more at the center of Asian policy, seen as a country that would be open to more long-term placements of Bulgarian machines.[79] More importantly, it was recognized as a potential source for equipment that was desperately needed but usually available only in the West.[80]

By 1974, however, the country was still exporting miniscule amounts of its priority goods, with only $43,000 worth of electronics in the first eight months of that year.[81] Despite some inroads by Elektroimpex, such as covering part of Bulgarian needs for cables through Indian enterprises, Izotimpex had not managed to capitalize on this in other areas.[82] Ivan Popov took it upon himself personally to rectify this and invited Professor M. G. K. Menon, the head of the Indian Department of Electronics, who came to Sofia in July.[83] Much closer cooperation was agreed on in research but also in joint production and entry into third markets. Menon was impressed with Bulgarian achievements in the sphere, calling them "imposing," and assured Popov that the Indian side would cooperate more fully so "we can build up inertia."[84] This only strengthened Menon's preference for Eastern machines over Western ones, which he communicated to Gandhi that same year.[85] This was part of his and other Indian specialists' conception of their country's socioeconomic landscape. Indian electronics policy was not only tied to protectionism but also to its own development dilemmas in the late 1960s and early 1970s, precisely when Bulgaria was trying to break the ice.

COMPUTERS VERSUS WORKERS: THE INDIAN SITUATION

The Indian government had its own particular problems when faced with computerization of the economy. In a labor-rich country which added around 1 million new workers per month in the early 1970s, any labor-saving devices were difficult to square with the need of finding gainful employment for a growing population.[86] The administration discussed such problems frankly in its internal reports, noting the slackening pace of development by the start of the decade; rising wealth inequality and monopolies; and accentuated unemployment, especially among educated technicians. The government also faced huge tax-evasion problems, preventing the giant rural infrastructure programs it envisioned (such as village electrification) from employing many of these people.[87] Much like Bulgaria, it also looked at its scientific bodies as a source of technological solutions to these problems, demanding that CSIR take an integral part in this effort.[88] Gandhi's government was looking for an almost Popov-like figure

> who should be able to visualise the critical role that science and technology have to play in the economic development of this country and, in addition, be capable of translating this vision into practical schemes of research and development.[89]

That man was Menon, who pushed for a much more active role of the CSIR and the Department of Electronics in the national economy. A 1972 ministerial report recognized the country's significant scientific capacity, which was sadly unorganized and lacking a true national coordination plan.[90] However, attempting to decide on such a plan in the pursuit of self-reliance faced the problems of labor resistance to automation.

Often modernization attempts were faced with unrest, such as in the case of the Indian Oxygen Ltd. Company, which tried to install a British ICL 1900A computer in Calcutta in 1974. The workers' trade union was appalled and wrote to Gandhi herself to state that it

> has warned the company of the inevitable labour unrest if the scheme for this full-fledged automation was not abandoned. The workers in the fold of the Congress-led union IOSKC have also expressed strong resentment against this sinister move on the part of the Company-management. The installation of a giant computer like this is bound to affect the interest of the entire work[force], endanger the security of their service, and substantially reduce the employment potential. At a time when the people of the country and the Govt. of India are seized with the grave and acute unemployment problem, how could this Company go in for full-fledged automation? . . . May I request you to take immediate steps so that the said Company is restrained from installing the computer, in the larger interest of the nation?[91]

This problem had already been seriously examined by two trade union activists and MPs with links to the independence movement, who were also experts in electronics.[92] With a foreword by the finance minister, and cowritten by a member of the government's own Automation Committee, their book argued that computerization in the developing world must proceed differently from that in more advanced countries. Workers' fear grew due to the "unknown and the displacement of labour," much like the Luddites reacted to mechanization.[93] Carefully recognizing that these machines were the future, the authors still sounded a warning that the computer-led increase in productivity can lead to "greater hardships for a large number of people, till such time the super profits are utilised to multiply industrial activity with the avowed objective of absorbing surplus labour."[94] In this period, at least, the government must be closely involved in computerization, so as to defend labor interests in the face of a profiteering management class. This was especially important in India, which in the these authors' view, was at the same time undergoing the transition from a rural and extremely backward society ruled by traditional customs to a scientific and individualistic society, and the government was not taking this seriously: "the challenge of a rapid change from a traditional, custom oriented society, to an individualist and cooperative economic oriented society required for the growth of modern technology, is still not reckoned with seriously."[95]

The criticism applied to other areas too, such as Indian experience of foreign technological cooperation. Gandhi herself commented that "our experience of foreign collaboration has not been a happy one," with imports that were meant to improve Indian capacities often failing as the foreign firms failed to establish home bases in the country. "The capacity to design and prepare engineering drawings is at the heart of any nation's industrial capacity," yet the country had let others do this task, and leading Indian engineers languished without work.[96] Cabinet discussions came back to the idea of domestic effects again and again: Did foreign firms train locals and did they aim to establish production in India? All those who set up factories must also prove how local know-how would be improved.[97] Proponents of closer regulation of foreign firms cited UN Conference on Trade and Development (UNCTAD) studies, which

> only reinforces the evidence which has been accumulating from many different sources over the last few years, that the activities of multinational corporations are inherently inimical to the self-reliant advance of developing countries.[98]

Increasingly feeling that the country was growing dependent on uncaring foreigners, the Gandhi administration passed the Foreign Exchange Regulation Acts (FERA) in 1973–1974, but taking until 1977–1978 to put these acts into full effect. Requiring foreign companies to reduce their equity in Indian subsidiaries to 40 percent, the high-profile casualties were Coca-Cola . . . and IBM. The American computer giant was heavily criticized for the fact that its 1952 entry into India did not result in the introduction of high technology, as it often brought in obsolete equipment that was second-hand, repurposed, and sold at its original prices.[99] IBM also preferred to rent computers to users, building up monopolies, and its response to FERA was absolutely brazen, wanting to maintain 100 percent of its export unit and 80 percent of all foreign currency earnings.[100] Dumping obsolete equipment in the country was not something that IBM was usually charged with, but the increased ire at its practices created the space for Eastern European competitors in India.[101] In 1974, the purchase of a Soviet ES computer for Roorkee University was considered to be justified, because the Soviets were offering a brand new system, with a price that included training and a spare parts package. The American counteroffer of an IBM 360/44 was in fact a second-hand Belgian machine that was no longer in production, so its lower price would come back to haunt the university as spare parts were scarce and would be extremely costly. The Soviet machine instead came with a full set of programs, and it used the same internationally accepted languages that the US machine did, so there would be no need for retraining—Comecon's IBM-cloning decision seemed to be justified.[102]

This turn to Eastern Europe in general was helped by the setting up of the Electronic Trade and Technology Development Corporation (ETTDC) in 1973, to explore what the Eastern market could offer India. The aim was to switch imports from hard currency areas to Eastern European sources, and the Department of Electronics (DoE) was to

> see that products which can be imported from Eastern European sources are imported, even if the public sector corporation wishing to make the imports initially indicates a Western source of supply [their emphasis].[103]

The Eastern Bloc benefited precisely because it didn't act like the multinationals. It was offering joint development, training, and reciprocity in deliveries. Even better, it came at a moment when India was kicking IBM to the curb but was not yet producing its own computers.

ENTER BULGARIA: THE 1970S

The first Bulgarian hit came in the year of ETTDC's creation, with the sale of an ES-1020 system to equip the new computer center of Jawaharlal Nehru University (JHU) in New Delhi, the country's leading university. The sale was finalized in 1974, complete with peripherals and programming packages.[104] Electronic successes such as this were seen as a key part of expanding trade with India, which had started in the mere thousands of dollars in the 1950s, and had reached $60 million in 1974, a significant number for Bulgarian nonsocialist trade.[105] The computer center was the thin edge of the wedge, as Bulgaria set out to study India's needs for mid-sized and large machines by 1980, and Izotimpex set up its first computers-only exhibit in New Delhi in 1976.[106]

In fact, as part of the 1976 Zhivkov visit to the country, it was only electronics that was called the "strategic" sector for further trade development between the two nations, in order to secure a "permanent placement" on the Indian market.[107] However, to do so, problems had to be overcome. The Delhi embassy was one of the most active in promoting the need for advertising and technical service reforms. In 1968, it wrote to the Ministry of Foreign Affairs to ask them to convince the Ministry of Foreign Trade to put aside more money for advertising, as competition in India was overwhelming; articles and advertising materials were to be of high quality and written by specialists and to reflect the latest technological advances.[108] Priority was to be given to the machine building and electric sectors.[109] The embassy also criticized responses to Indian firms and unions as "too formal," with no attached documentation, which was unacceptable for the market.[110]

The embassy continued to be a conduit for feedback on how Bulgarian enterprises were doing on the market by criticizing IZOT for its services and implementation problems with JNU's ES-1020 machine. Nothing was also being done about faulty ES-9002 text processing devices to work with the machine, and IZOT also didn't deliver English documentation. In fact, even the Russian manuals sent were for the wrong machine. The avalanche of criticisms continued: 250 memory packages had not yet been delivered, and Izotimpext still had no representative in the country, making all these problems impossible to resolve.[111] These issues were persistent problems for the industry into the 1980s, when the Mozambican deliveries also lacked English-language documentation, and they took a while to be fixed.[112]

These critiques had some result—IZOT did fix some of the machines in the same year, "upholding the prestige of your enterprise and of Bulgaria."[113] The embassy itself stepped up its promotion campaign, making Bulgarian engineering products and electronics the centerpiece of its magazine *News from Bulgaria,* which until then was dominated by cultural and political news.[114] The late 1970s saw a steady improvement in Bulgarian deliveries and services, with timely tests for the new disc drives and ESTEL teleprocessing system requested by the embassy, noting that the first sales to JNU had to overcome so many problems, not least stiff IBM competition:

Keeping in mind the nature of the good and its complexity, the Indians were hesitant to import them from Bulgaria. Many times it was stated that they are importing trial runs and singular items and the future orientation of India towards import from Bulgaria will depend on the quality of work during the testing period, the service quality, documentation, the data of the devices during their usage etc.[115]

The same report noted Indian feedback that criticized Bulgaria's inertia in not taking in enough goods the other way. Reciprocity was key, and in practice, this meant Indian cables but above all software. The second meeting of the Joint Bulgarian-Indian Committee on Scientific-Technical Cooperation had already identified Indian software that was fit for Bulgarian needs, as well as items such as oscilloscopes, computer relays, and other electronic instruments that Bulgarian labs could use. IZOT was ready to buy the first batch of software, costing around 40,000 roubles, seeing it as partial compensation for Bulgarian deliveries.[116] This policy continued in the following years, allowing IZOT to place its ES-5053 disc drives and ESTEL system on the market with timely demonstrations in exchange for $50,000 of software purchases.[117] In 1975, there were already orders for 250 discs of the ES-5053 type, and 20 for ES 9002 magnetic tape units—part of a general 77 percent growth in trade between the two countries during 1972–1975.[118] IZOT also vowed to study Indian firms' abilities to produce various components, elements, and applied software in order to expand purchases from the country.[119] The aim was "coordinating the redirection of import of some components from the 2nd line towards India," which would be cheaper and less problematic, and in keeping with the 1970s plan for an international division of labor between the two countries.[120]

IZOT's plans were ambitious, often immensely so, bearing in mind the poor technical services it often offered. It discussed production of magnetic tapes based on their own know-how but done in Indian factories.[121] Its discs and tapes were demonstrated in numerous cities and drew the attention of many firms.[122] Izotimpex deemed this interest sufficient to put India second only behind West Germany as a priority "capitalist" market for the 1976–1980 period. India always retained this curious in-between state for the Bulgarian trade representatives, who politically saw it as nonaligned but recognized its economic realities as a form of free market that required new approaches. The plan called for exports of 3 million levs in 1977 alone, with over half of them in peripherals.[123]

The embassy also kept pointing out to IZOT and other enterprises that the Indian market was saturated with foreign technology, often incorporated in domestic designs.[124] Purchases of Indian technology would thus be a back door to gaining access to embargoed Western achievements. This was one of the driving factors behind the State Security intelligence residence in India being taken over largely by the Scientific-Technical Intelligence (STI) Directorate by the early 1980s.[125] India was a rich source for the intelligence services to access Western technology and companies' expertise in a setting that was freer than the CoCom countries with their embargoes, restrictions, and counterintelligence. As was standard practice, the CSTP representative in major embassies was the worker in charge of STI activities in the area.[126] That individual in India by the early 1980s was Svetoslav Kolev. His monthly "economic information bulletins" highlighted Indian developments and companies that would be of interest. In 1983 he received the embassy's highest evaluation for his work, especially for facilitating sales of Bulgarian automation know-how, and was recommended for promotion to the rank of the embassy's economic advisor.[127] In fact, by 1986, he was the resident who took over as chief of the embassy when the ambassador was away on business.[128] The STI angle is one way to see the importance that the Indian technology market had for Bulgarian access to the science that interested it in developing its own sector.

The embassy was insistent on pointing out the developments of Indian science that could be helpful to Bulgaria. This was especially true in the "tropicalization" of Bulgarian machines, allowing them to function in difficult climates, and thus making them more competitive in the Global South, where ES machines often failed due to insufficient air conditioning and dust contamination.[129] The Delhi trade representatives harangued Bulgarian producers for replying to Indian requests with a curt "there is no tropical version," rather than resolving the issue. Bulgarians often did not respond to possibilities that the embassy presented, such as tenders for Air India's electronic system; while Computronix, the Indian firm chosen to represent IZOT in the local market, was not being sent enough prospectuses.[130] The embassy also harangued IZOT for underestimating Indian science, which would hinder the chimera of Bulgarian-Indian relations in the 1970s—the creation of joint hardware and software enterprises and entry into third markets.[131] It was a curious echo of the Soviet complaints that the USSR underestimated Bulgarian science in that same decade.[132] But here the Bulgarians were acting in the same haughty manner, buying Indian software in the smallest possible quantities, hampering their own disc sales in a country they had themselves identified as needing huge data-storage capabilities.[133] The ES-5053 discs were identified as compatible with Indian-produced computers of the TDS series, while another market niche was industrial control electronics, which India produced none of.[134] By the end of the decade, the embassy was thus noting that the possibilities were huge, if only the Bulgarians would learn how to act responsibly, and responsively.

By 1980, Bulgaria had also become the only socialist country besides Hungary to move its trade with India to convertible currency rather than rupee balanced payments. This was seen as the only way to increase trade and put it on a profitable basis, as there were constant balance-of-trade problems with Bulgaria not purchasing enough Indian produce in rupees. It was part of the plan to raise trade to around $200 million after the 1980 visit by Foreign Minister Petŭr Mladenov.[135] In the previous couple of years, IZOT had helped with the transfer and setting up of the production of disc drives in the range of 30 to 100 MB, and it had set up a dedicated engineering bureau in India.[136] Such developments and the IBM exit from the Indian market in 1978 had catapulted Bulgaria to occupy second place, behind the USSR, in electronics exports to the country by 1980,[137] and dovetailed with the electronics enterprises' wish to hold onto hard currency. The report praised the entrance of Bulgarian disc drives and CPUs into the Indian market, noting favorably that in the conditions of autarky that the nation aimed for in electronics, it was high praise indeed that Bulgaria managed to secure cooperation in the production of peripherals, as that was promised only to devices of exceptional quality.[138] However, this was no cause for complacency—Bulgaria was facing stiff competition in the open market from British and Japanese firms; its ES-1022 system was outdated and unlikely to win any new auctions; and due to lack of coordination, its socialist allies were in fact its competitors.[139]

The balance of payments was maintained only by the USSR, and Bulgaria would find it ever more difficult to maintain a good trade balance with the Indians under the new conditions. Continued presence depended on new pricing policies, full and proper documentation in English, newer computers, sales through local private firms and individuals, specialized exhibits, and "a substantial improvement in our work in the advertising and propaganda sphere." The Indians also were continuing, the Ministry of Foreign Trade noted, to insist on increased imports—the Bulgarian export/import ratio was 2:1 in the socialists' favor.[140] As the Indian government was facing problems that were unsolvable without computers, this was the best chance to solidify the position in the market.[141] However this was only possible if, as the embassy continued to note, there was unity between foreign policy goals and advertising realities.[142]

FROM PERSONAL COMPUTERS TO SUPERCOMPUTERS: THE 1980S

Bulgaria entered the 1980s with a renewed purpose and a heightened profile. This was thanks to the singular figure of Liudmila Zhivkova and her cultural diplomatic offensive. Her fascination with India was a cornerstone of her unique cultural policy that opened up the country to the world in particular ways, and her links with India were paralleled by a close relationship with Indira Gandhi herself. Her efforts resulted in the 1977 creation of a Bulgarian cultural center in Delhi, with a rich cultural program.[143] India was one of the last countries Zhivkova visited before her untimely death in 1981.[144] All this was a welcome boon for the electronics industry, which had matured by the early 1980s and could offer robots, PCs, computer-controlled machines for industry, and newer Winchester-type discs.

Buoyed by licenses, domestic work, and espionage efforts, many production problems had been ironed out, and by 1981–1982, Izotimpex had started a much more targeted and self-confident campaign in specialized Indian publications.[145] Much had been learned from contact with Western and Indian firms during its entry to the Santacruz Electronics Export Processing Zone (SEEPZ) in Mumbai.[146] SEEPZ was set up in 1973 and offered tax benefits and other incentives to promote foreign investment and technology. It was a place for Izotimpex to try out its new visual and advertising material before pushing to new heights (for it), such as participation in a three-page spread on Bulgarian engineering in the *India Express*.[147] The embassy also pushed for, and got, the whole November 1981 issue of the specialist journal *East European Trade* dedicated to Bulgarian economic development, coinciding with the country's gargantuan campaign to celebrate its 1300th anniversary. This issue presented the country as an advanced industrial state and a world leader in per capita production in electronic hardware (in fact, it was third).[148] It praised Comecon specialization as the right way to divide labor between states, and saw its cooperation as a model for all technological dealings between states.[149] This specialization allowed Bulgarian trade with the developing world to grow 37-fold between 1960 and 1980, with over 500 projects being realized in various

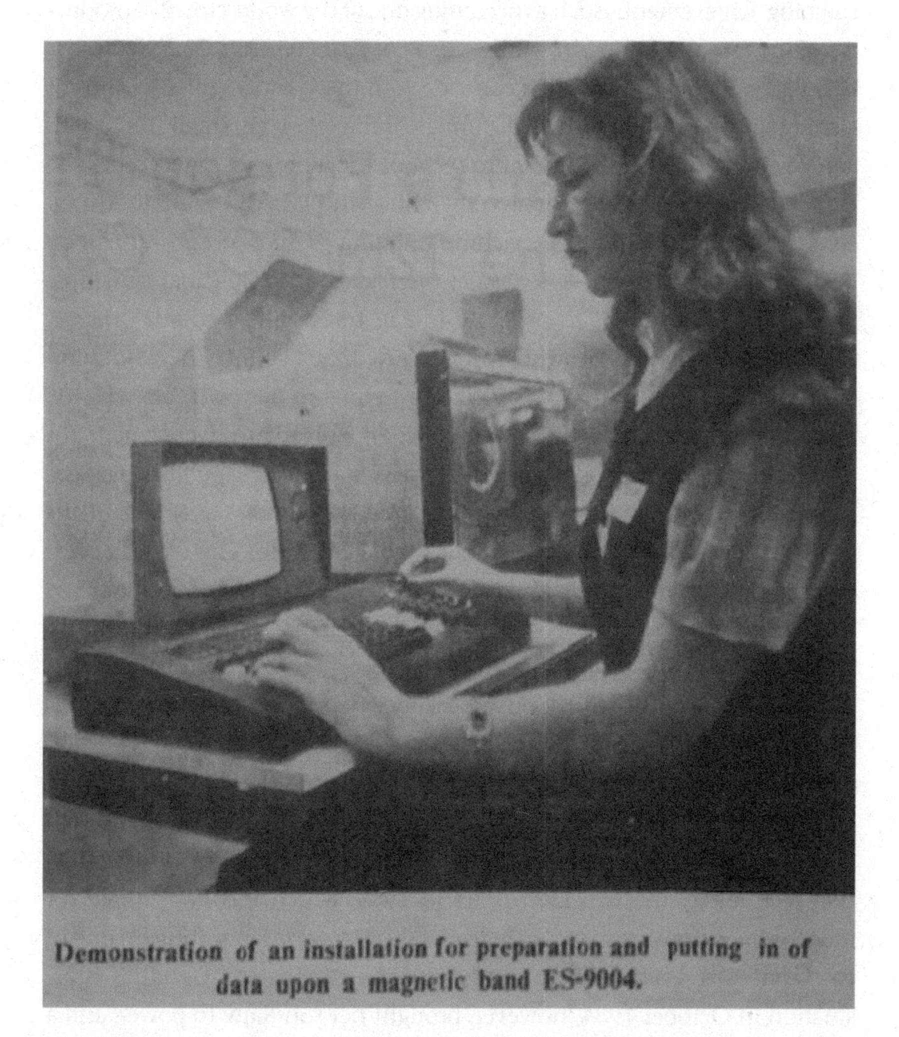

Demonstration of an installation for preparation and putting in of data upon a magnetic band ES-9004.

4.1 Bulgaria's new image in *East European Trade* (*Source*: Ministry of Foreign Affairs Archive, Sofia.)

countries, and 20 types of major electronics entering these markets. The mastery of the scientific-technical revolution by Bulgaria was the basis for such an advancement, especially in industrial engineering and electronics: Bulgaria was now a desired technology partner.[150] The rosy picture was the effect of the sector's growth over the previous decade, but also of course belied the quality of products on offer.

Press releases continued to accentuate on the most prestigious, eye-catching achievements, such as becoming one of the world's top five producers of industrial robotics, in another issue of *East European Trade* dedicated to Bulgaria in 1983.[151] Timed to coincide with the visit of Ognīan Doĭnov, it also sought to reassure Indian customers of the relatively small size of Bulgaria, which was no impediment to technological progress, as according to UN rankings, Bulgaria had achieved eighteenth place out of 36 industrial states in terms of export of machine building.[152] Such observations had been made earlier by the first resident ambassador to Bulgaria, Dr. Gopal Singh, who also sought to promote Indian trade with a country that at the time was still little known: "it is amazing that a country of 8.5 million people should have a foreign trade almost equal to our own."[153]

Robots were the chosen supermodels of Bulgarian advancement for export firms in the early 1980s. In 1983, *News from Bulgaria* published extensive coverage of new robotic machines used in Bulgaria, freeing humans from drudgery and raising the workers' quality of life. The coverage featured interviews with specialists who praised the quality of their work now that they were freed of menial tasks. Talking about the imminent entry of RB-232 robots to car-manufacturing factories, some stated that "if such assistants come to the workshop, especially where work is hard and monotonous, it will be lovely."[154]

Such measures were, however, unlikely to be too successful in the protectionist conditions of Indira Gandhi's India. Often the Indian side was to blame for failed deals, such as a large Bulgarian sale of floppy disc production technology to Rishi Electronics in 1983, which was not ratified for months by the government agencies, leading to delays in setting up one of the vaunted and hoped for joint production enterprises.[155] Gandhi's assassination in October 1984, however, brought her son Rajiv to power, and a liberalization of technology trade occurred. Cognizant of persisting problems with Indian backwardness, he launched six "technology missions" to increase literacy and communication in the countryside, as well as

EAST EUROPEAN TRADE

No. 246 SEPTEMBER 1983 Rs. THREE

1944
39 YEARS OF PEOPLE'S BULGARIA
1983

Secretary of the Central Committee of Bulgarian Communist Party, Mr. Ognian Doinov, with the Indian leaders during his recent visit to India.

4.2 Scenes from Ogni͡an Doĭnov's visit. (*Source*: Ministry of Foreign Affairs Archive, Sofia.)

pushing for the true development of nascent computerization programs started under his mother. He cut through debates about the utility of computers in a labor-rich society to lower import duties on components, allowing foreign manufacturers freer entry into the Indian market and encouraging the use of computers in offices and schools.[156] His view was that of many other educated Indians: The country had essentially missed the industrial revolution, despite Nehru and his mother's efforts, so its

Testing the Pirin manipulator at the 'Bereo' Robotics Combine in Bulgaria.

4.3 RB-232 robots as the face of Bulgaria in India. (*Source*: Ministry of Foreign Affairs Archive, Sofia.)

only chance was to participate fully in the information revolution unfolding now.[157] This was the start of the rise of Indian IT services, helped by the country's English education, low wages, and lack of competition from IBM that gave entrepreneurs the space to develop.[158] It coincided with the start of wholescale production of the Pravetz computers in Bulgaria.

Pravetz, the hometown of Todor Zhivkov, became the center of PC production in Bulgaria after repeated IZOT failures resulted in the creation of a separate Microprocessor Combine in the tiny mountain town an hour east of Sofia. It drew many of the best specialists from CICT and was headed by young, energetic managers, such as Plamen Vachkov. This development was concurrent with the embassy's continued criticism of IZOT's inherent conservativism in approaching the Indian market, not copying American or Japanese marketing or investment efforts. "Electronics is the most dynamically developing sector with the fastest change in production,"[159] and IZOT's history was burdening it with moribund practices. The Pravetz combine, however, was new and energetic, and it worked much more in the mould of the multinationals than IZOT ever did. It quickly identified the need for local partners, recognizing the huge potential for the delivery of thousands of PCs after Rajiv Gandhi's call for computerization of India's offices and schools.

4.4 Indira Gandhi visits the electronics manufacturing town of Pravetz, Bulgaria. (*Source*: socbg.com.)

In 1987, a partner was found, but the nature of the deal bears witness to the old adage that "bad publicity is good publicity." Adil Shahryar had signed a deal to import 10,000 Pravetz PCs to India, promoting his Priyadarshini Institute for Computer Aided Knowledge. He had already outsourced a project to Apple to create a typeface for Urdu, to facilitate computer education in that language, and the Pravetz systems were Apple compatible (of course, as they were copies of that computer). Shahryar complained that despite the DoE approving the sale before he signed it, and it being in accordance with the Indo-Bulgarian bilateral economic agreements, the DoE was now going back on its word, "sabotaging the prime minister's plan to computerize the country by defaming me, especially as he is associated with my institute."[160]

Shahryar went on the offensive, stating that this was the best technology available at the price, and was in fact in tune with the government's provision of $12 million for personal computers as planned in its own joint agreement. This fiery interview in the *Delhi Sunday Mail* was supplemented by an article recounting Shahryar's troubled past. This was a massive obstacle in a political landscape where Rajiv Gandhi was assailed by numerous corruption scandals. Just a month earlier, Swedish radio had blown open the Bofors scandal, where Indian officials were implicated in receiving kickbacks for buying Swedish guns. Amir Shahryar was closely connected to the Gandhi family and was thus under increased scrutiny. His excesses had already left a mark—he had stolen a car together with the late Sanjay Gandhi, Rajiv's brother, who had been groomed for succession but died in an airplane stunt accident in 1980. He had also spent three and a half years in a US jail for an escalating list of tomfoolery: fraudulent checks, currency counterfeiting, placing explosives on a ship in the Miami docks in a failed insurance scam, and a failed arson in his hotel room when he was trying to cover his tracks.[161] "Do famous juvenile delinquents never grow up?" asked the paper rhetorically, noting that his purchase of PCs was contrary to 1984 directives on PC import without a no-objection-certificate from the DoE, lacking in this case. The paper drily noted how he was received at Sofia Airport by Doĭnov and the prime minister himself, and how he had promised IBM peripherals as part of the deal. The journalist was astounded that a private individual with such a past could do these things and raised questions about the fact that

10-16 май 1987г.

Adil Shahryar's big byte in Bulgaria

by SUNIL SETHI

4.5 The *Delhi Sunday Mail* article on Adil Shahryar, May 10–16, 1987. (*Source*: Ministry of Foreign Affairs Archive, Sofia.)

the computer center would be named after Liudmila Zhivkova, insisting on this being evidence of close connections with the regime. Shahryar's defence, which holds some water, is that it was a return gesture for the naming of a Sofia school after Indira Gandhi. The paper saw all this as an obvious prelude to using government links to start a domestic computer empire, using both the Gandhi family connection and his father's name (the latter being the long-term chair of the Trade Fair Authority).[162] There was reason to believe this, as his ludicrously short jail sentence in the US was further reduced, and he was released in 1985 after India permitted the Union Carbide chairman, Warren Anderson, to return to the US after the Bhopal pesticide plant disaster of 1984.

Despite such virulent opposition and the doubts about corruption, the sale did go through, helped by the priority that was put on it by the

Bulgarian government throughout 1987.[163] The pressure and choice of such a close friend of the Gandhi dynasty worked, and the Liudmila Zhivkova center started work soon after, the showpiece of the Pravetz abilities in India. A sizeable article on the center and the Indian electronics industry as a whole was published in Bulgaria's specialized computer magazine, *Computer for You,* in September 1987—the only developing country to get an article in the magazine's run, and in fact the only one outside the US or Japan to get such a detailed treatment in its pages at all. The author praised Rajiv Gandhi's far-sighted policies of opening the Indian market to the world, noting that Indian policy since the 1970s was predicated on technology transfer in order to stimulate domestic production, but also pointing out that it was still doing badly in peripherals—a nod to Bulgaria's role in the market. The comprehensive article posed the same question, however, that Indian officials had been posing since the 1960s: What will happen to the already terrible unemployment problem once automation started?[164]

This personal computer saga was one facet of Bulgarian successes during this decade. The other was on the other end of the spectrum: the sale of the IZOT 1014-ES 2709 supercomputer in 1988, which demonstrated a different way that Bulgaria could exploit local and international politics and their intersection.

The supercomputer was the fruit of a 1984 project started by Stoi͡an Markov, the last head of the CSTP, and created by the main designers Vladimir Lazarov and Plamen Daskalov. It allowed the modular linking of different matrix processors into a network, overcoming the relatively weak power of each processor to create a machine capable of over 120 million operations per second—outclassed by Western machines, but unique in the Eastern Bloc.[165] It was finished by 1986, finding application in large-scale scientific research, oil field exploration, and the needs of the Soviet Strategic Rocket Forces and space program command.[166] It would be used to equip centers in China, Vietnam, and India, too.[167]

The case of the Indian decision to go for this machine, however, is an indicator of how, despite the computer being inferior to Western machines, Bulgaria could use other advantages to sell its computer. A 1987 Bulgarian embassy report noted that while the Americans, the leaders in supercomputers, were willing to sell such machines to India, they insisted on

4.6 The back cover of the September 1987 issue of *Computer for You*, showing the Liudmila Zhivkova computer center in New Delhi. (Source: *Kompi͡utŭr za Vas*, National Library, Sofia.)

military secrets protocols, demanding direct control over the end use to stop any potential information leak to the Eastern Bloc. The embassy notes, with open satisfaction, that this was unacceptable to the Indians.[168] Shankar Bajpai, the former Indian ambassador to the US, clarified in 1995 that "we were being denied a whole range of technology because it was considered either of dual use and, therefore, capable of enhancing our nuclear capabilities or that we might, willingly or unwillingly, pass it on to the Soviet Union."[169] However, as the former secretary of the environment notes in the same series of seminars, Rajiv Gandhi was very willing to consider alternatives when Washington denied them their newest Cray computers.[170] Ironically, it was the "free world" that proved more restrictive in its technological sharing than the totalitarian one. Bulgaria, coming from the position of technological backwardness, had no fears of its technology (often based on foreign expertise, anyway) falling into foreign hands Moreover, the political implications of such a possibility were nowhere near as dire as they were for Washington. At the start of 1988, therefore, an IZOT-1014 was demonstrated through the local firm Computronix, and a full delivery was completed in December of the same year.[171] Krasimir Markov was one of the CICT experts sent to install it and train local specialists. He shared how an American supercomputer then in operation in New Delhi (an older one had been delivered a few years before) was locked in a special room that only US technicians had access to, with Indian scientists able to only use terminals on other floors, buying machine time from them. "Our machine, they could do whatever they wanted with it! The Americans allowed them mostly to do weather prediction" Markov stated, and the freedom to use the computer for nuclear and defence research pleased the Indians more than the US machine's superior characteristics. "I read lectures to the Indian team for a month to train them—what they understood of my broken English, I have no idea, but they seemed pleased![172] In fact, American disbelief at even the possibility of the existence of such a thing as a Bulgarian supercomputer prompted the sending of the director of the computer lab at Los Alamos, New Mexico, to Bulgaria in the autumn of 1989, to verify its existence. Real as it was, its lucky creators were invited to a conference in Honolulu in the last months of the Bulgarian regime, and there was even interest in a purchase by the University of Minnesota.[173]

Despite the successes, by 1988, the Ministry of Foreign Trade was gloomy about where Bulgarian-Indian trade was going. There was little potential for expansion, as both countries had started to overlap in machine building. It evaluated India as still too protectionist, most of its technology as too low quality, leading to a low a balance of trade, with the 1988 figures expected to be $50 million of Bulgarian export against $20 million imports.[174] Indian trade was connecting to the West, and it seemed that Bulgarian goods were there mostly to fill gaps that other partners were not filling. Electronics, however, was the one area that did not reflect these gloomy predictions—RB-211 robots were being installed in the Bangalore and Mogoli factories of Hindustan Machine Tools in 1989, and 70 other enterprises were interested in implementing these machines.[175] The question remains open as to where trade would have gone, but 1989's tumultuous events put an end to Bulgarian abilities to offer technology at the same volume as previously. However, electronics and scientific cooperation remained the cutting edge of its outward-facing effort right up until the end. While other trade was lagging, India still sought its computers, and Bulgarian specialists were working on hybrid integrated circuits with colleagues at Chandigarh even as the regime was crumbling in the Balkans.[176]

The competition for markets in India and the Global South more generally changed the way Bulgarian computer organizations thought and worked. A whole new generation of younger cadres learned the rules of marketing, negotiation, and business more generally. In India, Bulgarian trade policy managed to learn from its earlier mistakes to be able to score some profitable successes in the 1980s. Turning away from dreams of a real division of labor, Bulgarian computer companies learned to provide better technical services, newer machines, glossier advertising, and maybe even the right type of bribes, in order to increase their market share. The nature of the electronics market in India was much more dynamic than that in Comecon, and even though many successes were scored in a period when Gandhi's protectionism had excluded IBM from the market, the advanced American and Japanese firms continued to be present and were a foe that the Bulgarians had to learn to fight. This tough schooling for the embassy, foreign trade ministry, Pravetz, and IZOT specialists, became a conduit of Western-style ideas and concepts, which had to be adopted to compete in the Indian market. One of these concepts was modern advertising, which fed back into wider economic thinking and was a catalyst for select Bulgarians to become part of the emergent transnational information economy discussed in the chapter 7 of this book.

LEARNING TO ADVERTISE

From early on, the regime's tangle with the Global South had an effect on another aspect of its thinking. Unlike Comecon, the developing market was a battleground of international firms, and did not operate according to the socialist economic bloc's politico-economic frameworks. Various embassies, amongst which the Indian one was a leading voice, called upon the Ministry of Foreign Trade to create a modern advertising system. Advertising in Bulgarian economic thinking had already entered through other Western-facing sectors, most importantly tourism. The need to attract Western tourists to the burgeoning 'red Riviera' became a conduit of new and glossier ways to market Bulgarian services. However, industrial design and the marketing of the goods that made up the vast majority of Bulgarian production and exports lagged behind strongly, and there seems to have been little to no cross-contamination between the tourist and industrial marketing sectors. As early as 1973, the Ministry accepted the need for a tripartite division in its advertising efforts. In socialist countries, they would use the full gamut of mass media, aiming to advertise the Bulgarian role in the socialist division of labor, its quick scientific progress, the intensification of its economy, and the price competitiveness of its products. The concentration would also be on heavy industry and machine-building products. The propaganda was aimed at the mass classes, not business strata, as its aim was to strengthen friendship between nations.[177]

In capitalist countries, advertising had to be flexible and targeted, avoiding standardized templates and reflecting the aims that the country had in each the particular market. The emphasis was to be on Bulgaria's growing technical abilities, with electronics at the forefront. Advertising was also a showcase of the country's normalizing relations with the West and would highlight the mutual benefit of increased links. Special attention was to be paid to the Balkan capitalists in Greece and Turkey, where the best Bulgarian products were to be sent to foster trade and thus regional peace.[178]

Finally, developing market advertising was aimed at showing a willingness to help raise local productivity and standards, from agriculture to high technology. Trade with Bulgaria was to show the road to socialist development and provide the tools and backing to break the shackles of Western dependence. Advertising here was to borrow the most from Western marketing, especially in visual terms, in order to deal with "the weak literacy rates of these countries' populations," which meant Bulgaria must use prints, films, screens, and cinema to create interesting and captivating ads.[179] These concerns were incorporated into the 1974 creation of a unified Bulgarian advertising agency for abroad, taking over myriad smaller agencies under the auspices of the Bulgarian Trade and Commercial Chamber. Coordination of advertising activities was especially targeted at the electronics and chemical industries; at the suggestion of the Ministry of Electronics, which was heavily invested in international exhibits for its sales, fairs would also fall under the auspices of this organization.[180] In this way, Bulgarian advertising finally got a dedicated umbrella organization in the early 1970s. It was to serve nuanced ideological aims, depending on which part of the world it operated in—from socialist brotherhood to peace to socialist development—but its practice was becoming more unified, more Westernized, and more modern.

With its increased footprint outside Comecon, the Ministry of Electronics informed the trade ministry that its paltry advertising budget was not enough when faced with Western competition: "US firms Hewlett Packard, Zinger, Bekman Instruments spend over 10% of their sales on marketing." Regime plans didn't factor in such sales utilization, so the Ministry demanded a special fund for "Market Development" to be set up, by increasing prices by 2 percent. This would allow for better advertising, joint firm investments, and the timely creation of prototypes for users to test.[181] "The market of electronics demands short delivery times and a fast reaction to the client . . . currently that is impossible as we depend too much on deliveries of some components from non-socialist countries," it argued, and hence the BNB must allow 10 percent of all capitalist currency sales and 2 percent of all socialist currency sales to be earmarked for this purse; while the Ministry should be allowed to move from "forward planning and inflexibility" to short-term plans that allow corrections, when faced with user demands and "just in time deliveries."[182] Twelve service bureaus had to be opened up immediately with this money, to ensure single sales turned into a proper presence in markets such as Spain, Greece, Turkey, and India.[183] The regime was also to allow the Ministry to partner with local firms so as to have local representation in markets it knew nothing about and to learn these local conditions.[184] This radical departure by the electronics industry meant it was the first sector of the economy that was allowed to use its own hard currency income for its own needs, rather than to feed its earnings into the state coffers, as Atanas Shopov (director of the Stara Zagora factory for disc drives) stated.[185]

The language of the demands is also revealing, utilizing terms such as "just-in-time delivery" or "manufacture," which were hallmarks of the new economic revolution of the late 1970s, after their development in Japan. Being at the cutting edge of technology meant that the industry was also at the cutting edge of business practices, copying these capitalist models in order to compete with them. The industry thus had pushed for reforms in financing and marketing beyond the Ministry of Foreign Trade's own plans. As in India, it was trade representatives faced with these local conditions that shouted loudest for American- or Japanese-style market approaches.

These battlefields, where Bulgarian companies cut their teeth in international competition, fed back into general marketing policies in the 1980s. New forms of advertising also targeted socialist users in new ways. Izotimpex paid special attention to visual and large-scale advertisements by 1982. Alongside the traditional participation in multiple fairs around the world, costing nearly 2.5 million levs, it also started producing more nuanced and picture-heavy brochures and catalogs.[186] Russian was not the only language utilized either, with English, German, French, and Spanish catalogs also being published in large numbers—the Stara Zagora factory printed 5,000 English copies of its catalogs: as many as its Russian-language ones.[187] Special, picture-heavy spreads were taken out in the Austrian journal *Made in Europe,* with 21 color photos.[188] German magazines that targeted the Arab world were also used, as were US ones targeting China.[189] Izotimpex also produced its own specialist magazines with ads in local languages for China, Japan, India, and the Arab world, with much success—76 such marketing publications resulted in 1,756 and 1,527 (for the years 1981[190] and 1982,[191] respectively) returned inquiries by interested enterprises through the forms attached to each advertisement.

The main change was the utilization of billboards by IZOT in the socialist countries—a novel thing in the 1980s, and one not explicitly needed, as deliveries were contracted through the government plans as part of Comecon. Yet these billboards could target users and enterprise directors directly, as well as being a Western-style technique that demonstrated Bulgarian prowess. Advertising displays popped up in places where those socialist technocrats—but also their Western colleagues—were most likely to see them: the Budapest airport, the Czech borders with Austria and West Germany, and 26 in Leipzig (where the famous fair was held) alone. Neon billboards were installed in Moscow, Berlin, and Prague.[192] Airports, hubs of the transnational business class of the rising information economies, featured prominently: Schonefeld in East Berlin and Domodedovo in Moscow got one billboard each, Sheremetyevo got two of them.[193] Izotimpex understood the connection between mobility and business, and billboards were also placed on the Moscow-Riga highway (linking two cities with important computer factories and institutes) and on all the highways leading into Berlin.[194]

The visual design of Bulgarian marketing material, including at fairs, was improving a lot by the 1980s, which was noted with satisfaction by Izotimpex.[195] It was a noted improvement over earlier advertising attempts, where representatives scattered electronic elements haphazardly on their stands, preferring chaos to an aesthetically pleasing display.[196] Until the mid-1970s, the firm was still delivering internally working but externally deficient machines, such as the earliest ES-1020 deliveries to India, described as unacceptable:

> The control panel delivered with the system was in a very poor condition. There were banal and unacceptable mistakes for a first delivery. Many technical deficiencies were apparent. The instruction panels and some control lamps had paint on them. Scratched and loose details gave a bad impression of the capabilities of the Bulgarian producer.[197]

Such problems had been overcome by the 1980s, as were the majority of service questions, such as delivery of the right type of cables, manuals in the right language, and most importantly, the provision of English-speaking technicians. Izotimpex had realized that its Comecon approach would not fly in the rest of the world and acted accordingly. A deal to deliver a computer center to Iran in 1986 called for a complete change of

4.7 Bulgarian computers became a common part of advertising the country's success, as seen in the India-targeted *News from Bulgaria.* (*Source*: Ministry of Foreign Affairs Archive, Sofia.)

control panels from Russian to English interfaces, and translation of manuals that were only in Russian—which proved impossible, however, due to the large size of the manuals.[198] By then, Bulgarians had understood that service quality and industrial design were as important as making a quick buck. The developing world market was a place of fierce competition, where Bulgarian firms had to innovate. There were no five-year price guarantees or political alliances to ensure the easy sales that had been the norm in the Eastern Bloc. As the regime wanted hard cash that came with deals abroad, it learned new ways of negotiation as well as advertising. The marketing tricks that Izotimpex had copied from their Western competitors in places like India were often driven by repeated calls by local trade representatives and embassy officials, who were adamant the old, crude efforts were just not good enough. By the 1980s, Bulgarian foreign trade, thanks to its entanglement with the developing world, had developed more modern and nuanced ways to capture the attention of a potential buyer.

In 1972, Dr. Gopal Singh wrote to a friend in Bombay industrial circles with a peculiar request. He was trying to help a Bulgarian man named Vesselin Stoinov to visit Rishikesh, the "yoga capital of the world," made famous by The Beatles' visit in 1968. He was interested in Indian spiritualism and wanted to travel to such centers for three months. Requesting financial help, he was willing to offer his services as an electronics engineer for as long as needed to Indian firms in order to repay this debt.[199] The paper trail vanishes, but one wonders whether this transaction ever went through.

Thus, although for some individuals, electronics was a facilitator for the soul, most Bulgarian computing in the Third World was a conduit of ideas as well as money, a symbol of prestige as well as business. Rarely were these machines presented as "socialist," for the story of the industry in India is not one of presenting a new model of development but of selling the tools of development in the best possible way. Nothing predetermined that electronics would become Bulgaria's leading edge into these areas except the actors at the middle and lower levels, who pushed for better sales and better techniques. It was also they who learned the skills needed, through trial and error and by observing their competitors. India was the opportunity to beat the West at its own game, and to improve Bulgarian marketing practices when found wanting. After 1989, such skills would allow their easier transition into the open market.

Computers were a great calling card and profitable commodity on the world market, but they were also *the* way to modernize and automate. They were, in everyone's conception, the future. In India, this idea caused problems, as automation enhanced the problems of unemployment that were exacerbated with each passing year. In Bulgaria, however, they held a promise that was tantalisingly simple for the party—boosting the command economy through better productivity of labor. Zooming out through the lens of commodity circulation, Bulgaria intertwined with the socialist and global markets. But a zoom back in shows how society and politics were also to be interwoven with the wires of the new age, and how computers became the shining star of the scientific-technical revolution.

5

AUTOMATIC FOR THE PEOPLE: THE SCIENTIFIC-TECHNICAL REVOLUTION AND SOCIETY

Luddites stalked the supermarket aisles and the copper mines of 1970s Bulgaria. The future was here, and it wasn't to their liking. There were no secret oaths like their British predecessors, just acts of localized resistance to the march of the party's dream. Machines were broken or burned because they were there, and because they threatened the skill set that many Bulgarian workers had developed. These skills didn't include their professional abilities only, and this isn't just a story of a fear of steel muscles replacing one's job. No, in the socialist economy of shortages and patron-client relations, labor practices included pilfering, stockpiling, and the black market. The introduction of computers and automation to the Bulgarian workplace was a momentous change meant to build a modern economy, but it also upturned the conventions of real existing socialism.

These isolated scenes were subjective responses to the party's quest for objectivity in information and quality in production. Its quixotic task was to find a nonmarket reform to an increasingly stagnating economy and to remove the "subjective" factor from production, which it blamed for the continuing poor quality of its products. Computers had been created as commodities to finance Bulgarian modernization, but they were also now tools to achieve the next stage in development once growth slowed down in the late 1960s. Bulgaria had created the basis of industry, something

Soviet-style development was very good at, but now it had to get the most bang out of the buck it had invested in its labor and enterprises.

Automation through computing and the constellations of machines it made possible also held a dream of the future. Through the 1960s, the BCP was finding that Marxism-Leninism had lost much of whatever mobilization capability it had once held. New unifying visions were required. One was the growth of nationalist rhetoric from the early 1960s onward, rehabilitating older narratives of Bulgarian history, and growing into a gargantuan cultural phenomenon under the auspices of Liudmila Zhivkova in the 1970s. But this policy could not offer a vision of prosperity beyond the balm of national sentiment. The party looked to build a gleaming vision of affluence and a new type of society, where the New Man would indeed be Bulgarian, the shining apotheosis of the nation's history, but he also had to be *New*–better than the old way of doing things—and *Socialist*—a distinct politico-intellectual being. Computers and cybernetics were at the heart of this new narrative, which fell under the umbrella term of the Scientific-Technical Revolution (STR). An obsession of both the BCP and Brezhnev's USSR, this revolution promised to implement the newest technology into everyday life in order to create a qualitatively new stage of social being.

"A book about Brezhnevites, however, is another story—that of a senile Cold War," Vladivslav Zubok and Constantine Pleshakov famously quipped.[1] Zhivkov was a Brezhnevite in the way he had been a Khrushchevite before, so it seems that this sentence would hold doubly true for gray, orthodox Bulgaria. But as we have already seen, its carving out of niches within the socialist family was just one indicator of its independence. This high-technology niche was reflected in a domestic, computer-imbued discourse around society and its path. Far from being senile, the Bulgarian drive to computerization reveals both the regime's dreams of automation and society's own strategies of life under the party state. The provenance of industry for small Bulgaria and Liudmila Zhivkova's peculiar vision for the country's psyche combined to make the social dimension of this technology key to the STR discourse that was increasingly the BCP's ideological program. Computers' march into society would intellectualize labor and free humans from drudgery. It would automate workplaces and streamline information flows. By the 1970s, this was a large-scale state program that did, contrary to scholars' dismissal of these proclamations as mere obfuscation of economic reality, aim at revitalizing the socioeconomic climate and create a new way of governance aimed at achieving the dual Marxist goals of a classless society and of a truly free citizen. Computers were more than a source of cash: They were the inspiration to build a future and a solution to specific economic problems.

This chapter thus explores the automation dream of the Bulgarian communists, as they introduced an array of computers, computer-controlled machines, and eventually robots into the life of the Bulgarian worker. Attempting to network the economy, they also sought to garner information about all aspects of their population. The chapter shows how pervasive this project was, how it ran into trouble, and how some sectors of society reacted to it. It thus blends the history of the technology with that of the political philosophy that drove it, and the history of labor's response to these encroachments—responses often shaped by the gendered or class position one had within the socialist economy.

STR AS THE NEW DOGMA

The dream of perennial growth was as inherent in communism as it was in its rival, but by the 1960s, the party had concluded that the extensive period had ended. The long-term plans had to change, and the Eighth and Ninth Congresses of 1961 and 1966, respectively, affirmed that next came "the movement from extensive to intensive development . . . on the basis of the widest possible implementation of the achievements of scientific-technical progress and the increased efficiency of social productivity."[2] The contours of this intensification were, however, up for debate. Market principles and the use of profit as labor incentives or tying wages to production levels were debated in various parts of the socialist camp. Some enterprises had moved to a self-financing basis in 1964, but subsequent discussions pointed out that this was counter to the fundamentals of central planning. The reforms were imperfect anyway, as directors were still unaccountable, and losses were covered by the state budget. Looking for profit, these enterprises called shortages or rushed production in the search for a quick delivery. The experiment was quietly set aside in the later 1960s, as differentiated approaches were tried.[3] The culmination of this was the 1968 July Plenum, when a New Economic Mechanism introduced a strong focus on prices and a scaled-back approach to quotas. Enterprises could now offer the State Planning Commission their own plans on how to fulfil the goals of the macroeconomics ones.[4] This, too, went largely unfulfilled, as the vast sectoral conglomerates of the DSO—such as IZOT—became the main economic bodies in the 1970s at the expense of individual enterprises.

The BCP was thus always looking for a nonmarket surrogate. Allies seemed to point the way. Both the Soviets and the East Germans had inaugurated ambitious reforms in the early 1960s, aiming at de-bureaucratization and devolution of responsibilities to low-level planners. While pricing and other instruments were key, they were also united by a newfound belief in the power of rehabilitated cybernetic theory: the application of data processing, operations research, and the perfection of information flows. This approach was palatable, as it retained the party in the process as the steersman of the modeling and the overseer of the processes rather than being a redundant body in the face of a rising class of technocrats—the very fear that scuppered the GDR's *Neues Ökonomisches System*.[5]

The panacea was STR, the object and aim of reforms. The price restructuring of the mid-1960s had also had this as its end goal, a "redirection of the economy" toward technological progress.[6] This revolution was seen as a qualitatively new stage in human development, the latest scientific achievements raising productivity exponentially and thus turning science into a productive force on par with other inputs. If it was harnessed correctly, workers' productivity and production quality would both skyrocket, and industries would be put on a "rational" basis. What this meant for the party was the minimization and eventual elimination of the "subjective" factor—a catch-all term for directors' and workers' failings, which were blamed for quality problems. My focus in this chapter is not a minimization of the numerous economic mechanisms the BCP toyed with, as cybernetic economics, a peculiarly Soviet science, was also strongly linked with the introduction of feedback mechanisms (such as profit and pricing) into the economy. Such thinking, however, had a closely related technological side that focused on the sinews of the economy and the flows of information between its different levels, which vested a belief in cybernetic machines (such as computers) as solutions to most problems.[7]

It is the technological vision that is less clear, and in the conditions of Bulgaria's electronic dawn, the implementation of the "objective" computer and robot with their perfect information usurped the core of ideology at the expense of the "subjective" and imperfect worker.

Thus, alongside the 1968 mechanism reforms, two commissions were formed to turn "science as a productive force" sloganeering into a reality. The first, overseen by planning tsar Zhivko Zhivkov, oversaw capital investment in new production so as to achieve maximum effect.[8] This Zhivkov had been a noted opponent of economic liberalization and was more amenable to increasing profit in new sectors as a solution to the financial problems of the country, a perfect conduit for Popov's promises of golden factories. The second commission was headed by Professor Popov himself and was to take care of the restructuring of production in light of the newest technological breakthroughs: in effect, the implementation of the first commission's decision. He also had a seat on the summary commission that finalized state budgets, meaning from its inception, STR's implementation was tied to electronics as the core technology.[9]

In October 1969, the Central Committee published its Decision 412, on the topic of technological progress and new steps in economic governance. Its prosaic tone noted that the country had left agriculture behind and could now turn to solving "social problems . . . easing the labor of workers."[10] This new economy was conceptualized cybernetically, as an interlinked organism, governed by processes of "modernisation and automatisation, the deepening of specialization and co-operation, the widening of industrial and economic links between . . . cells'—demanding 'perfect . . . forms and methods of control and implementation in full accordance with these processes."[11] Achieving these goals required a holistic development plan that used prognosis programs and modeling of economic and social processes, cells of a "full-bodied economic structure" in their own right. All this was predicated on the implementation of technology, above all "complex automation" and computing.[12] Machines would get computer-numerical controls (CNC), automated workshops would unite scattered production centers, and these processes would be continuous and free of the subjective factor—tiredness, mistakes, and the need for leisure. The vision was not far from today's dreams of "lights-out manufacturing," windowless robotized factories. Over a third of capital investment up to

1975 was thus earmarked for restructuring production and the introduction of automation.[13]

Machines would march into the national economy on the basis of ASUs—automated systems of governance. This didn't mean just CNC lathes in enterprises, but systems of pan-national importance, serving as capillaries for information, ensuring optimal governance and production at all levels.[14] This effort required 300 large computers in the next five years, to ensure the first level of automation of planning and operative control. It was envisioned that sectors reformed on this basis would raise productivity by a factor of at least two.[15] Automation, cybernitization, computerization became almost interchangeable terms in these documents, due to their "objectivity"—accuracy in measuring but also instantaneous control of processes. They allowed for the "conditions for objective evaluation of quality" by controlling everything from the inputs to the black box of a production process: Quality would thus increase throughout the economy.[16]

The ultimate aim was socialist governance in accordance with cybernetics, which was fully optimized thanks to STR. The machine on the floor was not enough if communication channels were not created to accurately provide information about the state of society. Dry as the language was, it was telling: "to ensure fully the normal functioning of all cells as systems and subsystems of a unified organism of socialist governance" through computers, network charts, mathematical modeling.[17] Enough accurate and objective information had to be gathered so as to be able to predict future developments. Hidden structural problems would be uncovered by the planners armed with these tools, so limited resources could be allocated to solutions optimally.[18] The first area this approach would be applied to was logistics, where a nationwide scheme of computerized governance was envisioned.[19] Ultimately, STR was to improve *flows* of information: between the machine and the product it was making; between the product and quality control; between the different parts of an enterprise and the different enterprises in a sector; between the sectors themselves; and between the sectors and the central planner, who could analyze this Everest of data and feed it back into new plans. A cybernetic organism par excellence, where control and communication ensured optimal governance—under the aegis of the BCP and not a market. Accurate flows would solve the problems of intensification.

This meditation on the 1969 document as the first clear conception of the party's approach to problems also introduces the defining topics of Bulgarian modernization until the very end of the regime. The triumvirate was automation, computerization (sometimes called "electronization"), and cybernitization, which all circled around the core problem of irrational usage of resources, poor production quality, obsolete technology, and the lack of accurate information about the economy and society. The workers were central, for they were the subjective factor that often led to problems, but they were also presented as needing freedom—the STR would save the human from unsavory work. Minimizing manual labor would free humans from drudgery to become a creative force, a concept that did unite automation with Marxist writings. Where ideology failed, technology would succeed, and it is unsurprising, given Popov's key role in late 1960s institutions, that such a road was taken—if the tools of the automated factory and computerized network were at the center of a solution, there would be no need for politically suspect fundamental tinkering of the planning mechanism. These proclamations would set the tone for the next 20 years, capturing much of the BCP's higher echelons' imagination and suffusing their language.

To a larger extent than for the Soviet case, where such obsessions also existed, the persistence of this language was also tied to the importance of cybernetic machines to Bulgaria's trade. The party sought increased world trade in the 1970s, and the dream of increasing living standards and labor productivity domestically was inextricably linked to that of raising the quality of goods to an acceptable world market level.[20] Cybernetic feedback would also streamline integration into the global economy, accurately predicting how much of a resource had to be sourced from abroad or tailoring output levels to the demand for Bulgarian goods abroad.[21] This science was indeed integrated into "regime speak" by the party, as Gerovitch has shown for the USSR; but in Sofia, it was not just because of a desire to conserve the economy but also to integrate it into the world, and the outsized importance of a growing, profitable electronics sector to the BCP gave an even stronger impetus to this language than it did for the CPSU.[22]

But the domestic goal was that of leaders in other socialist capitals—STR would boost the growth that was slowing down. By 1974, the party expected that intensified labor productivity would account for 95 percent of national income growth during the five-year plan.[23] The party would wait for Godot, however, until the end of the regime, as such a watershed moment never came. The slow introduction of ASUs and computers caused anxiety in many Central Committee reports.[24] Analysts railed against the fact that the scarce devices were being used on elementary tasks rather than on tasks of proper governance "which determines the optimal functioning of whole factories, ministries and economic sectors."[25] Factory reconstructions often ignored computerization, and ministries wasted machine time. A National Council for Automation was established in the CSTP to direct these processes, but it continued admonishing the economy's organizational weakness, where mistakes were repeated for years and hobbled "scientifically-based technology of governance."[26] The problem often stemmed precisely from the export-oriented nature of the industry, leaving scant machines for domestic users—the failure in the Bulgarian case was at least in part caused by the very success of the sector in its primary purpose.

The debate didn't change much in the 1980s, as Comecon growth stagnated. Andrey Lukanov, the key figure in foreign trade, reported on 1981 sessions of the body, which harped on about intensification and the need for priority areas to be engines of growth. "Intellectual factors" were an obsession for all his Bloc colleagues, who talked of concentrating on rational management, automation, and the end of as much manual labor as possible in favor of microelectronics.[27] Such further specialization was the basis of the economic plan unveiled at the Twelfth BCP Congress, held that year, where Prime Minister Grisha Filipov warned that that "if we allow modernisation and reconstruction to spill over into many places, we can just forget about solving our main tasks."[28] Electronics, as the premier STR sector, got a further boost in the party's dreams as it scrambled toward turning the country into the mini-Japan of Zhivkov's vision.

Thus the eighth Five-Year Plan, between 1980 and 1985, had at its heart the Automation-8 program, which had to raise productivity by at least 25 percent.[29] Capital investments were on a huge scale, even when compared to other Eastern Bloc countries—in 1982 alone, around 600 million levs were invested not just in IZOT but also in related developments, such as semiconductors or automation in metalworks.[30] Mechanized labor in metallurgy was to rise from 63 percent to 75 percent, and in the light industry from 56 percent to 66 percent. Special Problem Oriented Complexes were created for oil refineries, agriculture, warehousing, and the state investment bank, combining hardware and software in the solution of specific tasks. Most prominent was the focus on robots: They were to rise from 16 percent to 54.3 percent of all automated machines during this period, CNC lathes were to triple to 11 percent of all lathes, and CNC metal grinders—from 3 percent to 19 percent.[31] These scales and growth targets were huge, and wholesale automation was the byword of the BCP more than any other ally.

In the last years of the regime, other high-technology sectors (such as biotechnology) were to intrude into the dream, but even they were predicated on electronics and automation achievements, demanding clean rooms, qualified workers, and air conditioning: all aspects of the electronics industry's fraught birth.[32] In essence, genetics and other related technologies were just the latest feather in the STR bow that dominated Comecon thinking, even as the system was crashing. The thirty-eighth session of the body, in 1986, discussed the Bloc's plans up to 2000. The problems continued to be the same, and so did the solutions: radical modernization on the rational STR basis.[33] Advertising, user-supported networks, and better technical services were to transform socialist goods into globally desirable ones. Robotic automation and higher quality were closely connected in this dream of the new world.[34] The Bulgarian belief in automation was thus, until the end, a subset of the wider Comecon one. Yet nowhere else was the industry as important, and thus as domineering in the very content of STR language of the party.

Nowhere else did it draw in as much investment either, precisely because of its export-oriented nature, which always clashed with the dream of automation domestically. Yet the heyday of the automation dream, the period between the mid-1960s and the late 1980s, did bring real change to the economy and society. Automation, imperfectly and piecemeal, changed the nature of work and education. Despite its failure, the BCP dream was not just one based on economic surrogates and financial profit, but on the claim to be creating a new kind of citizen in the information age. It was a flawed but real attempt to bring about a Marxist vision of the future. Not taking these changes seriously would underestimate the giant effect that this profitable industry had on the Bulgarian landscape.

THE MARCH OF THE MACHINES

The first introduction of a computer-controlled automation system in Bulgarian industry happened in 1969, at the "Zlatna Panega" cement factory, one of the biggest such plants in Europe.[35] The cement industry was noted as one of the most automated in the world, so Bulgaria needed to catch up; yet the real reason is connected to the annual losses of over 1 million levs that were blamed on worker error and fuel consumption. All could be minimized if the constantly changing process of dosing and stabilization of chemical reactions, requiring precise calculations of over 81 equations with five unknown variables in three minutes, could be taken out of the hands of a person.[36] "Zlatna Panega" had plenty of automated or mechanized mixers and furnaces, as did many factories in many different sectors throughout Bulgaria. But for the first time, a Bulgarian ASU would have a computer heart.

This was just the beginning. The 1970–1975 plan was the first in which automated systems became a key part of investment, building on IZOT's growing importance. Cybernitization was the core, and automation was labeled a "structure-defining" sector of the economy, receiving 430 million levs of investments.[37] The plan called for the creation of 1,700 computers, 7,000 magnetic discs, 13,000 magnetic tapes, 270,000 changeable disc packages, and 17 million integrated circuits by 1975, which is the year the party expected computers, radio electronics, and related communication devices to make up 30 percent of all machine-building volumes. These devices would not just be sold but would automate, it was slated, the metallurgical, chemical, food, and textile industries.[38]

The quintessential sector of socialist modernity was of course steel, a staple ever since Stalinism.[39] The party noted that CNC machines and automated lines were proving their worth worldwide, while Bulgarian methods in the sector were woefully behind. The sixth five-year plan was to rectify this, with the first hughly automated systems being introduced by 1974 and the aim of widespread automation by the end of the decade.[40] Popov was a champion of this focus, noting that it would modernize the most widely used machine-building methods of the country. The knock-on effects would lower costs in other areas, too, affecting 70 percent of all machine-building.[41] He noted that this would "end . . . their dependence on . . . qualification of workers . . . or lack of working hands," which would in turn lead to "the highest form of complex automatisation of one closed cycle in a certain sphere of machine-building production—prognosis and planning, construction, technology, the production process, accounting, wages and placement."[42] Popov synthesized the party's thoughts in order to create the first complete cybernetic industrial process in the Bulgarian economy, with ASU input and governance at each level. IZOT would provide the digital control, Balkancar the hydraulic and pneumatic systems, and his old haunt of Elprom the power engineering solutions.[43] All the pieces were there—they just needed to be organized.

Popov highlighted this need to aggregate existing strengths but also to apply the *language* of automation. BAS and its key institutes were thus tasked with the machine languages to control the machines, as well as designing software for automated design and prognosis.[44] They would not go it alone, of course, as the USSR and the GDR would cooperate on automation, as they had been doing since around 1966.[45]

Other areas highlighted as prime contenders for computerization were logistics and internal trade, where the need was pressing: By 1971, over 10 million accounting operations daily were being done in the most "primitive" way.[46] Servicing, provisioning, and related logistical work that resulted in "huge physical and psychological stress" could also be mechanized and automated—the loading tasks in warehouses, but also accounting and stock-checking. Workers in this sector were noted as particularly overworked and were thus rude to or even cheated customers.[47] Automation would reduce the stress on the put-upon service worker, and the infamous socialist service rudeness would be gone. In effect, technology would humanize the person behind the till!

The CSTP reported that by 1972, over 3,500 automation tasks at the cost of 1 billion levs had been carried out throughout the economy. Some DSOs, such as shipbuilding, had fulfilled over 80 percent of their tasks, while IZOT had the highest success rating in implementation when calculated by profits.[48] Twenty-one enterprises were completing ASUs that captured the site's complete process, but all were still lagging. The ZIT factory in Sofia or the Neftohim oil refinery in Burgas performed well, but power plants in Sofia and Bobov Dol were falling behind.[49] This unevenness spread to the 200 enterprises that were to build partial ASUs, automating some of their processes: Only 110 were either complete or being completed, the rest were still at the idea stage.[50] The dream was unevenly applied but covered everything: even farms, such as the collective agro-industrial complex "Druzhba" in Ruse. Three DSOs had completed ASUs for their own administration—IZOT, Shipbuilding, and Elprom; seven more were being built, with various results.[51] Unsurprisingly, the best performers in automation were those already at the cutting edge, such as IZOT itself, or profitable and growing industries, such as shipbuilding. Other sectors were left with lesser specialists and were at the back of the queue for machines.

Enterprise ASUs were not the only ones being built, as 14 ASUs went beyond industry to be considered ones of national importance. They were to produce automated governing systems for the service sector but also for tasks common across many different areas of economic life. Areas such as financing, tourism, labor and social work, database processing, construction, and transport were among them, as were the two that lagged furthest behind: one for optimal planning of development and another for controlling research and development work.[52] The common characteristic of these ASUs was that they were to help automation within enterprises, but also links between them. If the enterprise ASUs were the mechanisms of the body's organs, in Weiner's cybernetic vision, the national ASUs were the feedback channels needed for holistic control. Thus the ur-ASU of national importance was ESSI, the Unified Social System of Information, which will warrant its own discussion later in the chapter.

The CSTP generously funded the expertise needed to fulfil these dreams. In 1972, over 900 workdays of specializations abroad were paid for—570 in the USSR, but also 25 in Japan and 30 in Denmark, for example.[53] Bulgaria was keen on learning how both socialists and capitalists organized their automation programs. The CSTP also oversaw the creation of the Unified System of Information and Means of Automation, which listed the 454 instruments and devices needed to create complex ASUs, of which 266 were being produced in Bulgaria. Yet by the following year, only 83.9 percent of the plan was implemented, with only 73 of 87 needed computer centers delivered.[54] Despite socialism's best effort, full automation required capitalist support.

The program kept growing, with partial or full ASUs at some stage of implementation present in 541 enterprises by 1974.[55] However, problems persisted, and they weren't restricted to just the lack of devices: Only 73 percent of computer machine time was being utilized, slowing down both calculations and enterprises paying back their costs.[56] At the current level of growth, it was estimated that by 1980, Bulgaria would have reached the levels of automation in West Germany or France . . . in 1971.[57] As usual, the Politburo's solution was *more*: more computers, more ASUs, more speed, more cybernetic models, more theory. The language became increasingly permeated by a focus on social governance rather than on just the machines, identifying the problem as "the cybernitisation of governance and the perfection of organisational structures" rather than just production of computers.[58]

Despite the slow rates, by the 1970s, automation was recognized as an embedded fact of the new society. Electronics was reorganizing the economy and were joined by new work in the social sciences, including law, where new reforms were envisioned to perfect social democracy, all aiming at "typification, unification, standardization in all spheres, the perfection of the system for the control of quality with the aim of achieving a maximum efficacy of material and human resources."[59] Everything from the plan accounting to material provisions was to be subject to sector-wide ASUs, requiring over 180 large computers, 700 mini-computers and 3,000 terminals by 1980.[60] The horizons the BCP chased in its documents were not in the USSR but in the capitalist world. The benchmark was not the Czechoslovaks but the West Europeans, as the party scientific theses of 1975 stated. Notably, these documents were drafted by people who had often specialized in the US, Japan, Austria, and the Netherlands.[61]

Doĭnov continued the Popov obsessions even as he usurped him. The 1980–1985 plan called for a further 243 ASUs for industrial processes, 322 for enterprise governance, 16 sector-wide ones, and 162 in design and research work. The expected effect was over 2 billion levs by 1990, as labor productivity would blossom, quality produce would roll out of the factories, and fewer resources would be wasted.[62] As chapter 3 showed, STI was at the core of procuring this technology, and the gargantuan plans of the 1980s account for its growing importance under Doĭnov.

Avtomatika-8, the National Program for Automation, was the umbrella for these ambitions. Over 1.8 billion levs were assigned for capital investment over five years, with many priority areas dedicated to upgrading key

ASUs, such as the ones in ZIT or Kremikovtzi, rather than building the first systems in underautomated areas.[63] Thus the perceived success in certain sectors deformed the economy as more money was poured into the already most modernized areas. Over 3,000 industrial robots were expected, as well as 2,100 metal-working machines with CNC.[64] A total of 480 enterprises would get some sort of ASU, with the largest ones again saved for the already automated enterprises that needed upgrading.[65] Altogether, capital investment rose by a factor of 3.5 over the automation spending in the previous plan.[66] Popov's plans were thus continued, but the solutions again largely exhausted themselves in throwing money at problems. Doĭnov's automation exhibited the gigantism that defined his industrial policy.[67]

The examples of what these ASUs and computer centers would do are numerous and abound in the archives. They were present even in collective farms—in Branishte, a computer helped with accounts, monitored volumes of milk produced, animal health, and their reproduction.[68] A couple of brief sketches of the computer center of DSO Shipbuilding in Varna and of Botevgrad's industrial and medical automation can shed light on how they developed, operated, and fared. The latter was the

5.1 Control room of the Beroe Robot Combine, Stara Zagora, Bulgaria. (*Source*: Central State Archive, Sofia.)

semiconductor-producing town about an hour's drive from Sofia and was the setting for some pilot programs. Due to its electronic heart, it was perfect as a testing ground—its high level of computerization and technically literate workers meant it produced 17 percent of the total production of Sofia district in 1980.[69] The semiconductor plant had well-developed automation in planning, logistics, accounting, material provisioning, and shopfloor governance thanks to an ES-1020 heart. The large "Georgi Dimitrov" chemical plant was also automating its administration and technical processes through a "Robotron" computer acquired in 1981. Even one of the milk-producing agro-industrial complexes had a computer center, equipped with a "Cellatron." The Chavdar bus factory was also automating, while much of the local district council's administration was done on ELKAs; even the regional hospital was introducing ASUs.[70] The CSTP thus concluded that the whole territorial system was successfully computerizing, but still in a localized way, not coordinating among the various systems. The local government was also not using computers to oversee the various ASUs, a key link between town and central state planning.[71] Computers also needed to be used not just in industrial processes but also in social provisioning and collecting complete data on local labor and financial resources.[72] Botevgrad was thus to construct a Computer Center for Collective Use, to put local governance on a rational basis and to connect it to Sofia's central decision-making.[73] The report astutely noted all this was well and good, but the problem "of who will collect information, who will move it, and who will be responsible for it, is not solved."[74] Harmonized development could be achieved by uniting the previously disparate systems, but the exact methods were still up for debate.[75]

The healthcare ASU was an example of the more general benefits electronization was supposed to bring beyond the factory. The Botevgrad hospital received its computer in 1982 and was to implement the Electronic Medical Establishment system by 1985.[76] It was the pilot program for automation of healthcare in Sofia, Plovdiv, Varna, Pleven, Ruse, and Stara Zagora, the country's biggest cities.[77] The program involved both advanced electronic diagnostic equipment and automated governance of the hospital's activities.[78] SM-4 minicomputers were used to create a central information system, and local terminals and microprocessors would carry out four specific tasks: "hospital activities and economics" (accounting and administration), "para-clinic," "poly-clinic, "statzionar" (three systems connected to the gathering of general and medical information of patients in emergency rooms, for doctor visits, or in the wards).[79] This system was envisioned as improving diagnostics but also rationalizing medical labor and freeing doctors and nurses from superfluous activities. Mundane tasks such as appointments would be automated by gathering patient information at entry; this data gathering would also allow for "mass prophylactic examinations of the population."[80] Automation was the path forward for industrial and medical workers alike, and the hospital ASU is a perfect illustration of the party aim of transforming all spheres of Bulgarian labor.

The logic behind it was not really health but, as always, increasing productivity. Automation of clinical laboratory analysis plus automated classification and return of information would cut down labor time by 30 percent. New programs analyzed blood sugar, urine samples, cholesterol, and creatinine levels—all justified as they would double laboratories' productivity.[81] Six bed monitors linked to a central processing station would look out for ten different types of cardiac problems, take patients' pulse, register ECG patterns, and connect to an alarm if anything went wrong, allowing for less direct observation and more time for other tasks.[82] The equipment and methods were Bulgarian, and the innovative (for that time and place) systems were again aimed at rationalizing labor by combining disparate processes on one screen and one printout—an ultimate effort to simplify the overwhelming information flow of modern life.[83] Success here would, it was expected, spur more production of such specialized equipment for the "intensification and intellectualization of labor of highly skilled medical personnel."[84]

The archives are replete with examples of ASUs and computer centers. Some were industrial, such as the "Sigma," a joint Soviet-Bulgarian plan to automate car production, which was implemented in 1981 in the Lovech Balkancar factory. It integrated both horizontally and vertically, encompassing and connecting different parts of the enterprise shop-floor's (and also the whole plant's) governance with regional and state planning. It compared targets against performance in real time, it accounted for movements in warehouses, keeping track of shortages, as well as overviewing design and research, cadres, and qualifications. Through its ES-1020 and ES-1022 hearts, the ASU created the full feedback loops envisioned by experts in order to make the plant fully "rational," resulting in numerous economic effects on savings and elimination of wastages.[85]

But most of the holistic work was concentrated in computer centers. Concentrating on one particular center allows us to trace various issues in the final chapters of this book. The computer center of the DSO "Shipbuilding" in the coastal city of Varna was created in 1968. Its mission was to design programs and ASUs with the aim of "achieving higher economic effects" in the sector. This included both enterprise programs and optimization of the governance of the DSO as a whole.[86] Throughout the years, the Varna Shipbuilding Computer Centre would become the heart of the industry, designing everything from its first program in 1969[87]—a way to choose elements to be used in ship construction that also fit the desired price ranges—to full ASUs for factories in the DSO in Varna, Ruse, Burgas, and beyond.[88] Its archives are full of catalogs of the programs it implemented, numbering in the hundreds. It is worth noting the international scope of these programs—apart from in-house designs and those designed by Soviets or Poles, it is full of IBM packages and those coming out of the laboratories of the Massachusetts Institute of Technology. Between 1969 and 1978 alone, over 50 IBM programs were acquired and used in Bulgarian shipbuilding, for example, local proof of the porousness of the Iron Curtain highlighted in chapter 3.[89]

The center was well equipped with Western machines. Its dealings with both the DSO leadership and CSTP highlight the power that local technical experts had when defending the interests of their professions. At first, the center was offered the ZIT-151 machines, which it rejected: The Fujitsu copies were specialized for economic work rather than engineering projects. If the DSO desired a real ASU-designing center, the experts argued, it should cough up the money for the right machines with the best specifications. An ES-1020 was subsequently offered, with the full range of IZOT peripherals—this computer was also rejected, but some of the peripherals were kept. Ultimately, the center's experts argued that they needed a Western machine that was fast and reliable—the ES was not yet proven. The advantages highlighted were that IBM machines also came with many packages already designed for shipbuilding. Ultimately, over 2.5 million levs were freed up by the DSO and state to pay for an IBM 370/145 machine, imported through Vienna.[90] As was often the case, then, local centers could get the machines they desired, even when their purchase required scarce Western currency. Like many others, the Varna computer center was thus equipped with a Western machine rather than an Eastern one, the latter seeming to go for exports or to second-echelon and enterprise centers. It must be noted—shipbuilding was a key industry for the regime, so the center did have the clout it needed to get what it wanted, too, unlike a more provincial center tied to a less strategic industry.

The center's activity was also highly international, another proof of the porous nature of technological work in the Eastern Bloc. Some of the main technologies used were plotters and specialized shipbuilding programs from the Norwegian firm Kongsberg. Money for their devices, as well as specializations there, were often demanded—and received—from the DSO and CSTP.[91] Close cooperation developed with SENER, a Spanish engineering firm with cutting-edge expertise in shipping engineering. In 1976, the Bulgarians implemented its "FORAN" software for CAD/CAM (computer-aided design and manufacturing) ship design, applicable to all stages of the development of a vessel. This program would remain the backbone of Bulgarian shipbuilding design until the end of the regime.[92] It was acquired for $165,000, which was negligible, given its huge effect later on—and a massive hit as the Romanians paid $1.5 million for the same program, having failed to develop or acquire much of the auxiliary software before.[93] Once again, it seems, STI had done its work well in preparing the deal for the computer center.

These international links are worth noting as further proof of the transnational networks of expertise that Bulgarian engineers participated in throughout the socialist period. The catalogs of programs indicate an increasing reliance on in-house programs, but many of these were designed either after copying or inspiration by programs by IBM, SENER, Kongsberg, or Japanese firms. Western software carried out much of the design and automation work in Bulgarian industry and was the basis of other Bulgarian programs developed later. Varna's engineers trained, learned from, and designed parallel to Western programs, dissolving the simplistic dichotomy of capitalist and socialist technology. In effect, participation in export-oriented industries (such as shipbuilding, computers, or the amalgamation of both) meant that one was simply part of a universal technological family. What else is one to make of the center's 1983 library catalog—650 volumes of Soviet and Bulgarian technical manuals, overshadowed by over 1,200 IBM manuals, 150 magnetic tapes with IBM programs, volumes on Spanish FORAN technology, foreign elements catalogs, and subscriptions to Western magazines alongside Soviet and East German ones?[94]

By 1987, the center employed 196 people with a high level of training.[95] Throughout the 1980s, about a third of all staff were sent out on specialization or refresher courses at various Bulgarian and foreign institutes and lecture courses, keeping them abreast of the latest developments.[96] Women predominated, from the earliest period of the 1970s when they made up 63 percent of the staff, to the 1980s when much of the software was designed wholly or partially by female software specialists.[97] Most workers were young, in their late 20s and early 30s.[98] Both the gender and age ratios were significant, and we will come back to their significance in the next chapter—computer labor in calculation centers were often the provenance of women or youths. Even in years the center labeled as difficult, such as 1983, it fulfilled an immense number of tasks—one nationwide task, three specific enterprise tasks, and 53 automation tasks at the DSO level (such as accounts and labor databases), establishing terminal teleprocessing links between Varna and Ruse factories, and more beyond.[99] Its input cut down design times for 25,000-ton vessels by a factor of more than 7.5; its automation of design for ship's electrical systems had effects in the hundreds of thousands of levs per year in saved labor or resources.[100] The list of achievements is long, but the abovementioned ones above are worth highlighting: Plenty of Bulgarian computer centers, despite the shortages of equipment or problems in planning, *did* rationalize labor and create fully functioning ASUs or tangible effects in their sectors. The tasks ranged from mundane teleprocessing to designs of whole ships, the full gamut of modern industry in each sector. Due to the nature of their work, thousands of specialists in such centers learned Western technology and software, and they worked hard to fulfil the directives of their enterprises and DSOs, implementing STR in their sectors.

By the mid-1980s, there were 143 computer centers of territorial, DSO, or ministerial level (rather than just enterprise ones); more than 850 large computers had been installed, fulfilling key tasks of the automation programs of the party, and with the microprocessor revolution coming, the possibilities of automation would—in the regime's mind—grow exponentially. Over 1,500 sites in the economy were serviced by computers in some capacity, over 60 percent of them in direct material production. Over 15,000 specialists were employed in the computer centers alone, all possessing higher or specialist secondary education.[101] But the automation dream did not rest on single, industry-specific centers, toiling away on local tasks—the computer centers were never envisioned to be, and never could be, isolated.

NETWORKS

The crux of the dream was, however, very literally a crux. Socialist cybernetic governance was predicated on the center possessing accurate information on social and economic factors at all levels. This was discussed and agreed on after the Ninth Congress intensification proclamation, and the 1967 CSTP report stated that

> For the perfection of governing work, planning and scientific-development and project-construction work, fast and precise computations are key, as well as accounting and statistical records. The fast and precise completion of calculations, accounting and statistical operations at minimum expense can be solved through the building of a unified system of computational centres [their emphasis].[102]

Computer centers were not only to exist in enterprises, ministries, or institutes but also at a territorial level. The first regional computer center (RITz) had started construction in Ruse, Bulgaria, in 1965, as the first base to study mathematical programming and apply it to a region. It became a core center for designing software that would address the tasks all such centers would face, such as calculations related to accounting and planned production. Its specialists were quickly sent out to specialize in the USSR, GDR, and Czechoslovakia in 1966, where they familiarized themselves with linear programming in industry and agriculture.[103]

The RITz got its machine in 1967, the first provincial computer center to be equipped with one—a British-made ICL 1904.[104] It is interesting to note that together with an Elliott computer system sold to the Devnya fertilizer plant the next year, such early British successes in Bulgaria and Eastern Europe more broadly led the British to envision themselves as masters of the market, before ICL was overtaken by IBM—not least due to the decision to base ES computers on the American and not on the British system.[105] Some of the 57 staff were also sent to Britain, to train on the machine, and by 1970, they would help implement COBOL-based information-processing systems, ones for accounts, labor, and wage databases, used throughout the Ruse province, such as the locomotive factory or tractor stations on collective farms. "Optimization" programs for specific tasks (such as optimal usage of chemicals in metallurgy or the feeding of animals in AKPs) became the bread-and-butter of the RITz. In May 1970,[106] due to its primacy among such centers, it hosted the first conference on ASU implementation and was hailed by the CSTP as cutting edge.[107] In 1971, it was renamed as a territorial center (TCC), as part of a nationwide consolidation, taking on more administrative tasks in the region rather than just industrial ones. Gradually, it incorporated regional statisticians and created a Scientific Centre for Territorial Planning, facilitating party and state plans at the regional levels. Its British machine was no longer sufficient, and it became truly Bulgarian with the delivery of an ES-1022 and SM-4s after 1975.[108]

Ruse was a forerunner, but from 1968 on, other regional centers started operating fully equipped, too. The first was in Gabrovo, becoming also the first provincial center to be equipped with an IBM machine.[109] The "Bulgarian Manchester" was a symbol of pre-socialist industry, but during these years, its smokestacks got new neighbors—a technical university and an IZOT factory, "Mehatronika." By mid-1972, there were 91 computer centers, including 28 TCCs (one in each region), yet 55 of them had computers.[110] In the first nine months of that year, they had only used 76.2 percent of machine time, which was considered a good achievement: 62.5 percent for data processing, 30.6 percent for program testing, and 6.7 percent for other tasks, mainly training.[111] Performance varied widely—one center used 97.4 percent of machine hours, while the Ministry of Education's center managed just 26.9 percent.[112] But gradually, priority tasks such as economic planning increased in number, with routine statistical processing falling.[113] The TCC system grew and matured, becoming integral to local authorities' fulfilment of socioeconomic development priorities. The huge variation in usage, however, never went away, with some (such as Ruse) remaining champions, while centers in poorer regions or overlooked ministries often lagged.

The dream seemed to be getting nearer every year, with many caveats. By March 1975, there were 114 computer centers, of which 60 percent had one computer, 17.5 percent had two, 4.5 percent three or more. And although 18 percent still had no computer, being computer centers in name only, this percentage was down from the nearly 40 percent three years earlier. Concentration was still problematic, however: 45 percent were in Sofia, and another 30 percent in seven large district centers. The CSTP sought to slow down the creation of new centers to ensure that all centers had machines, and it also defended the doubling up in major cities as "redundancy."[114] Creation of these centers had proceeded at a breakneck speed in the last five years, but demanded complex construction work that was sometimes not met—better buildings with regular cooling and reliable climate control meant that some, such as the TCC in Vratsa, took over four years to be built. By the end of 1974, the country had produced 183 large computers, of which the ES-1020 types made up 35 percent, and 22.5 percent of the computers were the older ZIT-151. Yet even with the 88 to be built in 1975, gaps would remain. Only 7 percent of machines delivered in 1972 had memory capacities over 64 KB, and not much had improved, so peripherals were to be prioritized, even though they were of course the prized export good. Indeed, 53 percent of machines installed were imported, another glaring example of the tension between the party's computerization plans and its export policy.[115]

Not all was doom and gloom, of course. Over 9,400 highly qualified specialists were working in these centers by 1974. Numerous university and postgraduate courses were producing graduates, and more had to be done only in terms of practical training.[116] At least 76 enterprises had dedicated computer centers that were automating industrial tasks, not just administrative ones.[117] Thirty universities and institutes would have complete centers by 1975, increasing machine time available from 179,000 hours in 1972 to 345,000 in 1975, leading to a 230 percent increase in data processing, and hourly productivity would rise. The average use of a machine was 14.1 hours a day, a full 90 minutes higher than the Soviet average, with the Gabrovo TCC a champion at 94 percent of machine time used—but the Vidin TCC, in one of the country's poorest regions, achieved a paltry 39 percent, highlighting the continuing problem of uneven development.[118] Real economic effects had been felt, despite this unevenness: 24 major engineering projects saw their duration cut by factors of four to six; 5 million levs were saved by the Ministry of Transport in 1974 alone.

The effect was calculated at between 12 and 14 million levs per year and would only go up with better training and machines.[119]

Thus a fledgling but unconnected matrix of computing power existed by the mid-1970s. The ASU and computer deliveries would continue at a growing pace into the 1980s, but the Politburo recognized this moment as a watershed for deciding the future of development. In December 1974, Doĭnov reported on the future direction of economic growth and emphasized the need for minicomputers rather than large centers to allow for a cheaper and deeper penetration of computers into more locales. This would help create a hierarchy of automation, which was needed for cybernitization: "small and medium enterprises and organizations, thanks to minicomputers, linked with powerful computational centers for collective use, achieve the ability to automate the governance of production without possessing large machines."[120] Apart from smaller machines, there was also a lag in software, which would allow more tasks to be carried out but also "makes the computer needed, marketable, effective, and sought on the world market."[121] Bulgaria also needed an expanded and modernized communications system, with radio-electronic equipment based on the Crosspoint system with over 1,260 relays.[122] Thus the focus was not only on more personal computing but also on the communication between nodes in the system. Another report at the meeting highlighted that the world was on the cusp of a communications revolution, with new laser-fiber and cable developments. Bulgaria had to ensure its connectivity both within the country and to the outside world—only 22 channels existed between Sofia and Moscow. To pique Zhivkov's economic interest, it was suggested that the state could become a hub between Europe, Asia, and Africa.[123] Yet it was the domestic effect that was highlighted:

> I feel that we can solve all this [bad integration of information and communication networks] on the basis of completely new means of communication and information activity . . . and if this material gives more space and importance to this question, I think that it will help in the fuller solution not just of the question of automated systems but the whole realisation of governing activity of our leadership.[124]

This importance, together with the fact the country was now producing ESTEL teleprocessing systems and the focus on new effective telephone exchanges birthed the next dream in the plan: a network that would enable the transfer of information.

This dream was Unified System of Social Information (ESSI), part of the party's thinking since the 1960s, when it became clear computing would allow for the processing of a huge amount of information. The planning had been consolidated into a DSO Machine Processing of Statistical Information, which then became the "Avtomatizatsiya" in 1974, while the work on creating a national network of collecting and processing social information was taken out and put under the Council of Ministers in a committee called KESSI (Committee for the Unified System of Social Information) in 1977.[125] ESSI was envisioned to collect many things: local plans, enterprise production and accounts, but importantly, it would have a major subsystem for citizen registry: the Unified System for Civilian Registration and Administrative Services for the Population (ESGRAON). Its original envisioning in June 1974, in the midst of so many computer-center births, was "to improve the technology and culture of administrative services for the population and to perfect some parts of the informational processes in the local and central organs of government."[126] It created registration cards for all citizens, containing a great deal of information: education and work status, convictions, financial obligations, property, car ownership, and the Unified Citizen Number (EGN). The latter, the citizen registration number that every Bulgarian citizen has to this day, also coded in its 10 digits the date and place of birth, and gender. The CSTP admitted that this task was hugely complex.[127] It would require over 300 million levs in research and implementation, including the abovementioned better communication systems, and would have a great but unquantifiable effect.[128] ESSI was a gamble for a party that liked its projects to be predictable, but it also had hitched its wagon to the cybernetic engine.

ESSI was to be a "rational computer network."[129] As the skeletal structure of nascent computerization was in place after 1975, it was at the heart of that year's scientific theses that called for the creation of over 3,000 smaller terminals that would encompass as many offices and enterprises into "multi-machine networks."[130] The BAS Cybernetic Institute took the lead, releasing reports on large multi-machine networks and processing in 1976, a topic it cooperated on with the Soviets. It studied existing systems, such as ARPANET (Advanced Research Projects Agency Network) and TYMNET, but strangely, didn't talk about OGAS (National Automated System for Computation and Information Processing)—the Soviet failure was glossed over and of little interest when compared to networks that operated successfully and exchanged data over great distances in the West.[131] The Institute's definition of ESSI is worth quoting fully, revealing its techno-utopian scope:

> [A] pan-national, automated information system for the gathering, processing and storage of data for governing, planning and accounting of the social-economic development of the country, which has as an aim the complex informational servicing and mutual interaction of all organs of social governance. Its aims are the ensuring of the needed prerequisites for the solution of given tasks in achieving the maximal economic and social results with the optimal usage of natural, material, human and other resources of the country. ESSI is an integrated system, which uses the informational base of automated systems of governance (ASU) and all automated information systems (ASI), created in the country, including the systems designed for the servicing of citizens.[132]

Each TCC was to house a territorial database with information on the material, natural, and human resources of the region, which both local party planners and enterprise managers would resort to. But each had to be linked to national centers and databases, accessible to all organs of governance. Mini-machines and channels for data transfers that went parallel to the main one would ensure autonomy in the event of a breakdown of the network—unnamed in the report but always hanging over this was the biggest possible threat of nuclear war.[133] The experimental communication system was to be built by 1980, with the first echelon to be connected encompassing Sofia, Plovdiv, Varna, Stara Zagora, Burgas, Veliko Tŭrnovo, and Vratsa, ensuring that both the TCC and local enterprises' ASUs were connected. Then the national ASUs, such as the pension database, would be connected.[134]

The system required computers of at least the third generation, capable of 250,000 operations per second, which the ES 1020 and ES 1030 range was capable of. Modems of at least 4,800 bits/second as well specialist software was also required—all these were baked into the modernization programs of all TCCs.[135] The fourth iteration of ESTEL was also explicitly based on the IBM System Network Architecture and used the internationally accepted Synchronous Data Link Control algorithm, making the Bulgarian network compatible with international standards. User access was improved through the new ES 8501 terminals, again based on IBM machines. Databases were developed that would help in tasks beyond mere management—Horizont, for example, would predict the development of advanced economies, so that enterprises could tailor their trade policies; while Sirena would ease research work by compiling all dissertations and scientific implementation papers in the country. Moreover, the national network would be connected to the Moscow VINITI (the Soviet Institute for Scientific and Technical Information), greatly expanding available knowledge.[136]

The Soviets themselves were interested in the Bulgarian system, noting its facilitation of "timely and accurate statistical and other information and analysis for the development of economic, social and cultural life in the country for the needs of social governance at different levels," as well as the standardization of programs and documents to be used nationally.[137] The Bulgarian system was late to the game—after OGAS, after APRANET—and that was its advantage, avoiding the pitfalls of the Soviet dream and taking advantage of Western developments. It took a gradualist approach, an echelon at a time, and had the advantage that in Bulgaria, computers were less tightly tied to the military. Its functions were also more targeted at social governance, the computers becoming a tool for "reading" the population rather than for planning the economy. One of the standardization methods that most struck the Soviets was the EGN (*Edinen Grazhdanski Nomer*), which allowed every citizen to be entered into the national database, with the number encoding the date and place of birth as well as gender. It was just one among many variables that the system was to encode, putting massive strains on the limited capabilities of Bulgarian computing in the late 1970s. In a report on the heart of the system—the National Information Calculation Centre—the head of KESSI, Dimo Balevski, stated that the system was already tracking over 500 indicators of development per day, recording both social and economic effects of the Plan.[138] But much like the Soviets had found, the complexity of the task in a modern, industrial society was greater than the technological capabilities—the center required a computer capable of 1.5 million operations/second and 1,200 MB of storage. At the time (late 1979), it was exporting over 3,000 proccessing hours to other centers, as it only had an ES 1022: "the lack of machine time destroys work rhythms and destroys

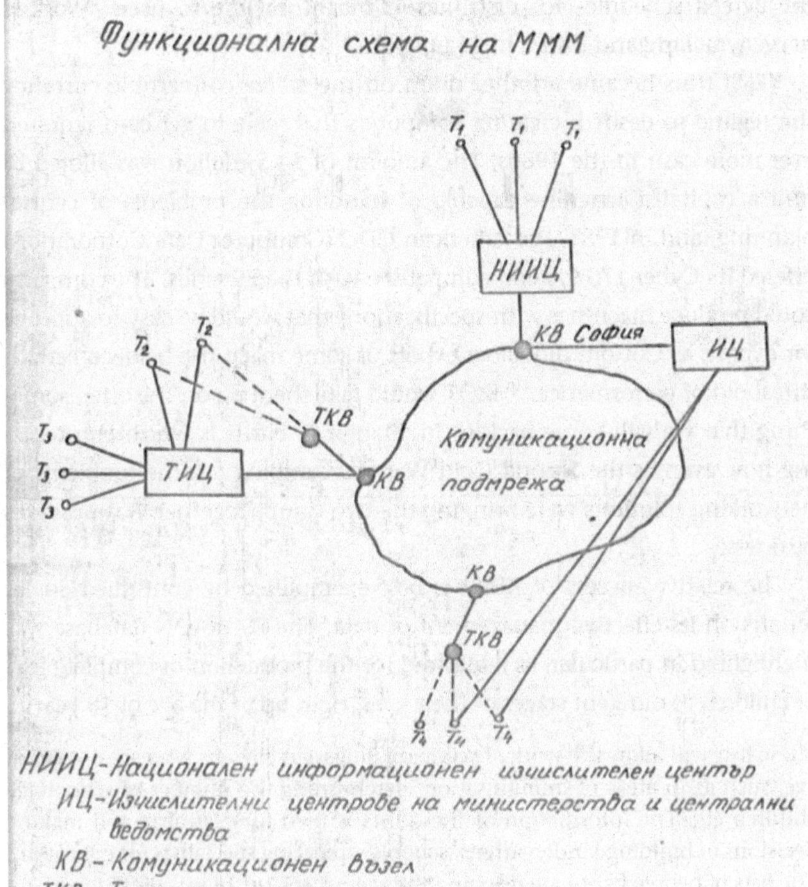

5.2 Map of the network. T1 through T4 are the terminals of local party and state organs as well as enterprises. They are connected to the central TCC, while the territorial administration was to have a secondary, backup connection to the national network through a dedicated terminal. T4 (enterprise terminals) were also to connect to the central computer center for the sector (usually housed in the corresponding ministry). Through the Sofia interchange, the network was to be connected to the national databases. There was thus a lot of back-up and overlap built into the system, to ensure safety—while not voiced, the specter of nuclear war must also be factored into this network redundancy. (*Source*: Bulgarian Academy of Sciences Archive, Sofia.)

the agreed schedules for distribution machine time to users. Work is nerve-wracking and under huge pressure."[139]

KESSI thus became another drain on the scarce convertible currency the regime so desired: creating computers that were to get cash required ever more cash in the 1980s. The amount of $4.5 million was alloted to find a capitalist machine capable of handling the problems of central planning, and in 1981, the American CDC (Computer Data Corporation) offered its Cyber 170 system, compatible with the ES series. The company could produce machines with specifications that would be easy to approve for export, as CoCom did allow export of some machines below a certain threshold of performance.[140] IZOT would take them up on the offer, something that we will come back to in chapter 7, but it is worth emphasizing how even as the Second Cold War was ramping up, the problems of networking solutions were bringing the two camps together. Business was business.

The relative success of KESSI is best exemplified by continued Soviet reports on its effective management of data. The ESGRAON database was highlighted in particular, as it allowed for the production of complete lists of children at different stages of their lives, right up to the age of 18 years:

> These lists will allow the work of servicing Bulgarian citizens who are not yet of age, such as in areas of immunisation, determining the number of school-age children etc. The information of these lists is used for planning and making decisions in building kindergartens, schools, preparing specialists for education. The lists of people [who] are reaching the ages of 16 and 18 is useful in work on passport issuance and the distribution of labour resources.[141]

ESGRAON was used to determine army conscription lists, and it also worked at the other end of the life cycle, drawing up lists of those about to retire. It presented users with clear visual data in the form of charts and graphs on so many topics—from gender to education level to unemployment lists—that it made Bulgarian social governance vastly easier than the Soviet type.[142] The system, together with a related system on "clerk services" operated since 1978 after pilot runs in Ruse and Burgas. Over 150,000 households in the country could pay taxes and bills without the usual long lines at tills, but through databases, working to achieve the aim of "maximally freeing citizens from their direct participation in services, saving them time and raising the social productivity of labour."[143]

Many of the largest cities were part of the system by 1980. The pension system had been computerized, as were the labor histories of 4 million workers. Citizens could now request certain documents or administrative services through the phone, saving precious labor hours. Bureaus for such "complex administration" had been set up in Sofia, Pleven, Ruse, Gabrovo, Vidin, Blagoevgrad, Veliko Tĭrnovo, and other cities.[144]

KESSI decisively made servicing society easier, but more importantly, it made that society much more transparent for the state. We are used to thinking of real existing socialism as having been invasive and all-knowing in its Stalinist excesses, but it was in the supposedly drab but consolidated 1970s and 1980s that the Bulgarian state actually came to know the most about its populace. At the moment that backwardness seemed to have been overcome, and the Bulgarian social landscape was to all intents and purposes stereotypically "modern"—urban, industrial, literate—it also became highly readable to a party that always demanded more information. In 1983, the Politburo still complained that KESSI was not serving it with enough information.[145] With most Bulgarian electronic products exported, while the administrative economy ballooned, KESSI remained underequipped—in 1980, there were 1.4 million workers in services and administration, more than double the 1960 figure.[146] Productivity in the sector had risen just 4 percent over these two decades, while industrial productivity had grown tenfold; 80 percent of time was spent on meetings or travel between them.[147] The state was producing the tools of the information revolution, but it was not harnessing them for its own purposes. KESSI's solutions continued to be large-scale—the world of centralized computer centers and planning. What could really change workplace productivity was the entry of the revolution into offices and on an individual's desk. The early 1980s were thus a key juncture as the party latched onto the next stage of automation: the personal computer.

MICROPROCESSORS AND ROBOTS

The Intel 4004 started it in 1970, the Intel 8800 perfected it, and the Altair turned it into a true commercial breakthrough, with the "it" being the PC revolution. In 1977, the Apple II came out, making the company a household name and catapulting its cofounders Steve Jobs and Steve Wozniak to public stardom. Companies like IBM, established and colossal, were slow to pick up on the trend of small computers aimed at individual users, but by 1981, they too entered the fray with the powerful 16-bit IBM PC. Paired with the MS DOS operating system, created by Microsoft, the last Cold War decade was dominated by a new type of computing. By 1982, *Time* magazine had made the computer the machine of the year. The information age had truly arrived.[148]

A domestic microcomputer was also at the heart of the eighth five-year plan. The ITCR was tasked with the creation of the first Bulgarian PC on the basis of the MOS 6502 processor (in its Bulgarian variant SM603), and the first 50 machines were created on an experimental basis in 1980. This was the IMKO-1 short for *Individualen MikroKOmpiutŭr,* although the joke was that it stood for the Bulgarian abbreviation of "Ivan Marangazov [the leader of the development team] Copies the Original."[149] The examples were sold to various organizations, who reported its ease of use and workability. The CSTP thus developed a program up to 1985 to bring this new technology into serial production. A report on the microprocessor revolution highlighted the immense capability of PCs to intellectualize labor of all workplaces, as well as to enter the home. In 1981, over 500,000 units had been sold worldwide, and by 1984, it was expected that the figure would exceed 5 million.[150] The CSTP decided that much like the ES, the PC must be based on the latest foreign developments, with dedicated video monitors and at least 64 KB ROM as well as BASIC-based software. Research also had to start on 16- and 32-bit machines, as the latter would have the processing power of an IBM-370, on which the latest ES-1035 was based. The microcomputer could usurp many of the functions of its more complex cousins, while it could also attract a whole new generation through both games and education.[151] The Soviets and Hungarians had also started making these machines, but they weren't commercially available. The PC was not just another niche but an extension of the 1970s automation discourse, which pitted socialist against capitalist usages:

> the possession of a personal computer doesn't have the characteristics of a purely commercial usage; here the personal computer serves for the easing, intellectualisation and efficacy of labour at the workplace, for scientific activity, education, prognosis, planning and many others.[152]

The IMKO had already demonstrated Bulgaria's ability to be at roughly the world level, ahead of its allies, and "this, as a rule, creates the preconditions for an eventual acquisition of profitable specialisations to Bulgaria's benefit."[153] Much like in the late 1960s, the BCP had its eye on the massive Comecon market: "as illustration of the reach of these markets we can mention, that in terms only of regional agricultural centers in the USSR there will be a need for tens of thousands of personal computers in this five-year plan alone."[154] The plan thus envisioned for a multi-front attack: 8-bit computers that could be attached to a family's TV would invade the socialist home, while more sophisticated 16-bit machines would automate the office.[155]

The future was here, and it had to be prepared for. Classrooms would be equipped with computers, and children would be taught BASIC. Blagovest Sendov at BAS was already developing innovative methods for computer-based education, the so-called "second literacy." The DKMS would engirdle the country with computer clubs and push students into software development. The ultimate aim was for the PC to be the final stage in automating Bulgarian life, even down to your bedroom. The CSTP report is worth quoting at length:

> The realisation of the aims of the programs will ensure the mass entry of personal computers into the spheres of education, professional circles, scientific circles, schools and universities, trade and construction organisations, in the many hobby clubs, in open spaces for games and work through PCs, and many other places. The effect of the entry of personal computers into our country will reflect en-masse in all spheres of social life. The wide distribution of personal computers will constitute the real manifestation of the all-encompassing intellectualisation of our life, and in the end will lead to a massive and geometrically multiplied economic and social effect in all areas of the national economy and social life. In many cases, with the help of personal computers it will become possible to solve economic and technical tasks through cheap technology at the workplace of specialists, the engineer, the economist, the doctor, the economic director or supervising worker.[156]

The Bulgarian worker finally had the chance to be the Marxist Renaissance human. While being a capitalist revolution, the PC had a distinctly socialist flavor, too, stemming from those earlier automated dreams. If the kernel of ideology by the 1960s was becoming science as a productive force rather than Marxism-Leninism, then the microprocessor was even better suited than the old computer, locked up in a secretive computer center that serviced esoteric plans and churned out numbers. The PC was a personal ASU, much better suited to the STR ideology, as it could enter your home and be part of your uplifting as a worker. It could ease your work in the factory through CAD software; but it could also ease your personal accounts or let you relax by playing computer games. Together with the other facet of the 1980s automatic dream, the robot, it freed you, so you would be a controller rather than a cog.

Thus in 1982, a PC section was created at CICT, and the ITCR developed its own laboratory, too. In only a few short months, the Marangazov-led team created the IMKO-2, a complete Apple-II compatible machine, which entered the market as the Pravetz-82. From 1983, this machine entered serial production at the Instrument-Building Factory in the town that had been a village—but one where Todor Zhivkov was born. Under his patronage, this town would become the high-technology mountain home of the PC. Printers (based on EPSON licenses), monitors, and floppy drives also entered production.[157] The following years saw the range of machines and peripherals increase: The 16-bit machines of the Pravetz-16, analogs to the IBM PC, entered production in 1986, when the factory became the Scientific Production Combine for Microprocessor Technology. IZOT also produced its own series, the 1036 and 1037 machines, but the talent was increasingly pooled at Pravetz. The old production organization was slow to adapt, a parallel to IBM's lag. It was too vested in the ES series, its mainstay, and was already facing problems in minicomputer implementation, let alone microcomputers.[158] The best ITCR and CICT cadres were pooled in Zhivkov's hometown, helping it produce 5,000 units in 1984 and approach but never fulfil its capacity of 100,000 units per year by the end of the regime.[159] It produced desktop PCs but also the Pravetz-8 series, which could be plugged into the TV—cheaper and easier to use, it was the first step toward home automation. Apart from the talent, the money was also earmarked for Pravetz, especially scarce Western currency.

The other face of the last automatic dream was the robot, the steel muscle to be controlled by the microprocessor brain. Robotics was item two on the Doĭnov plan for technological modernization presented in July 1978—item one being microelectronics. He waxed lyrical about human's ancient

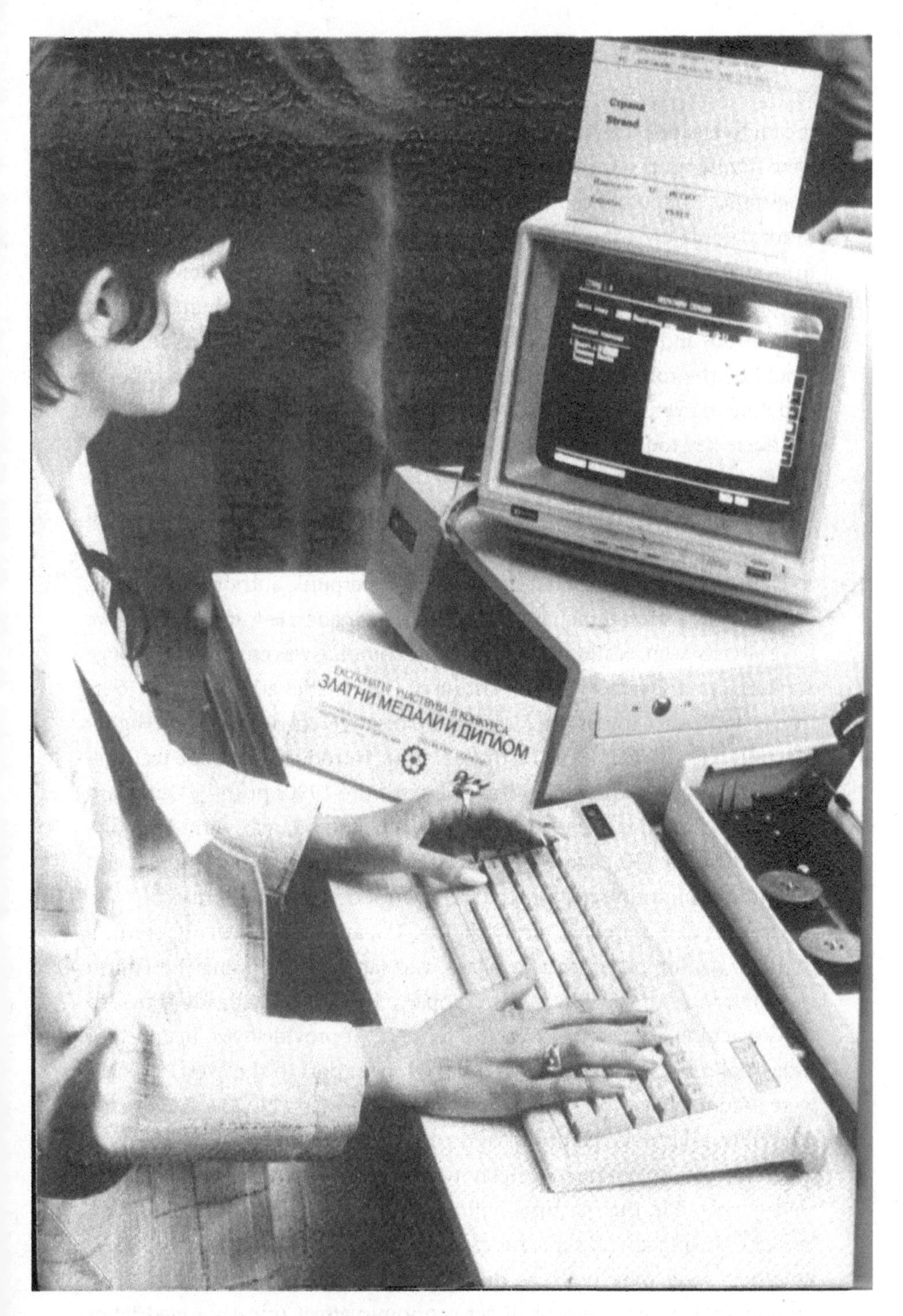

5.3 The Pravetz-16 computer. (*Source*: sandacite.bg.)

dream to create a machine that could copy human functions at work. We were finally on the threshold, due to the range of innovations in microelectronics, pneumatics, hydraulics, and precision machine-building. More completely than any ASU, this combination would free humans from manual, repetitive labor. Robots were a qualitatively new stage in the organization of labor, as they broke the monopoly of the human factor. If the 1960s and 1970s were under the flag of the computer, then the 1980s would be the robotic decade. Globally, entire factories would be robotic, and Bulgaria would be no different.[160] The vexing problem of quality was of course key too: "robotisation will introduce changes in the role of the subjective factor in production quality. It will no longer be determined by the psycho-physical and physiological abilities of man, but by the stored programs and capabilities of the machines."[161]

Doĭnov might have made a passable cyberpunk author, but he was undoubtedly a master manager. Robotics were made a key aim of STI, as we have already seen, while a "Beroe" robotic combine was created at the huge Stara Zagora disc factory. The Institute of Cybernetics added Robots to its title in 1978, too, under the directorship of Angel Angelov. VMEI (Higher Machine Electro-Technical Institute) Lenin introduced a robot technology center in 1979, preparing the cadres for the 1980s push.[162] Old guard party functionaries were bludgeoned into compliance with Doĭnov's ambitions by a 150-page collection of articles from the world press and scientific community demonstrating that robots were the future.[163]

Licenses were purchased from the UK, US, and West Germany, starting in 1978. An important source, again, was Japan—continuing the Fujitsu cooperation from the 1960s, the company FANUC had already provided CNC-machine licenses in 1973, and by 1979, it provided two licenses for industrial robots, with some earkmarked for export to the West.[164] These were used to create the RB series of machines—110, 211, 231, 232A: used in welding, lathing, painting, and other industrial tasks.[165] Serial production was implemented quickly thanks to the investment, and 3,000 were expected in the machine-building sector by 1985, with 422 million levs of CNC-metalworks machines and robots to be produced by 1983.[166] By 1982, there were twice as many robots as there had been in 1980, producing 4 million levs of direct economic effect (mostly saved labor costs), with the aim of having over half of all "mechanical manipulators" computerized by 1985.[167] Once again, Comecon specializations ensured Bulgarian profits. In 1982, Moscow noted that while Beroe was still underfulfilling its orders, it had implemented more types of robots and with far fewer failures in delivery than in the Soviet industry.[168] In places like India and Zimbabwe, Bulgarian robots were used as more proof of national advancement. They were the next step of the ASU project, automating the factory while the PC automated the office. The party, forgetting its promises of full employment and in pursuit of the panacea for quality, was busy identifying workplaces that should be robotized beginning in the late 1970s: 180 in the "Madara" factory, 40 in a plant in Plovdiv, while the car and truck Balkancar plant was specifically targeted.[169] On the Branishte collective farm, 25 people worked in animal husbandry before the 1986 installation of a computer: afterward, only three vet technicians were needed.[170]

A 1983 report to the CSTP by Vladimir Lazarov extoled the new machines' capabilities: "as a result of their 'intelligence' there is now the possibility for the decentralisation of governance functions, and automation can enter areas which were until now technically impossible or economically unviable."[171] Humans would now be intellectual workers, far removed from harmful production "and through digital displays will have access to the technological progress."[172] Even as Gorbachev came to power with radically new ideas for economic reform, the BCP plenum held onto STR applications as the pinnacle of success: "this is the stage of the powerful development of . . . microcomputers . . . new generations of computer machines and robots, of biotechnologies, of the miracles of informatics."[173] It was also the cusp of cybernetic self-regulation discourse being weaponized as a tool to attack bureaucratic deformations, maybe the ultimate proof of Gerovitch's thesis that the party subsumed the language. But also, in Bulgaria, cybernetic and informatics discourse allowed the young cadres to turn it into a language of potential anti-party reform. But first we must say a word about the impact on the worker. Rationalizing labor and raising productivity remained the aims—computers and ASUs remained the means. But whither the people?

5.4 Advertising brochure for an RB 251 welder robot. (*Source*: Scrapbook presented to Angel Angelov on his birthday, Petûr Petrov personal archive.)

RATIONAL CONTROL AND THE MASSES

There was one little problem—the "subjective" factor that was to be eliminated through rational control and automation. Behind the dry official phrase lay the socialist worker, made of flesh, blood, and desires. The party's obsession with automatic control centered on removing this pesky ability of the worker to get in the way of a streamlined, recordable, and observable workplace. Workers didn't always see automation as helpful, despite the slogans proclaiming that it was all socialists' duty to introduce the latest work methods into their existence. Being automated away was an obvious fear, but so was the anxiety of being melded into human-machine interfaces.

Some of the problems were the continuation of the mundane—computers could not take away the attitude to work that was infamously summed up by many as "they pretend to pay us, we pretend to work." Even the computer centers or cybernetic institutes themselves were not immune to the foibles of the socialist worker. In the Varna shipbuilding computer center, a report complained of too much walking around in the corridors. People gathered there to smoke and chat, a breach of both discipline and the center's air control. "Also there is no case of when you walk into any of the rooms and not finding a mess of people who are sitting on the desks or drinking coffee in large group chats," the report continued.[174] Carelessness even sometimes led to security breaches, as in when February 1975, two engineers and a porter were punished for allowing a technician from IBM-Vienna, an Indian British-educated man named Ishu Mirchandani, to record system error information onto two tapes that he took out of the center.[175] At Izotimpex, a 1970 report also complained of workplace attitudes:

> We have observed and report that in our enterprise there are workers whose feeling of responsibility for the quality and timely fulfilment of tasks is at a very low level and that these workers have decided to use their time here as a temporary requirement. These same workers wander from room to room, don't take on their tasks, solve the problems posed incompetently, make mistakes, [and] become the reason for harming the interests of the enterprise and the state.[176]

This was, of course, not distinctly socialist—walking around drinking coffee and not caring about work is a reaction to the capitalist workplace as much as it is the socialist one. Introducing computers and "rational planning" without other incentives or restructuring of the workplace could never achieve the results that the party wanted. Nedelcho Vichev, who worked in the Varna computer center, recalls a little cartoon drawn for him by a Japanese programmer who visited from Hitachi to help in the implementation of software in the 1980s. "He drew a little cartoon of this cart being pulled uphill by two figures, which he labeled Vichev-san and Avram-san, surrounded by many other figures cheering and waving flags. It was me and my colleague Avram [Avramov] who did all the work, that's what he was saying!" The rest of their colleagues just observed.[177] Most things don't change.

The regime had also positioned itself as a champion of women who had finally broken free of centuries of oppression and were active agents in the building of communism. The astute reader will note both the presence and absence of women in this story: present in the pictures and statistics, absent from the narrative. Mar Hicks has brilliantly demonstrated the importance of female labor to British computing power, and how its mismanagement and discrimination was directly linked to the eclipse of the UK by other competitors.[178] Women were present in university courses, worked in the computer centers and ASUs, and were the majority of the workers who actually put together the machines on the factory floor. Yet problems, of course, persisted—not least because automation added rather than subtracted from female labor. *Zhenata Dnes* (*The Woman Today*), was the regime's second biggest publication in terms of circulation and was often candid about the problems in this sphere, where the dream of automated labor ran against anxieties about the traditionalist view of women as mothers and caregivers.

The BCP's own pronunciations at the 1985 congress, when discussing STR's entry into the family, was typical: "the woman and mother has another . . . task: create the moral and psychological atmosphere which will encourage children in their active development as personalities with a contemporary education, a discoverer's will and ability to create."[179] An article by a male BAS academic also meditated on the woman's role in the computer world. He acknowledges that most of the work in the industry and computing centers is done by women as "the delicate women's hands help us create the most fine, sensitive elements, schematics and devices." But those hands are being underutilized in a pressing sphere—software—which is well suited to female labor, as it requires no physical labor and can be done at home, and is thus easily combined with family life now that PCs are being produced: "her participation in programming will probably strengthen the family, will raise the birth rate, and will solve other social problems too." She can program her computer-controlled kitchen, or car, and these skills are easily transferable to the future automated shops, where they often work. As a mother, programming at home will also be a great benefit for her kids.[180] Women's labor was thus once again to plug the gaps in the industry, but it would also solve social problems, as she would produce not just programs but more kids of the future.

When the writer was a woman, the conclusions were no less insulting. "A woman who cares for her appearance . . . would not be too happy to wear clothes designed by a computer. What does it understand of fashion? Turns out, however, quite a bit" starts an article by a female engineer. She cites the usage of computers in all areas—from science to banks—but the focus is on how women must overcome their "cyber-phobia," not just because of the future of jobs but so as to not look like dinosaurs to their children:

> Let's not forget that at the previous century's end our curious great-grandmothers were not afraid to learn to read and write together with their children and grandchildren. Why can't we, then, modern parents, without fear and anxiety, start mastering . . . computer literacy together with our children—and from our children?[181]

Even when encouragingly reporting on the possibilities of computing for easing labor, such as usage in the state DSK bank by mostly female tellers,[182] the magazine was thus preoccupied with showing how computing dovetailed with a woman's role as a mother or a worker who had to help fix the economy's failures. Sometimes even it slipped into easy stereotypes of what women were interested in, such as running an article on the rise of computer-aided romantic matchmaking in the US and Czechoslovakia.[183] But some journalists were more willing to delve into what the computer meant for women, giving them a more candid, direct voice. An article called "Everyday Next to The Computer" was centered on interviews with two Sofia workers: Krasimira Mateeva, a programmer, and Violeta Furnadzhieva, a hardware specialist. Both were graduates of VMEI-Sofia, married to colleagues, and in respected positions. Both state that there is nothing special that makes women better suited to the work, and in fact when they give birth (assumed to be certain), they are even given permission to not care about the job so much—the baby is central. Asked whether they have hopes for a future working from home: "No! . . . the question of time persists. Also, working in a collective has a social meaning. Imagine being left alone at home for a year. Would you be able to last surrounded just by the kids?" Mateeva muses how her kids were growing up lonely, with a latchkey—"this is a problem that exists ever since woman went to work for the first time." Both women love their work, seeing deep meaning in it, often staying late at night—something their husbands understand—and getting so lost in their work that once Krasi was asked by a visitor whether Krasi (whom he didn't know) was there, to which she replied "no!" as she was so taken in by her work. However, they see little compatibility between their roles as a mother and a specialist—even if they could get a computer at home, to keep programming, a couple of years out of the institute would be too much in such a fast-moving sphere. Finally, even though they are happy their kids are growing up around computers, they do worry: "if a person is not connected to nature, he is lost, he turns into a robot."[184]

The emancipated woman in Bulgaria suffered the same sociological double shift that women everywhere in the developed world experienced—her job, however highly qualified, added to her expected role as a mother and caretaker, rather than replacing it. This is a universal story, yet the Bulgarian obsession with the industry and the high ratio of qualified female

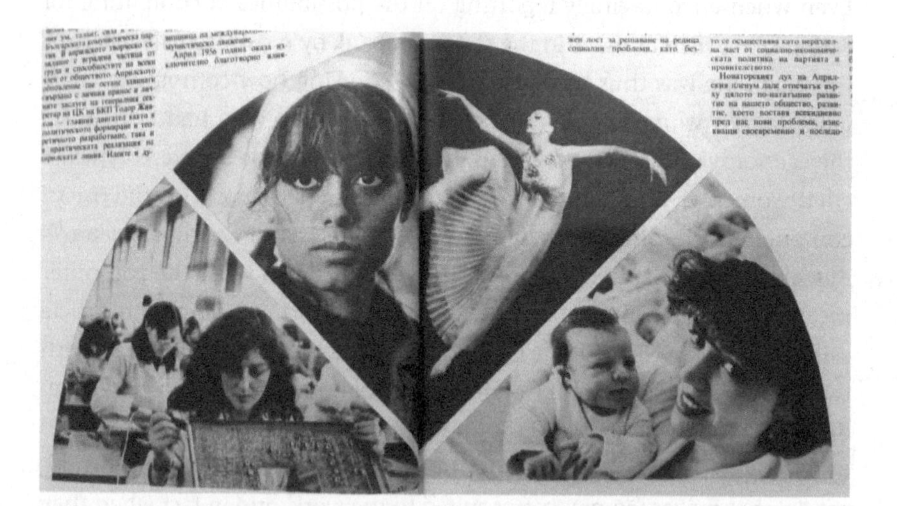

5.5 Four faces of the modern woman—electronics worker, farmer, ballerina, nurse. (*Source: Zhenata Dnes*, Regional Library, Varna.)

engineers made the contradictions even more glaring—the computer was a possibility, an obsession, but also something that due to motherhood meant that women fell further behind once out of the fast-moving industry even for a little while. They experienced this as the mothers to the supposed next electronics generation, as discussed in chapter 6, but also as workers facing the march of automation in their jobs.

These issues of the whole labor force were even clearer in two examples that have been uncovered in archives and by interviews. Needless to say, the traces in official documents are rare and scattered; but historical evidence lives on in various ways. The first example, a testimony to the value of repeated interviews with the same interlocutor, is that of ASU "Astra," a project led by Petŭr Petrov and Vasil Sgurev in the ITCR. Astra was a system for governance of industrial transport in open-air mines with the help of a specialized computer. Driven by the self-initiative of the two young scientists, it started as a side project to their official duties—a curiosity to see if they could fix a modern industrial issue. A single, large, purpose-built computer was situated at a command post, linked by cables to various radio receivers and magnetic sensors within a quarry or mine. These receivers gathered data about the location, movement, speed, and loading status of large trucks that undertook the loading and unloading of ores at a site, which were fitted with radio transmitters to communicate this information to the receivers. The system thus allowed for the accurate gathering and transmission of data regarding the mine's logistics, with the command computer able to rationally distribute the trucks along optimal routes and areas. Its sophistication and performance were praised by Soviet scientists in 1972.[185] Once completed, it was implemented in the "Medet" copper mine in a 1973 trial. By then, the ITCR had recognized its innovation, backing the project, and promoting both Sgurev and Petrov.[186] The system won a gold medal at the Plovdiv Fair of the same year, for a "significant contribution to the development of science and technology," leading to Soviet purchase inquiries from 10 organizations and further inquires by another 38.[187] It was the subject of a short, prize-winning documentary, "The Horizons of Astra," in which the trucks were shown in operation against a background of upbeat electronic music.[188] Use of this system was estimated to save 400,000 levs per year by 1976, when it was upgraded, and its sale price was a profitable 1.2 million levs.[189] It was the perfect poster child of Bulgarian automation—the fruit of young scientists' efforts, brought to fruition by institutes that recognized their talent, successfully implemented, and with export potential.

But the archives belie the difficulties of implementation, which were anything but smooth. Petrov recalls the physical struggle of installing the various receivers throughout the huge site, done in winter, knee deep in snow, back-breaking physical labor as opposed to the glossy dream of seamless automation.[190] But his most vivid tales are reserved for the clash between technology and workplace culture. The old way of doing things at "Medet" involved UHF radio links between drivers and a command post staffed by female dispatchers, who were responsible for directing trucks to sites and recording each trip. Due to friendships and sometimes workplace romances, each dispatcher played favorites. A favored driver would be sent to areas where the loads were consistently heavier, allowing him to fulfil his quota quicker; or their tallies would get a couple of extra ticks—"phantom" loads. Less popular drivers, of course, suffered conversely. As the system became the arbiter of truth, the personal networks started to break down.

5.6 Science in the field: installing "Astra" in the "Medet" mine. (*Source*: Petŭr Petrov, personal archive.)

But so did the system itself. Petrov checked all the modules, the cabling, even the trucks' transmitters—everything was in order. But what he found when he went back to "Medet" in late 1973 were smashed sensors—the intermediaries that relayed information between the trucks and HQ. Drivers claimed they accidentally bumped the sensors during their workday due to their placement. During those weeks, the mine thus operated according to old routines. Petrov recalls sending a young member of his team to the military factories that made tank armor in order to get custom-made boxes. Once each sensor became a mini-tank, the accidents mysteriously ceased, and the mine was successfully automated.[191]

The first computerized supermarket in the country was also struck by mysterious ailments. It opened in 1977 in Sofia, sporting a central computer connected to electronic tills, a showpiece of how the regime's technology would revolutionize the infamous socialist customer experience. Cash flows would be tracked, as would stock, alerting management when resources were running low. But it very obviously threatened the petty theft and misappropriation that was the hallmark of the shortage economy, and on the very first night after its installation, someone poured water over it. The hope was that the whole complex system of keeping track of even the last penny that couldn't be misplaced into someone's pocket, let alone goods that could disappear from the warehouse to be sold elsewhere, would break down, but the computer was waterproof. By the end of the week, another primal element was tried—fire—and this time the sabotage succeeded.[192] As Albena Shkodrova notes when recounting this tale, the socialist shop assistant was unpopular, widely seen as dour but also corrupt, so it was no wonder that the rationalization of his labor would also be opposed.[193] Another issue was that the computer's rationality ran counter to a much bigger problem—the shortage and supply problems of the Bulgarian economy, especially its consumer sector, which was dominated by limited variety and intermittent supply.[194] Even if sugar was running out in the supermarket, what could a manager do if there was none to be found in the supply chain? The computer's precision required the rest of the network to function perfectly, which it did in the models—but not in reality.

In 1984, sociologist Stoi͡an Mihaĭlov, who would become secretary of the Central Committee, stated that there was lying in the country.

Networks such as KESSI amplified the untrue, subjective information that was entered into them, circulating it far and wide. Documents were liars, entering fictive achievements into the databases. The people filling the documents were liars, due to socialism's conservative attitude toward quotas and plans—unchangeable and thus unachievable.[195] KESSI was actually an exercise in subjectivity. It never allowed for true horizontal connection between users, and it didn't truly address the center-user relation either. All it did was codify it and cover it with a veneer of "scientific objectivity." The subjective factor, with its fictitious numbers of economic achievement, were reified by the information flows.

As early as 1983, psychological and social barriers to automation were observed, and it was noted that increasing volumes of technology did not equal success.[196] Other observers talked of fears of unemployment, cutbacks, fear of risks, reluctance to retrain, and the lack of self-confidence in the face of new screens. Workers felt that the only value of this technology was to showcase the party's success to the world, while in fact placing huge demands on the workers themselves. Many felt the quality was lacking anyway, and the money could be better spent.[197] Seymour Goodman cites this as an extension of the New Soviet Man's reality—a human who has no rights but a tiny piece of power: to mock, to steal, to bribe, to work poorly. This view ignores the real ability of ASUs to become wardens of the workplace, to automate work and be methods to decrease worker's tactics of everyday resistance—the very act of destroying systems indicates that they knew well what the effect was. But the BCP's dream of automation was too successful, maybe even more so than the Soviet, in convincing the party that it had hit upon a solution—the importance of the industry and the money it brought into the state coffers were taken for a parallel success in its economic potential. The fundamental challenge of massaged statistics was not addressed, and in fact, computers reified a false picture into reality through the very process of computerization—the party had made the screen the final arbiter of rationality, so when the number was displayed, it had to be true. The Politburo had tried to harness the computer to socialist progress but had only succeeded in adding another barrier to true reform. And while many workers broke the machines, others were inspired, leading to a veritable intellectual revolution that was more varied and complex than the party could ever envision.

6

THE SOCIALIST CYBORG: EDUCATION, INTELLECTUALS, AND CULTURE

If Bulgaria's per capita production of peripherals or robotics was impressive, it was unrivaled in its per capita production of new robotic laws. Throughout the socialist period, two new laws appeared thanks to Bulgarian authors—and a satirical one, which came on the back of another 95 unwritten laws in the same vein. To foster this industry, the party had also created the conditions for a deep fascination with these new tools to emerge. This was most obvious in the human dimension of it all—thousands of electronics specialists who were trained to research, tinker, and think about the information age. Many labored in the Academy of Sciences (BAS) institutes, such as the Institute of Technical Cybernetics and Robotics (ITCR), IZOT's own Central Institute of Computing Technology (CICT), the universities, the computer centers of ministries, and State Economic Unions (DSOs). Increasingly, technicians and other intellectuals started wondering—what was it that these tools would bring to Bulgarian society? Computers and cybernetics posted huge questions that weren't just engineering riddles, but social ones, too—to be a person in the information age was to be a fundamentally different type of human, who related to labor and to the world itself in novel ways. The new horizons were both utopian and troubling. Over time, the language and issues burst out of the institutes' doors and onto the pages of philosophy journals, books, popular magazines, and science-fiction literature. The language of cybernetics permeated not just the computing sciences but also sociology, philosophy, pedagogy, and psychology. Finally, it burst into the minds of the children of the 1980s, who faced the new currents in the classrooms and in their dreams. Bulgarian computers and robots exited the factories and entered the last socialist generation's thinking

Cybernetics as a language of intellectuals and regime was the subject of Slava Gerovitch's groundbreaking work. He posits it as a precise language employed by specialists in opposition to the increasingly empty Marxist-Leninist rhetoric of the post-Stalin era. It was an attractive "dissident" language, a "cyberspeak" that was all-encompassing but also provided definite rules and thus provable claims. Eventually, the regime subsumed it within "newspeak" and its own official discourse.[1] In Bulgaria, however, cybernetics remained a tool of communication that even within the regime's confines—and the Bulgarian regime screamed cybernetization even louder than the Soviet one did—was a field on which competing claims and visions could be mapped. Even in late socialism, when it became the content of the BCP's future-oriented philosophy, it was thus not just "newspeak," and in fact it was increasingly used as a cudgel with which to beat the party on the head.

The computer was protean, bursting out of its disciplinary institutions and boundaries, as it entered almost all spheres of science. Ksenia Tatarchenko's innovative work on Akademgorodok's computer center makes this claim clear: Disparate fields were brought together by the computer, new dialogs emerged, and new things had to be made sense of. Tatarchenko's call is to see the computer as a locus of many competing fields of knowledge and technology, as it collapsed distances—both geographic and disciplinary. This chapter takes on this challenge, seeing the computer as a "trading zone" that Peter Galison also encouraged us to investigate. As he pointed out, the computer was a zone par excellence, allowing us also to simulate outcomes, being able to create other realities and possibilities for all sciences and beyond. Bulgarian science, too, became increasingly computerized, a new discourse for different intellectuals to converse in.[2]

In Bulgaria, the trading zone was even wider, as the debate progressed from institutes to philosophy journals but also into education—both in schools and computer clubs. Finally, and understandably, it permeated the imagination through literature, too, in novels, short stories, and comics. The regime was proclaiming that it was building a science fiction society, so it was only natural that the genre would become another key indicator of the regime's mindset and reactions to it. This chapter follows these strands to draw three important conclusions. First, cybernetic discourse in Bulgaria flourished due to the regime's backing but remained vital and not beholden to party theses only throughout the period. This was at least in part influenced by the peculiarly Bulgarian cultural politics of the 1970s. This culture was centered on the vision of Liudmila Zhivkova, which was both official and "spacious" enough to allow cybernetic thought to go down more esoteric avenues—away from governance and toward discussions of creativity. Second, these clashes were a real space for articulating a future vision for Bulgarian socialism. They became powerful political tool as they both tried to square the circle of the Party's program and delved deep into the practical problems of the socialist information age. The computer offered possibilities for humanity to progress or digress, and thus became a prism for reformist thinking, too. Third, the spread of computer education and discourse created a lasting social impact on the last socialist generation. This happened not just in schools but also even entertainment. Both through their discussions of the party's dream and by influencing ideas in the classroom or computer club, the technical intellectuals had a bigger impact on society than mere philosophizing. The practical question of how to integrate human and machine in the factories and office became a wider humanistic concern, as it was envisioned entering the lives of more and more people, including the most impressionable: children. The cyborg was real, and how to make it socialist and Bulgarian was a real and vexing question.

This chapter thus continues themes explored in the previous one: What was actually meant by the "future" when technical experts talked about it? By tracing the discussions of intellectuals but also the reactions of children and the reflections by writers, a picture of ambivalence emerges. The computer, this chapter shows, was both a promise and a curse for many people. Taken together with chapter 5, we thus see a picture of the local story of how Bulgarian socialism tried to harness cybernetics, created the tools for "legibility" of its population,[3] and how people reacted.

These two chapters thus show how the conditions of the Bulgarian cultural landscape—of both work and thought—made this both similar and different to the general anxiety about cybernetic control.

TRAINING THE TECHNICAL INTELLECTUALS

The head of the ITCR, Nikolaĭ Naplatanov, complained in 1965 that Bulgarian universities were not training enough graduates ready for electronics work. His institute had to train those who were sent to them, taking up valuable research time.[4] This was unsurprising, given that the Higher Machine-Electrotechnical Institute (VMEI)-Sofia had just opened its Computer Technology faculty, and that the 1964–1965 school year was the first in which semiconductor classes were taught at the university or at a Bulgarian school (the Sofia Technical School).[5] These were the first steps toward rectifying a generally poor technical level of many specialists that a Central Committee report noted, and which had called for the Ministry of Education to help the progress of the Scientific Technical Revolution (STR).[6] The birth of Bulgaria's computerization upped the demand for specialists that far outstripped the supply. Stop-gap measures were introduced, such as retraining talented mathematics students to work in the computer centers until the first specialist cohorts graduated. This hunger for specialists meant that even ex-political prisoners, such as the anarchist Georgi Konstantinov, were employed: He had spent years in prisons and labor camps after blowing up a Stalin statue in 1953. Released in 1962, he was allowed to study mathematics, and upon his graduation in 1969, was hired at the Computer Centre of the Ministry of Internal Trade. He even authored part of the ZIT-151 manual, and headed a section in the center—all despite being the subject of one of Bulgaria's largest secret police files.[7] The usual political considerations often were ignored in the name of the industry—in another case, from the late 1970s, the son of another anarchist ex-inmate was allowed to study for a doctorate in Ilmenau and then work in the sensitive Pravetz factories.[8] Progress demanded cadres, and talent trumped politics.

By the 1969–1970 school year, courses were offered in operations research, complex automation, and automated governance systems at various universities, enrolling 1,220 students in automation courses alone.[9]

Over the 1970–1975 plan, 193 million levs were invested in higher education, twice the previous amount. Much of this money went to equipping higher education institutions with computers and automation devices, while doctoral studies in the sciences were tied to national development plans, with topics chosen from CSTP-approved lists.[10] Admissions to these courses were increased; five new degrees and specializations in automation were created in universities and four in technical schools. Between 1969 and 1971, over 6,300 students went through such courses, higher than the 4,300 planned.[11] As admissions were tailored to specific industrial needs, the students flooded into the courses as IZOT was opening its factories.[12]

Many students received privileges reflecting their importance to the economy. In 1972, the rector of VMEI-Sofia complained to CSTP that too many students were being requested for the seasonal agricultural brigades that were key to the regime's ailing agriculture. Certain specializations should be exempt from this disruptive, useless work, he argued.[13] Military service was also reduced or made lighter, with graduates destined for electronics courses kept in units based in or near cities, so they could do their coursework there; or they were posted in relevant units, such as signals or electronic warfare.[14] By 1984, those accepted into electronics, automation, or biotechnology courses were fully exempt from service if they agreed to work in their specialization for at least 10 years after (and undertook a four-month military course, rather than the full two years of service).[15]

By 1979, the expansion of computer education was part of a wider expansion of higher education, which made Bulgaria a world leader in students per capita. Over 720,000 workers in the economy had higher or specialist education, and over 80,000 per year were going through some kind of further qualification.[16] Practical training was emphasized, with placements in industry, with the aim of "intellectualizing" labor and "satisfying their information needs" through familiarity with the latest technology.[17] As the numbers of the specialists working in the Varna center cited in Chapter 5 show, the gender ratios were also astounding: women became the majority of students in Bulgarian universities, including often in the computer specializations—a phenomenon also testified to by the numerous photos of such work.

But if the future was to be automated, everyone had to be familiar with automation, not just those who chose to study it. The party had discussed

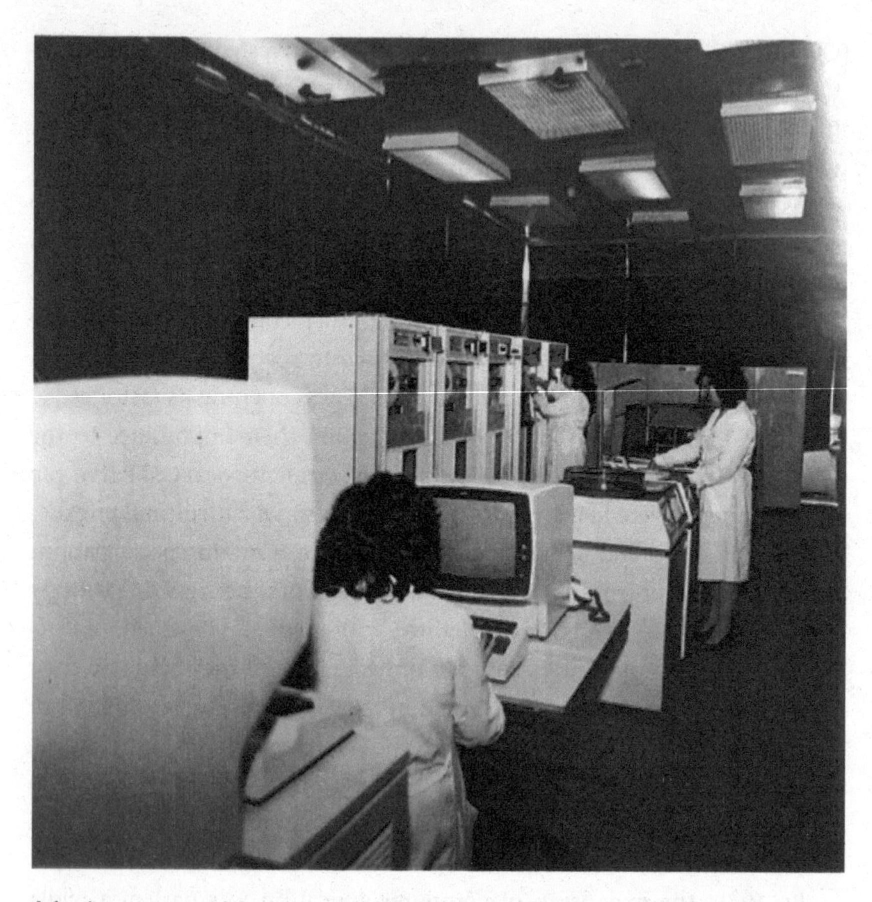

6.1 A computer training center in VMEI-Varna, c. 1980. (*Source*: socbg.com.)

the introduction of SM-4 computers and terminals in the late 1970s, but the PC revolution truly allowed the implementation of the STR to the classroom, with potentially immense repercussions for pedagogy and the economy, too. In 1983, the Sofia Electronic Technical school received 18 Pravetz'82 machines—and so the first computer classroom in the country was established. The computers came with education software and BASIC—the computer language all children had to learn throughout the 1980s.[18] Within a year, there were 300 PCs in Bulgarian schools, and children were learning Pascal, Logo, and other languages. The eleventh grade was restructured to include classes such as "Iintroduction to Cybernetics" and "Automation of Production." Everyone would start computer classes by the fifth grade, and by 1985, it was planned that over 3,000 PCs would be in schools.[19] The number is unlikely to have been reached, given that the Pravetz factory was still producing smaller runs at that time.

The plan was utopian: Each Sofia school was to have a classroom with 20 networked PCs, and even the smallest school was to have at least 5 computers by 1987. The total expected was 105 large computer classrooms and 98 smaller ones, equipped with any educational programs that would "automate" education and produce the next generation of specialists.[20] Architects teamed up with pedagogues to design the new classrooms so that teaching would "open up towards the environment and integrate with it and use the school . . . as a centre for out-of-class education, as an education-cultural centre with universal functions." The new education process required precisely designed spaces, desks placed a particular distance apart, chairs of a particular height. Students were turned into cyborgs—"integration into the school environment is effective only if it ensures the optimal functioning of the system

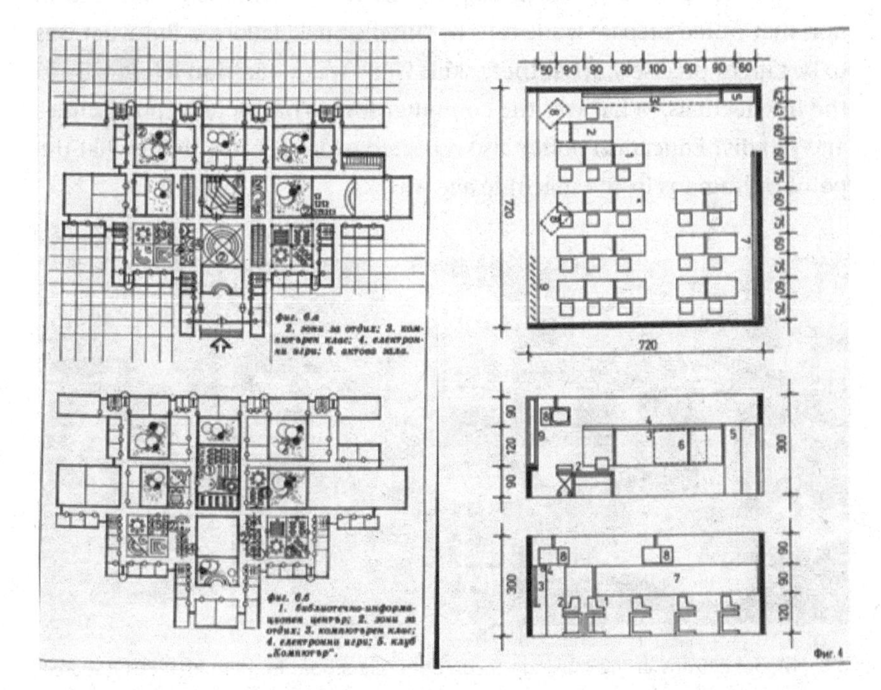

6.2 An ergonomic computer classroom (left) in a modern school, with the computer lab as part of an integrated system with information centers and a public computer club. (*Source: Kompiu͡tŭr za Vas* magazine, National Library, Sofia.)

'Man-Machine-Environment' and in accordance with ergonomic viewpoints."[21] Schools were to prepare them for an economy where they would again be expected to be part of human-machine systems.

Progress was slow but real, as the Soviets observed: The design and usage of Pravetz PCs was good and was stirring up deep discussions on education.[22] In 1986, the town of Pravetz opened up its own technological center, probably the best-equipped computer school in the country, tied directly to the microprocessor combine, a sign of the growth that computerized education was expected to create.[23] Such education was only the first step of a general program of computerizing most knowledge acquisition, a new age where the machines would be part of a person's life in all spheres: at work and at home.[24] This was STR's cutting edge, and these tools were expected to unite education, culture, and science into a new integrated production unit.[25] Zhivkov had posted the question of what was "the most valuable thing that our society has—the person as a creator of all material and spiritual goods in Bulgaria" in 1984.[26] The answer was education that would prepare workers to be "intellectual" laborers. But what was to be taught precisely, what these skills were, was a question left mostly to the intellectuals. What was the computer for? What were its possibilities and pitfalls? Education policy also reflected wider debates about what the point of humans in the machine age was.

6.3 Model citizens in the making—a computer classroom in the mid-1980s. (*Source:* socbg.com.)

THE CYBERNETIC TOWER

Actually, there were two towers—IZOT's huge CICT, and BAS's growing ITCR—that were the beacons of computing and cybernetic thought in the country. The former was more concerned with the production of the lucrative items of the industry, receiving huge funding to do so. The latter also produced the tools—it was the birthplace of both the Bulgarian microcomputer and the RB series robots—but it also researched the application of these goods to the economy, the theory and methodology of game theory, modeling, ergonomics, bionics, and planning. It was thus the ITCR that was the cauldron in which cybernetic thought bubbled most fiercely.

The institute started in 1959 as a BAS section on Automation and Telemechanics under the directorship of Denyo Belchev, working on both the theoretical and implementation problems of automation.[27] Initially it focused on the energy sector or the elements base, housed in just a few rooms in the scientific city outside Sofia's center, starting with just three workers. In 1961, the institute completed its first long-term plan, right down to 1980, but the going was slow, as it was hard to find the right cadres. It was transformed into a Central Automation Laboratory in 1963, when the directorship passed to the energetic, ambitious Naplatanov. Petrov paints him as a committed communist with impeccable credentials. He had been a partisan in the 1940s, while his father, who had emigrated to the USSR, was inserted back into the country by submarine, captured, and shot by Bulgarian police in 1942. Naplatanov graduated from the Leningrad Electric Technology Institute with a degree in electrical automation, and he specialized in Dresden and East Berlin in the 1950s. By 1961, he was an associate professor in VMEI (which he would head between 1968 and 1970), a young and promising scientist that the short-staffed BAS took notice of. It was under his leadership that the laboratory became an institute in 1964, thanks to his managerial ability and political background, allowing him to sell his ideas to the CSTP and BAS. As soon as he was made head of the laboratory, he had invited Boris Sotskov, a leading Soviet automation academician, to help plan his new institute on the Soviet model. Thus, by the mid-1960s, Bulgaria had its own promising cybernetic institute with Naplatanov as its head and Petrov as scientific secretary.[28]

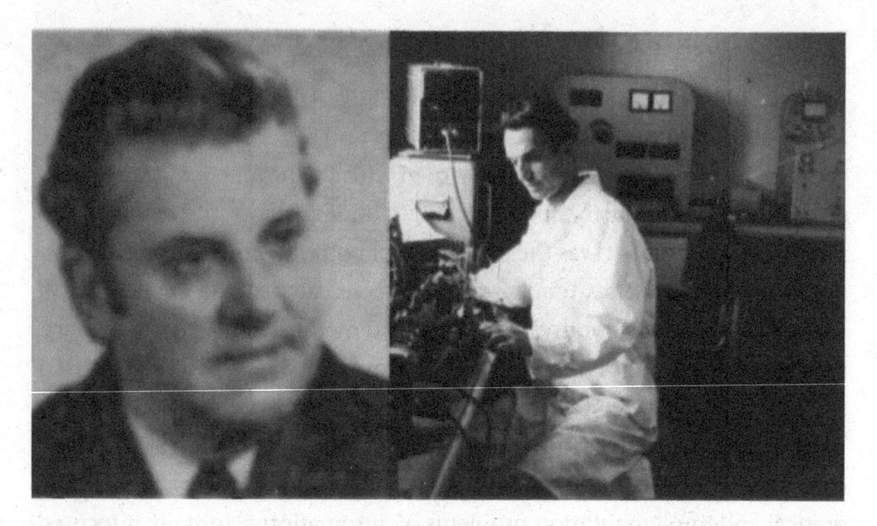

6.4 Nikolaĭ Naplatanov (left) and Petŭr Petrov (right). (*Source*: Petŭr Petrov personal archive.)

The institute had an acute need for workers, as there were just 17 in 1964, as well as specialists in English to create an information bureau.[29] It was stressed that there was an "absolute need for world level information," and translation of Western literature was a priority plan task in the early years.[30] Its model and aspiration was the Institute of Cybernetics in Kiev, run by Victor Glushkov, the towering figure of Soviet cybernetics, and operating with a staff of 3,200. Naplatanov demanded 135 extra staff as soon as possible, so as to cover the key areas of research, including his personal favorite—bionics.[31] Despite the usual administrative squabbles, the energetic director got most of what he wanted, and by 1972, ITCR operated with 168 workers, of which 56 were scientific workers (a term denoting those with doctorates who were in various stages of their post-doctoral career); 44 others had university degrees or were doctoral students, 47 had specialized secondary education from technical schools who engaged in scientific assistance, and 21 were administrators.[32] By the end of the 1970s and Naplatanov's directorship, the numbers were higher, and the institute included a workshop that built prototypes and small runs of the institute's designs.[33] It had its own computer center, freeing it from the crippling need to buy machine time from other BAS branches.[34]

Angel Angelov was the next capable director, as the institute added the "R" (for robotics) to its name as part of Doĭnov's reorganization of science for the next stage of automation. He was elevated to the office after discussions with the scientific secretary Nikolaĭ Iliev, who was also Doĭnov's eyes in the institute, highlighting robotics as a field for the future.[35] Angelov had been educated as an electrical engineer before specializing in semiconductors in Moscow in the 1950s. He worked in industrial electronics throughout the 1960s, headed CICT in 1968, and was the longest-serving member of the Council of Head Constructors (SGK), which dictated the bloc's electronics policy. He had also held deputy directorships at both IZOT and CICT.[36] The wealth of experience at the height of Bulgarian science demonstrated that his ascension to ITCR was a new stage in the institute's importance, which immediately was felt in the creation of the "Trial Base"—a production wing with 350 workers that manufactured the first PCs. By 1989, ITCR employed over 1,300 scientists, assistants, and specialists—a proportionally huge element of Bulgaria's but also the bloc's electronics landscape.[37]

The makeup of the institute and BAS in general reflected the gendered aspect of scientific work. Despite all the regime's proclamations about upward mobility, there were obvious ceilings. A 1988 article in *Zhenata Dnes* talks of "reaching the middle of the pyramid" regarding BAS work: Women made up 50 percent of the academy yet only 3.8 percent of the professors, and not a single full academician. Anonymous scientists are quoted stating that to advance, you have to show yourself to be superior rather than equal to the men.[38] In a compendium of interviews with the highest-ranking members of the industry, published in the 2000s, only one woman was featured: Snezhana Khristova. Key in the development of the calculator, and reaching high positions within CICT, her interview doesn't touch on her experience in such a male-dominated field. In general, the ITCR's relative lack of women at the highest positions was a reflection of a wider reality of the Bulgarian electronics industry.

The institute's devices were thus embedded in wider realities, and the two strands of the institute's thinking—theoretical and applied—increasingly merged. Even seemingly mundane problems, such as the modeling of water resource systems of the Iskar river, raised very concrete questions about governance. The designers noted that the "rational usage of resources . . . is now regulated by administrative and not economic means . . . the production of one ton of steel takes two times the water that the norms require."[39] For a cybernetic system to work, precise information was needed at every level, stymied here by enterprise managers' fudging of numbers. To automate the river's usage, the computers allocating resources needed accurate data about usage and shortages in factories, farms, and towns. Those managers who sought to fulfil their quota by insuring themselves with extra water—ergo, all of them—hindered the objective governance envisioned by the party. Information accuracy was an overarching concern of any cybernetician, whether building a simple mechanical tool or modeling a system. In the Iskar system, current information was based on "non-economic criteria . . . [and] due to this they can't be used as a template for the creation of models in real situations"—the criteria for what the system was to achieve were "parachuted" in by higher organs that applied general solutions to every problem, incongruous with the realities on the ground.[40]

All technical systems of this nature had political consequences. The water system contained many variables, depended on the environment, generated statistical noise, and was influenced by human choices—it was constantly changing. Centrally decided parameters of usage did not apply well to chaotic systems, where the informational structure was related to function—a truly functioning system of gathering, processing, and acting on data was inseparable from changes in the governance system of the waterworks.[41] The ITCR's intellectuals ran into the problem of cybernetic information in a centrally planned economy: These governance system problems could be resolved in dialog among the technicians, state, and industry, but in the logic of the Plan, the issues had to be streamlined through new technology rather than fundamentally changed. The blockage of local actors reacting to local problems that could not be predicted by the general plan can be traced throughout socialist history, exemplified best by Peter Palchinsky, who advocated for local autonomy within the confines of the central plan, and was executed in 1929, just before the Industrial Party Trials in the USSR.[42] The ITCR engineers realized their social roles and the inherent political implications of their analysis in the 1970s, much as Palchinsky had in the early days of Stalinism.

At the base of cybernetics was the information loop, a constant feedback mechanism that showed how the system's behavior reacted to its environment. If the whole structure was to respond to this fast-changing world, it required that accurate information.[43] The Politburo's dream contained this kernel but was much more explicit in the ITCR's work. Naplatanov's own research on bionics, optimizing human operation in a technological system, bore this out. So did the institute's 1972 prognosis on development up to 1990, putting the onus on global information systems that would subsume ones such as Unified System of Social Information (ESSI), as well as placing robotics, automation, and human-machine interfaces on bionic bases and applying biological methods to control and information processing tasks.[44]

Such issues were Naplatanov's forte. His ergonomic research from 1975 on aimed to optimize operators' actions within a governance system where the operator's contact with the outside world was only through indirect measures, such as screens and gauges. In this "Informational Model System," the operator acts on information rather than on the environment itself. The operators were thus just a link in a closed system—whether on the production line or in administrative work—and their interaction with the data, which had to be sufficient to facilitate decision making (but not overwhelming), was to be streamlined.[45] Further wide-ranging research pushed toward neural modeling and biological information processing, as Naplatanov was convinced that only through such research can the next leap in computing power be achieved. The task was to study how biological systems operated in conditions of environmental noise, isolating the key information needed for survival—highly applicable to the complex socioeconomic world humans inhabited in 1976.[46] On such a biological basis, visual recognition algorithms could be introduced into human-machine systems, freeing the operator from even more mundane tasks.[47]

It was increasingly apparent to Naplatanov's team that the quantity of information was ever-growing, threatening to bury the human operator. Using the metaphor of a pilot in a fast jet flying over a changing terrain, Naplatanov stated that

> the quantity of information that a pilot receives from separate instruments is so big that he can't assimilate and decode in such a part of the second as is needed for the control of the supersonic flying machine. This requires a search for ways for optimal congruence between the system of information presentation and governance.[48]

This was true of all operators. A worker could now know the ratios, timings, and chemical balances of an industrial process instantaneously, but what was useful? Scaled up to the social governance tasks of ESSI, the problem multiplied almost infinitely. What was the optimal usage of information and its optimal quantity? Naplatanov sought the solution of this universal question, inherent in all cybernetic systems, in the "conalogue"—the Contact Analogue Indicator—a means of transforming data from a quantitative to qualitative form, allowing operators to orient themselves quickly.[49] The complicated environment—natural, industrial, social—was simplified into an analog presented to the operator, easing the operator's orientation within the process being governed. The conalogue would create a full but simplified picture of the general situation. Factors such as how long eyes lingered on particular instruments or parts of a control board, the tracks they made over screens (a "road map" of operator vision) were considered to create a picture of how humans build conceptual models of events and keep track of information.[50] Naplatanov thus could model behavior, making certain assumptions, such as that an interruption of a gaze indicated daydreaming or a mind going off track. Thus, the need to design the optimal control system led to engineers having to take on the task of workplace psychologist too, considering how humans interacted with machines and what the machine was *for*.

These concerns dovetailed with the other big ITCR project of 1976, preparing for ESSI's construction. ESSI would integrate information gathered on all social levels, from the ASUs in enterprises to ministerial systems: a

6.5 Conalogue: How to combine quantitative and symbolic information (left) and the operative visual field of operators (right). (*Source*: Bulgarian Academy of Sciences Archive.)

database of all available information gathered in the state. It was the first visualization of how this system would work. Maps linking and expanding territorial computer centers that would be hubs for particular areas' information suggested that ESSI was around the corner.[51] This weighed heavily on Naplatanov's mind, as it entailed an exponential rise in data for human operators. His 1977 report on the seemingly separate area of robotic control indicates his concern, as he pondered the social and psychological impact of perfected robots on the operator, not just at the control panel but at the social level as robotization brought a new stage in automation. This new horizon for Bulgarian society demanded new methods of human-machine interaction, through natural language interfacing and machines based on human physiology. Only then would automata help economic intensification.[52] Robots' social function was to "free man from the need to carry out unqualified work and so give workers the ability to move towards highly-qualified labor, which expands their knowledge." The new labor would thus be intellectual. Nature would become more knowable, as robots would allow research into areas inaccessible to humans, spreading their mental capacities into hitherto unknown areas.[53] Naplatanov was also clear this development had an impact on a state predicated on full employment—worker numbers would fall, and wages would be reduced. A qualitative jump in technology entailed one in social development, too.[54] His bionic interests made the ITCR wield massive power within Comecon science, with a 1975 pan-bloc meeting in Varna adopting the definition of bionics as the merging of technical and biological sciences into the study of "systems of task-oriented behaviour."[55] The merger, rather than mere application of biological examples to engineering solutions, was born out of Naplatanov's own obsessions in these years, while of course highly influenced by his own knowledge of Western bionic debates. For him, cyborgs were already here, rather than yet to be created.

As the vanguard of automation, it is not surprising the ITCR was closely involved with philosophical questions. Designing tools for engineering labor led to discussions of creativity "which by its nature is hard to formalise, creates special difficulties in designing the very systems of automating design."[56] The automated workplace demanded programs, algorithmic languages, and graphically displayed information, meaning that engineers had to consider not just technical specifications but also the whole process of interactions between all elements, including the operator. This meant a concern with the information that the system used, fed often from above or horizontally from other cells, which could broadly be termed "the rest of the economy."[57] Cybernetic thinking meant holistic thinking—the office couldn't just be an office—and it hung over every project: from ESSI to designing the right medical questionnaires to allow easy computerization of data for systems such as the Botevgrad one discussed in chapter 5.[58]

This informational obsession was not just Naplatanov's. Cybernetic theory was a Comecon-wide concern, as questions of optimal distribution of resources within hierarchical systems were of interest for every socialist planner. Ivan Popchev's team led research on ways to control the variables within such fluid systems, rejecting both full centralization and full decentralization and using water resources as their main example. In the first case, the users' interests were not taken into account, with governance reduced to solving of tasks set by a distant center; in the second, everything was subordinated to the user, and the analysis of a local situations would never lead to optimal results for the system as a whole. Cooperative government was the solution, creating a "game situation" where both sides affect resource and task distributions. This solution meant the periodic resurfacing of the need to distribute resources, including information, at all stages of exchange, implementation, and evaluation—a feedback loop par excellence.[59]

In all this discourse, the kernel of actual reform of the economy by truly empowering users—the enterprises—was evident. It ran counter to the regime's expectations that the tools of information would lead to nonreform solutions. Naplatanov's and his team's research often led to conclusions that were not in line with a top-down application of technology, instead concluding that this new age required new economic organizations and certain decentralization. "Socialist democracy" and cooperative control can thus be seen as correctives: the party's attempt to integrate the intellectual impact of automation into orthodoxy. Bulgarian scientists were particularly preoccupied with the modeling of social processes, noted their Soviet colleagues, who commented favorably on results in modeling demographic, social, consumer price, economic, and even musical questions.[60] The party's obsession with applying automation to all tasks had naturally thus led the engineers to pose politico-philosophical questions, which now spilled onto pages read not just by engineers.

INFORMATION-AGE PHILOSOPHY

The rise of the industry, coupled with the party's desire for scientific governance, meant that the cybernetic debate took over the pages of the premier philosophy journal, too. *Filosofksa Misŭl* was published monthly between 1945 and 1991, aimed at not just philosophers and students but also at neighboring disciplines, too. It often published translations or commissions by other socialists and by Western philosophers, too. The journal was the main venue where the state of the field could be observed. Its stance was often marked by the figure of Todor Pavlov, the country's premier Marxist philosopher, who headed the Institute of Philosophy until his death in 1977 (which also terminated his 10-year tenure as Politburo member). An early opponent of cybernetics, he toed the Stalinist line: "even the most complex robot does not assimilate, sense, remember, think, dream, fictionalize, seek."[61] But as Soviet science changed with the times, so did he, and cybernetic discussions spread onto the pages, allowing multiple disciplines to be infected with the bug.

The journal's publications were a bellweather for the regime's acceptance of cybernetics and often reflected the concerns of the newest stage of automation. Cybernetics was positively introduced through translations of Soviet luminaries, such as A. Berg or V. Pekelis, while Bulgarian authors jumped straight into the effect that computerization had on people.[62] Throughout the regime's history, this remained the case, with Bulgarian cybernetic thought particularly centered on the social, philosophical, and cultural aspects of the new age, more so than on its economic or technical aspects. *Filosofksa Misŭl* was the "trading zone" of historian Peter Galison's nomenclature, where multiple disciplines engaged with the growing influence of the industry on life. As soon as cybernetics was allowed back onto the philosophical stage in the 1960s, writers such as Trifon Trifonov called for improved psychological investigations into the new, more complex workplace, where the psychological burden was increasing. "Repetitive actions lead to the formation of conditions such as apathy, boredom, slacking," so new labor psychology was needed if the regime's automatic march was to not result in mental damage.[63] Shift work would have to change to screen out fatigue, and workers' psychological characteristics now had to be taken into account when assigning jobs.[64] The psychologist would have an increased role in the cybernetic age, considering each person holistically, so as to forge the perfect workplace collective—only then would the technical progress the regime demanded actually transpire, as it was currently threatened by the reaction to its social aspects on the part of workers.[65]

At the start of the regime's proclamations, the journal also grappled with what the "informatization" of life would mean beyond the party's theses or the workplace. The more we know, the more unknowns appear to us, an article stated pithily.[66] The country's course toward a networked economy and automation thus raised questions of social governance, where information would be used to maintain homeostasis, a tool of social consciousness that could create durability in the social system. To achieve this tool, "good" information was needed—the regime wanted information, but the question was to determine which was the useful, accurate type.[67] Western social consciousness was full of "bad" information and noise, as narrow class interests created competing datapoints and wasn't able to solve its inherent contradictions. "From a social cybernetic viewpoint, the more promising system will be that which can most rationally use the streams of social information in its governing processes," and science would judge capitalism harshly on this count. This was the kernel of the socialist difference—its social information brought unity to all members and allowed systems such as ESSI to objectively solve all problems. The amalgamation of social data allowed for charting future roads, as you could draw "objective" conclusions from the dataset.[68] Socialist cybernetics was superior because of centralization, and the political nature of the regime—and the regime was superior because of socialist cybernetics' predictive powers! It was a perfect closed loop.

The ITCR's scientists weighed in on such issues, because making the tools of the information society was closely connected with considerations of how these tools were to be used. In the 1970s, the journal hosted more and more of the engineers' own philosophical musings, which moved away from general considerations to advance models for this new social governance. Nikolaĭ Stanulov was the most prominent ITCR contributor to the journal in those years. In 1973, he meditated on cybernetic social governance and the role of feedback, without which it there is no "comprehensive information about the governed object and all other outside disturbing effects."[69] Each cybernetic system required positive and negative feedbacks, the former allowing the system to exit its programmed situation by amplifying variation and thus tending toward destruction, the latter keeping output within the input parameters. Mass competition among workers, development and fulfilment of plans, rising labor productivity were all positive feedbacks that completed tasks that meant progress was possible.[70] Each social system, like a biological one, also exhibited change in a temporal sphere, moving from the past to the future. In a technical sense, which Stanulov was familiar with due to his specialization, a human operator could monitor such changes in real time—the same could be applied to social systems. A "governing mechanism" was the key, a regulator of all information—and thus the party needed a General Theory of Social Governance.[71] A primitive society had no need for governance, as the human was both object and subject of it. With growing complexity, subjects and objects become independent parts that have a dependent relationship as subsystems that can create larger systems. But unlike technical systems, where someone from the outside creates the system, people *formed* the social system. Stanulov critiqued Norbert Wiener's ideas of the cybernetic nature of class relations as wrong: The discipline's father saw capitalist societies as practicing regulation rather than socialism's governance, and he felt that was superior. But Bulgarian socialism was just as well based on regulation, which was just another name for feedback! Socialist society purely used feedbacks and had separate elements that allowed for information to circulate—the party, the judiciary, the workers—as long as you defined who was the governing and which the governed part.[72] Relations between the levels were possible through legal or political proceedings, which acted as inhibitors, or through scientific prognosis, which allowed certain levels to "surge" progress.[73]

Stanulov's deep engagement with technical sciences thus translated into a novel model of how socialist society operated. He placed the most importance on social consciousness—the changing amalgam of temporally changing social activity and views—which could regulate the "social being" (society) through the state's regulatory mechanisms, such as law or the plan. The changes of social being, the day-to-day experience of life, would feed back into consciousness—the social views—allowing consciousness to retain dynamism and regulate its conclusions once it observed the positive or negative effects they had on the social being.

This mechanism was a way to get the true engine of change—humanity—the tools needed to act on the world.[74] There was no governance without information, which was independent of action, or the transfer of energy within the system. Information could affect action overtly or covertly, a dialectic system dependent on the presence or absence of deviation in the governed object. The primary importance for the receiver in this system was the pragmatic nature of information—its usefulness in regulating the system, allowing social consciousness to be an active governing actor. To do so, the system needed subsidiary, mediating systems that would affect the political and legal realities.[75] This lengthy meditation on Stanulov's theory is borne out not just from its complexity but because it is the fullest expression of an attempt to take the BCP's theses seriously and work out the political philosophy behind them: What *are* the translations between cybernetics as a technical system and as one for socialist governance? The

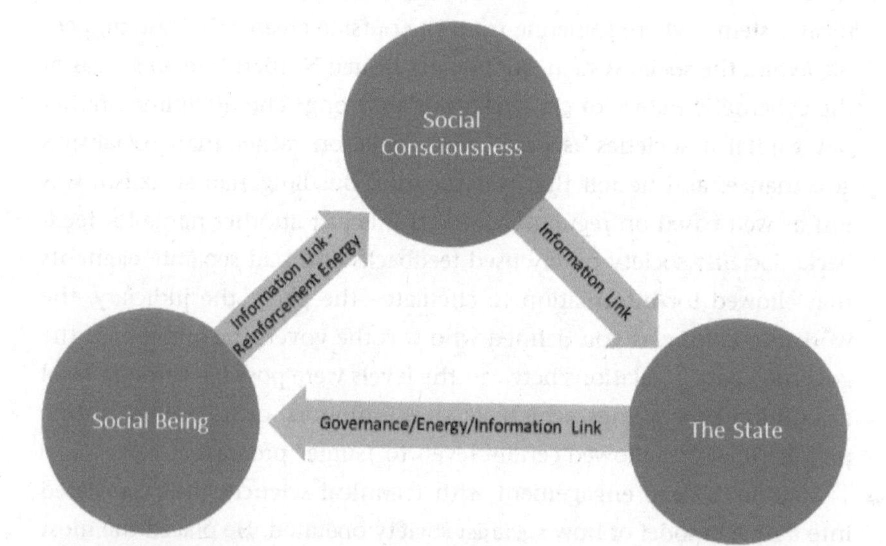

6.6 Stanulov's "social consciousness" model of governance. (*Source*: *Filosofska Misŭl*, no. 1 (1973), p. 49, National Library, Sofia.)

theory contained both an orthodox cybernetic understanding of who was governing—humanity itself—and a more radical placing of consciousness as governor. There were distinctly socialist elements—the proletariat was in effect a substitute for humanity, as its historical dynamo. This idea was fully in line with a state that consciously presented itself as walking the road to communism, hence the state could appear as a mediator for the force of progress, enacting the "will of the worker." Yet this consciousness was not a disembodied historical force—it was a real existing force consisting of all people who made up socialist society. With the advent of ESSI or an insistence on the existence of socialist democracy, the state could discursively be maintained as an expression of the dictatorship of the proletariat. In effect, Stanulov's model was a way to show how cybernetics would work better in socialism than in capitalism, and even better in Bulgaria, which was building the feedback loops that allowed the different levels to communicate accurately. Yet this scheme was, of course, false—the elements were not independent, objective mediators: The judiciary was never independent, for example. The scheme thus created the space for a cybernetic questioning of how well the socialist state was reflecting the popular will. Subordinated to a social power but also a temporal one—the road to communism—the state was just a link in the system and could be judged by cybernetic principles on how well it facilitated information, governance, and the ultimate aim of the system. The dry, complex philosophical discussions the regime encouraged in order to put its own theses on a sounder intellectual footing created the space to be attacked on the grounds of the regime's own promises.

In the same year as this model was published, Stanulov's salvo continued with an overview of the general philosophical questions cybernetics was still to answer. "Information" was the key one—the carrier of functional properties of material bodies, and thus something that occurred in the inorganic world, too. This was a broadside against orthodox Marxist philosophy at the time, prevalent in the socialist world and exemplified by people like Pavlov and his "reflection" thesis—information for them was just the function of living beings, narrowing cybernetic purview to society and not encompassing the whole universe. This obscure debate is worth highlighting to show that cybernetics was truly a revolutionary language that could undermine all aspects of the regime's previous foundations. And even when cybernetics became the party's political language, it didn't necessarily lose this radical effect that it could have on aspects of socialist society, as Slava Gerovitch has argued. Stanulov highlighted that modeling the human mind mathematically led us down a road where eventually a machine could do everything we could—so we should see ourselves as the most perfect cybernetic machine.[76] Cybernetics could thus infuse Marxism with new power, helping it uncover the true basis of the universe, with a new logical analytical power that the old guard (such as Pavlov) lacked.[77]

The social and philosophical aspects of cybernetics thus found their rightful place alongside technical issues in the 1970s. A 1975 book has sections on automation of passenger lifts alongside articles on feedback in enterprises or Stanulov's quixotic dream and continued revisions of his ideas. In the ITCR, scientists confidently defended their ability to model and thus uncover more and more about the world, allowing them to engage in debates beyond engineering—computers were allowing them to simulate reality.[78] This was also the decade that Naplatanov edited the nine-volume bible that all students in the sphere had to read—*Foundations of Technical Cybernetics*—solidifying this view for the new generation: Cybernetics was not just about simple technical controls but about control and "knowing" more generally.

The computer's emergence in Bulgarian material life was thus accompanied by cybernetics' growing hold on the Bulgarian mind. Cultural, psychological, and social aspects and prisms thus traded concepts back and forth with the technical and political sciences. For some, the new power of information meant that the human personality could now truly flourish to develop a multifaceted complex nature, capable of vast analytical leaps when it was armed with the power of computers and robots.[79] Philosophers ruminated on the implications of the STR—humanity's core, that which set the human apart from the animal, was formed and determined by social relations, the relations of particular forms of matter that are in hierarchical relations to one another.[80] Labor was the most important such structure in humans, connecting our biological and social aspects by making the former serve the latter. This was the basis of all other human structures.[81] The massive changes in our labor that had started with the Enlightenment were now accelerating beyond belief through computing. Humanity is now pushed beyond mere production into the role of governance—brain matter, not brawn, was now our tool of production. As humanity changed nature, we changed our own nature, too.[82] Pollution, noise—including information overproduction—meant that humas were becoming different, not always for the best. These were common tropes among many—the human was now an intellectual laborer. But only in socialism could this development come to its full fruition, as the means of production are owned by the people, so there was "true freedom of information," unlike in bourgeois countries, where "noise" abounded and unclear information, key for governance, circulated.[83] The BCP's dream was taken seriously, with all its utopianism and anxieties, by the country's intellectuals, who saw themselves as being at the crest of the new technological age's wave.

The debates in the 1970s also became somewhat swept up in the changing cultural politics of the country, initiated by the culture minister and leader's daughter, Liudmila Zhivkova. Her own utopia centered on the multifaceted personality, a new renaissance human driven by beauty and aesthetics. Often esoteric and influenced by Eastern mysticism and theosophy more than Marxism, her ideas were interwoven with a cultural revival that was also centered on the Bulgarian nation and its history. She is best remembered for eccentricity but also for artistic patronage, especially of the grand vision of Bulgaria's role in world culture, but her obsession with children's education, aesthetics, and the development of human ability meshed together with the automation discourse.[84] After all, the party's STR dream was also to create the intellectualized laborer, a creative governor of production. The journal *Filosofska Misŭl* printed articles that supported Zhivkova's thesis, such as that by Soviet philosopher Victor Afanasiev, who argued that even a self-learning machine would bend to the human will, lacking the emotional comprehension of the world that was a key tool in human creativity.[85] This period was possibly the most "Bulgarian" of all social cybernetic discourse during the regime's history, as articles laid out how the unfolding STR would help develop not just socialist morals but also the "aesthetic" culture of the modern personality that was desirable—Zhivkova's language thus had an imprint on cybernetics, too. Science would unleash this new creative power and with it would come new moral responsibility, based on the higher social consciousness that would emerge with the information age. The higher amount of information allowed more freedom, more flexibility in norms, and thus more individualism. But unlike in capitalism where this individualism leads to alienation, the Bulgarian path of socialism would allow individualism to proceed along harmonious paths toward both rationality and aesthetic personality traits.[86] The computer would help create the socialist human, but also a *Bulgarian* socialist one.

But by the 1980s, as the economy struggled to implement the electronics policies, the debate shifted. The country's primacy in producing the goods did not translate to "scientific" governance. As exports were prioritized, computers were lacking in the country or broke down often when they were delivered. The gap between the dreamy promise and the reality grew even more as the Pravetz PC was created in 1982, itself an indicator of IZOT's shortcomings in innovation.[87] The PC promised automation in the office, schoolroom, but also the home—the fuller intellectualization of life was even more possible now.

Yet over a decade of rich debate and actual production had made the technical elite more confident of its own abilities to contribute to the world's store of cybernetic ideas. Under the leadership of mathematician Blagovest Sendov, who had Zhivkov's ear, the country was experimenting with new approaches to education. These included unification of subjects, such as the natural sciences, into one to facilitate "integrated" approaches to knowledge, and eventually the computer classes mentioned above. The "Sendov method" was being tried out in 27 schools beginning in 1979, with computers at the core. "By entering the information century, the object of education is changing," Sendov held, as the computer would be a weapon for the student's brain and a continuation of thinking.[88] Knowledge acquisition was to be structured in such a way as to reveal to the student the commonalities between hitherto discrete subjects. Syntax and morphology, for example, could be taught through mathematics; social sciences were to be combined so as to move away from traditional focuses on war or politics and uncover the dynamics of social change or how ideas arose.[89] Computers would teach coding and decoding—skills applicable to literature as much as to maths. The "student-armed-with-a-computer" was the new object of education, as creativity would be unlocked through integrative knowledge.

Sendov shared his ideas with the world at a 1984 international seminar in Plovdiv. In his inaugural speech, he highlighted the imminent second computerization wave, which would impinge on all social practice. The computer was better than us at remembering, a traditional educational aim, so it would be used as the encyclopedic repository, while "humans, with its help, will have access to information and the ability to integrate it creatively."[90] In 1985, Varna hosted the International Conference on Children in the Computer World, organized with UNESCO. The journal summarized the discussions as proving that Bulgaria wouldn't repeat previous mistakes in "upgrading" the workforce, instead learning from others' failed attempts. Sendov stated that "computers can democratize our imagination" as long as we took care to present knowledge in an imaginative, integrated manner.[91]

The journal's articles in the last decade thus presented a utopian image that the regime often approved of, with intellectuals demonstrating how human creativity—the crux of cyber-society—was being fostered by Bulgarian methods. Yet the journal also hosted critiques—something in socialist administration was holding back the promise made in the 1960s, which hadn't yet materialized. Philosopher Ana Krŭsteva overviewed bourgeois cybernetic theories and concluded that the new type of human fostered by the machines couldn't be a slave to hierarchy. The tools had indeed allowed "creativity," but also "symbolic" transformations—computerized labor was abstract and called for a particular type of creativity. While bourgeois informational concepts were also flawed, ignoring social relations, Krŭsteva's main assault was on hierarchy as impeding this creativity.[92] For another intellectual, Vladimir Stoychev, it was too much to expect this flourishing—the regime was setting up unfulfillable expectations. The cybernetic human could be a Salieri rather than a Mozart, but that was no bad thing. The Italian broke down big tasks into smaller ones, and through perseverance mastered them, rather than wait for the one-in-a-billion chance to be the Salzburg genius. Salieri was the paragon of the cybernetic model of creativity, as opposed to the capricious gifts of Nature: "one is marked with God's spark, while the other one creates his talent alone and to some extent manages to equal himself with the Salzburg magician."[93] Salieri was no sad figure but the model for the creator in the machine age, who perfected himself through constant labor. The famous Pushkin tragedy that framed Stoytchev's discussion was a projection of conflict between cybernetics and a faith in the "muse." A contemporary laboratory for algorithmic machine music would bear Salieri's name—after all, he taught Beethoven and Liszt. Modern genius was thus the ability to choose correctly: Mozart might be the dream, but Salieri was the "real, achievable variant" of creativity: "The creator who makes us believe that even people with conventional creative abilities can reach, with much labor and perseverance, the heights of perfection."[94]

Moderating expectations was thus key, as was recognizing that this new human was possible but was not transpiring in Bulgaria. Naplatanov, too, highlighted this in 1986—the computerization of governance could lead to "hybrid intellects" of artificial and natural intelligence, and real flourishing was possible if the two united in a cumulative nature. The question was how to furnish socialist humanity with the needed environment and technology to do that.[95]

"This question is about the most valuable thing that our society has—the person as a creator of all material and spiritual goods in Bulgaria," said Todor Zhivkov in a 1984 Central Committee discussion of an ambitious program that had the aim of raising the intellectual and creative abilities of Bulgarian citizens.[96] But while the regime elevated this question even higher in the consciousness in its last years, the journal's discussions became more and more critical of its realities. Some observers tentatively argued that increasingly, not just in the West, the information workers were becoming the dominant social class. Export and import of information was becoming as key as material production, and capitalists were winning at this—socialists had to fight to put this international information order on a democratic basis. Whatever the international situation, though, we must not reject all Western analyses of the information age, for it seemed obvious that information did add value to products, so a rethinking of the labor theory of value was needed.[97]

As preustroistvo's (the Bulgarian version of Soviet perestroika) wider reforms became part of the Bulgarian landscape, this voice of the information worker was increasingly raised, often mouthing Western concepts of the information economy while exposing the regime's economic shortcomings. Cybernetics retained, in Bulgaria, a reformist or even radical tint that could be turned against the regime rather than just subsumed by it. In the twilight year of 1989, a specialist issue of *Filosofska Misŭl* on informatics was published, with eight essays. One, by Petŭr Mitov, stated that a society closed to new ideas was doomed to die. A monopoly in science or politics prevented free development. The microscope and the computer had to take their place alongside the hammer and sickle if socialism was to strive, shearing the latter of conservativism. Information allowed people to be linked simultaneously and instantaneously, and it could create veritable, true, mutualist networks. To accelerate and rescue socialism, recognition of the rise of info-commodities that were free had to happen. Marxism was a powerful analytical tool of capitalism, as it had been 150 years ago—but not a recipe for the current, transitionary period. Information had to go where politics would take it, and in Bulgaria—as in the West—it had hitherto served a dogmatic political culture. The new reforms could stop Marxism's deformations only if the bureaucratic authoritarianism was combated by the cure: free information. Bulgarian socialism had promised big changes but delivered little, leading to passivity and confusion among the populace. But now the information age, which collapsed time and space, promised a really free society.[98]

Others also attacked the regime, this time on economic grounds. A wide-ranging essay eviscerated the labor and workplace organizations that created socio-psychological barriers to social computerization. In the automation age, the psychological burdens were not reactions against the technical system but against the social change associated with it—the relations between people rather than human and machine. Attachments to old work methods was normal, if conservative. Routine work was automated—but this put new responsibility on the worker. Workers feared that they would be reduced to their technical functions, or that fewer jobs for workers would be created. Those who embraced the computer faced other problems—a new type of office that isolated you, running counter to our proclivity for contact. Instead of fostering it, this led to a *loss* of creativity, as people retreated into themselves. Humanity is surrounded by machines and is dependent on them. The human also had to know the machines' "algorithmic languages" and the working of their "thought process." The specialists who knew these codes were feared, resented for their mastery over a machine that subordinated you as a lowly worker. The darkest fear of bourgeois society was true here, too—that the machine surveilled you. It uncovered your smallest imperfection, exposing you as a fraud. Even if you loved the computer, you risked losing a sense of reality by falling into dream worlds. An increasingly anxious society was the result, proliferating mental illness, as human emotion and connectivity was hampered. The regime's computerization lacked coherent applications. It was more present as an idea than as reality. Not only would more computers have to be delivered, but more psychologists, too. To create new ways to communicate with the machine, programmers had to work with psychologists to create more natural languages, rather than the formal ones that predominated. And finally, information was fetishized. Data were considered to be "right" without reference to human components or actual reality. Decisions hide behind the computer's authority, the ultimate guarantee of science. A job was only well done if a computer was involved. Bulgarian computerization had lacked the desired results, because it had lacked the consideration for human needs—the regime loved the prestige of the technology but in fact hampered creativity with its own programs.[99] The regime had a dream and that was exactly what it had created—unreality.

Briefly, it is worth mentioning that the regime did take workplace conditions in computer centers seriously and commissioned studies to find out whether these conditions were within norms. A 1989 study of the Varna Shipbuilding Computer Centre concluded that everything was within normal noise levels, screens did not emit harmful electromagnetic radiation that could harm the workers, and the climate was pleasant for work.[100] Thus some steps had been taken to at least pay lip service to the recommendation of workplace psychologists and health authorities.

The final essay was a Soviet broadside against informatics as a control tool. In it, A. I. Rakitov stated that an individual armed with a computer was the total opposite of state control over the individual. In a society without democracy, "information technology can become not the basis for freedom but the instrument for total coercion and control." Without glasnost and freedom of information, computers could be used to deepen social disinformation and backwardness. Once again, it seemed that old socialism had run its course—a new type was needed in order to truly unlock the promise that was made over 20 years before. Without such reform, socialist countries would become informational colonies of more advanced countries.[101]

The journal's pages thus reflected the full spectrum of the machine dream—from 1960s rehabilitation to 1970s utopian application to all spheres to 1980s disappointment and morphing into a tool of critique. Computers offered the possibility of both objective social governance that the party demanded and the full-blossomed individual that Zhivkova's cultural policy spelled out. But by the later 1980s, the computer had also become a prism through which to view the regime's failures—rigid and unreformed workplace organization, lackluster productivity, and conservative approaches. To these shortcomings were added anxious fears of psychological damage and calls for freedom of information, which were unleashed fully under the Bulgarian variant of perestroika. The aim had always remained to create a socialist cyborg, not a capitalist one; but also a particularly Bulgarian one. This was partly due to the intersection with Zhivkova's policies, and partly due to the successful reach of cybernetics to most disciplines through the pages of the philosophy journal. This concentration on the human dimension also explains the country's particular concern with the psychological failure of the cybernetic dream. Of course, it wasn't just in journals that this idea played out, as both the state and its intellectuals impacted the last generation of socialist citizens, who became the object of the future in an attempt to overcome the conservativism of their elders through an introduction to computers from the earliest age.

A COMPUTER FOR WHOM?

In 1984, children's lives were invaded by two new institutions—a network of computer clubs and a magazine called *Kompi͡utŭr za Vas* (*Computer for You*), both run by the Dimitrov Communist Youth Union (DKMS). The magazine's first editorial in 1984, before a 1985 redesign and launch as a monthly magazine, proclaimed that nothing in modern Bulgaria was done without electronics. So the magazine would bring young people the knowledge to enter the world of "her majesty—Electronisation": new programs, developments in the clubs, and from around the world. Children were encouraged to write back to the magazine with requests or concerns, so as to turn their "work with computers into a calling and duty of the young generation."[102] While the leading, dry, article was penned by the DKMS's head, the first issue set the tone for the magazine's existence with well illustrated and accessible articles on computers as second literacy, such as interviews with kids who spent their time at a computer club rather than at the beach. Bulgarians and East Germans on holiday, they are enthusiastic acolytes that eschewed fun in the sun now for the future inherent in the computer a few meters away. The journalism was candid, covering how those with fathers in the Merchant Navy wanted computers from them rather than prestigious stereos. They knew Bulgaria produced computers, but the journalist asks where they are: "there are French cosmetics, whisky—presents, cigarettes, and whatever colourful rags you want, but there are no computers."[103] The magazine would always feature the shortages and the regime's failures in the sphere alongside its educational and practical articles.

The magazine introduced many tropes, such as "UMKO" the computer (a play on IMKO, the first Bulgarian PC, and "Um"—the word for "mind," or "clever") who helped the youngest with the new world. Lectures were printed throughout its run, as well as the most important thing—code. Complete programs were offered to subscribers, from simple ones that allowed calculation on a home computer, to engaging ones, such as programming a computer to become a musical keyboard; to play chess; or a

6.7 The "Microcomputer Kids" and how computers would help at home (the screen reads "Put more soap in the water!")—the magazine's views of the future. (Source: *Kompiu͡tŭr za Vas*, National Library, Sofia)

number of other games, such as space battles. The magazine was very well aware of its audience and its interests.

Alongside the launch of this magazine, the first computer club opened in Sofia in late 1984, meant for all ages who wanted to use its Pravetz'82 machines. From the start, the club suffered from a lack of adequate software literature, with users bringing in their own notebooks to copy programs from manuals, and the computer lacked sufficient peripherals, like printers.[104] Yet by 1985, there were 28 computer clubs and a further 350 were being organized. A software enterprise—"Avant-garde"—comprised of students and teachers in Sofia and Plovdiv schools and universities was tackling the program problem by creating 60 computer games and 20 educational programs.[105] By 1986, there were 1,500 Pravetz'82 PCs in the network, together with IZOT 1013S microcomputers and ROBCO training robots that taught children how to control robotics through a PC. Children were producing software in Sofia, Ruse, Haskovo, Sliven, Kystendil, and other towns through self-teaching aids for languages like Basic and Pascal; these materials had sprung up out of the youth movement itself.[106] Bulgarian computer clubs, showcasing domestic prowess by teaching foreign kids, opened in Moscow, Leningrad, Kiev, Kharkov, Havana, Pyongyang, Hanoi, and Addis Ababa by the end of 1986.[107] By 1987, around 530 clubs operated throughout the country, at least one in each of the 28 regional centers, but also 306 clubs in factories, schools, universities, and youth centers.[108] Those in Sofia or Varna routinely had 18 PCs or more, while smaller ones had just two or three.[109] Time was limited, as many machines needed constant repair or the clubs had to balance teaching children alongside adults who wanted to requalify for these new sectors—yet the network was widespread and afforded many people their first taste of the fabled future.

The magazine covered the clubs extensively, often using them to display children's experiences. A questionnaire given to 200 students who attended clubs revealed that 66 percent of them had a family member who worked directly with computers. The youngest children wanted to use the PCs two hours a day, while teenagers wished for at least four hours. Unsurprisingly, games were among the most popular activities to do on computers.[110] Other articles highlighted how fast children learned to use the machines—an interviewed adult recalled it taking him three hours to

6.8 Children and adults together in a computer club, late 1980s. (*Source*: socbg.com.)

land a spaceship in a simulation game, while a middle-schooler watching him got it right in under a dozen tries.[111] The magazine was both part of and a testimony to the computerization of Bulgarian society, especially its youngest citizens. Other magazines, such as *Zhenata Dnes* (*The Woman Today*) also noted children's excitement. Kindergarteners are quoted in an article extolling parents to also learn about computers: "I want the comrade (teacher) to take me to the little computer every day"[112] Older children were also noticing how their parents' lives were changing, and were preparing for that world—"my mum told me about the factory in which she works, about the machines they create . . . electronics became my hobby and my dream is to orient myself professionally towards them."[113] Parents, however, are concerned with how exactly these dreamers are interacting with the computer, in a fashion that is universal: "space battles, treasures, monsters, to win you hate to kill, to kill . . . yes, 'pretend,' but pedagogues and psychologists need to finally say something about this."[114]

The role of computers in childrens' lives also dovetailed with articles in other publications that wondered about what the machines would also

6.9 The universe at your fingertips, as well as the socialist computer dream: graphic design helped it capture an audience, while it ran advertisements for the domestic industry. (*Source*: *Kompi͡utŭr za Vas*, National Library, Sofia.)

do to the primary caregiver, who, for all of socialism's feminist fanfare, remained the Bulgarian woman. The first instances of computing entering the pages of other magazines are from the mid-1980s, when the article "Getting Closer to the Computer" exalted the new technology, as was usual: Every citizen now had to contend with the revolutionary entrance of the machine. This was especially true for adults, who are portrayed as going to classes, where, by the time a mother learned to draw a basic house, the children had programmed a whole company of paratroopers. Overall, the article focuses on women as teachers in the new computer clubs, but also on children—in many ways calling on women to learn about the new technology due to their position as mothers, rather than as citizens who should use the machine for its own sake.[115]

Another 1985 article also focused on how computers would transform home and family life. It starts with a dream of how you come back from work, the lights turn on automatically, dinner is just warmed up, your VCR has recorded your favorite program, and you turn on the PC to book theater tickets. Comically—and candidly—it also tells you which shop is currently stocking coffee, an admission of the shortage of such goods.

This dream of an automated home is what an electronics couple—Valeri and Valentina Zhekovi—work on in their spare time. They are both specialists in the Pravetz microprocessor factories, graduates of VMEI-Varna, and having fallen in love over a shared interest in the computer. Their daughter is born amidst their work and is featured as the future—taking her first steps in a PC exhibit at the Plovdiv Fair. The article is a showpiece for the modern couple working on the future automated home while raising the next generation of cyborgs.[116]

But the article also reveals the burdens. Valentina doesn't just program at Pravetz, but is working on a thesis at home, while taking care of little Aneta—the grandparents are far away, and there is no mention of any real help from the state. Valeri seems absent from that part of domestic life—despite working in a similar sphere at Pravetz, he is the one traveling to different factories to demonstrate the programs' abilities. Read against the grain of what the editor wants to portray about the future, the article is a reflection of the double shift of the Bulgarian woman: taking care of the home and taking care of the party's technological revolution. It is little wonder that her dream is of a fully automated home.

Meanwhile, *Kompi͡utŭr za Vas* editorials served another purpose, ready to point out the state's failings in the field they loved. One asked from the first page of every issue whether the reader could buy a Pravetz 8D home computer, supposedly available everywhere. Acknowledging that schools and clubs were prioritized, it slammed traders for not being clear about available stocks. The lack of software literature was attacked too, a failure laid squarely at domestic industry's feet. "We know that the floppy discs that we produce are shoddy and work (chattering away like a clapper) only in exceptional circumstances. We don't know how long it will be like this," it bemoaned, stating that the editors were uneasy facing their readers—extoling things unavailable to most of them. As compensation, it offered 50 percent subscription discounts.[117] In 1988, the magazine was still complaining about the same things, running an investigative piece on how easy it was to obtain a home device. Until recently, it found, only one store in Sofia stocked them, where kids would bring sceptical fathers to buy computers to plug into their TV or telling them they could use the home tape player to record programs.

The father is even more of two minds. The tape player thing is good—the house will finally rest from all that disco . . . and heavy metallurgy [a jab at heavy metal]. But what about schools? What about the university entrance exams?

But the journalist defended the kids, as this year 2,000 Pravetz 8Ds had been made available for the home market, at 420 levs each. He extoled the computer as a long-term purchase and even a disciplinary tool—a father is quoted as locking it up when his son misbehaved, for immediate results! Yet the computer was difficult to find in Sofia and almost impossible in the provinces, as traders ignored its potential. Despite promises, the internal trade network was still failing the new generation's obvious demands.[118]

By printing code, the magazine explicitly tried to offset the industry's shortcomings. It ran competitions to give away Pravetz 8D computers and even whole Pravetz'82 desktops, which for some was the easiest way to acquire these machines.[119] Its humorous articles and caricatures didn't shy away from poking fun at some fears about computers, wrapped up with pointed barbs at the political climate. In 1985, before preustroistvo was announced in Bulgaria, and before a loosening of censorship, it could run a picture that made fun of fears about how computers could aid DS spying, without any repercussions. The magazine's reflections of the

— . . . Написал донос срещу мен. . . !

6.10 "He wrote a donos [denunciation] against me!" (*Source: Kompiûtŭr za Vas*, National Library, Sofia.)

hopes and dreams inherent in the regime's language existed along with cutting, truthful reporting on that same regime's failures. Its importance as a genre that reflected both the march of computers in Bulgarian society and their shortcomings can't be understated. The magazine's very popularity reflected the image that the computer had in young, hopeful minds who were to be the regime's cyborgs of the future. That same mental space was also increasingly occupied by another cybernetic dream—science fiction.

THE NEW LAWS OF ROBOTICS

Bulgarian science fiction's heydey was undoubedly during the socialist period, often breaking with the "reds in space" genres that simply projected Marxist utopias into the future. The biggest boost came in 1979, when the "Galaktika" series started publication, presenting Bulgarian readers with the best in domestic and Western science fiction. Classics by Bradbury, Herbert, Le Guin or Clarke nestled alongside Eastern Bloc classics like those of Stanislav Lem or the Strugatsky brothers. One hundred and one volumes were published by 1989, a prized possession for many.[120]

But some Bulgarians had started grappling with the effects of computers and robotics before that series began publication. Li͡uben Dilov was maybe *the* towering figure in Bulgarian science fiction, with his 30 books selling over 1 million copies altogether.[121] While studying in philology in university, he encountered the genre, becoming a fan of Clifford Simak. His first book, 1958's *The Atomic Man,* was the closest to an orthodox Marxist science fiction, showing how a capitalist wakes up in a world where communism has won and his reactions to it—in Dilov's 1979 edition, he resurrected the original story he had wanted: The man who wakes up was a Bulgarian who had escaped communism, and now he was back in it, thus turning the book into a meditation on the national psyche, too. His crowning works, however, include 1974's *The Road of Icarus,* which won an award from the European Science Fiction Society in 1976 and is considered one of the best socialist science fiction novels by Arkady Strugatski himself.[122] Heavy on philosophical and interpersonal themes, it tells of *Icarus,* a generation ship made of a hollowed-out asteroid, which takes humanity's best to explore the universe. The protagonist, Zenon Belov (referencing Zeno of Elea), is the first child born on the *Icarus,* a true citizen of the stars. Strands of the story concern love and coming-of-age. Yet there is also a deep description of the society, and the philosophical question of its engagement with machines—the topic of Zenon's dissertation, which he isn't very interested in.

A turning point comes with the trial of a scientist who creates a cyborg child, programming it to play and learn, convinced that it is our propensity for games that allows us to innovate and become individuals. This act is against Icarian law, in which robots are allowed to be helpers rather than mimickers. The little cyborg also exhibits brainwave functions identical to the creator—an attempt at cloning in its way. The child is killed, but the debate centers on what to do with the scientist, who is supposed to be frozen for a decade and to have his brain altered before release. Parallel to this story runs a debate about allowing the young to conduct recon flights on their own, rather than use automatic probes. *Icarus* is at a point where it must decide whether it can change its rules, a fool's errand in Zenon's eyes.

But his father, an influential figure due to age and earthly achievements, takes his side, criticizing society as cruelly unchanging. "We are the end of our wisdom . . . the drama of Icarus is that it is fit to burst with scientists . . . Einstein himself stated that we scientists are conservatives by nature and only circumstantial impetus can make us sacrifice our sacred positions," he bemoans, arguing that while drifting farther from Earth, this society created nothing new—for two decades it has just identified Earthling's mistakes about the universe. All of *Icarus* aimed at preserving a world rather than knowing the real one: "If for a civilization this can be defended, for a research team it is deadly." *Icarus* knew nothing real, as it looked out at the universe through apparatuses that deformed reality. Icarians had been waiting for a massive jump over "the gaping abyss of contradictions between our new knowledge and old views," but such a revolutionary leap could happen in the mind of a single person, not a collective. One scientist could expect to carry out but one revolution in his lifetime—Zenon's father himself shouldn't be here, as his achievements were done and dusted. Icarian heads were full of chaos and needed an empty-headed fool to be born: and now society wanted to freeze him.[123]

Dilov defends individual innovation against a staid collective, warning against rule by experts who were great at maintaining social functionality but not so good at allowing it to leap forward. Icarians don't progress meaningfully until a few outliers, including Zenon who is born in this society and is not satisfied by a novelty that is his whole life, shake things up. It is a powerful story of generational and social conflict, exploring science's limits and supporting the power of curiosity and spirit. Dilov also tackles a technological anxiety—what is it to be a human among machines? Where do they meet and what sets them apart? He thus invents a Fourth Law of Robotics, to supplement Asimov's three: "The Robot must, in all circumstances, legitimate itself as a Robot."[124] This was a reaction to an increasing wish to give robots human qualities and appearances, often copying animal forms. Zenon muses on humans' interaction with machines that start from a young age, the child possessing power over the robot from the start, undermining the child's trust in the machine that she depended on. Humans need a distinction from robots, assured of their power and that they weren't lied to. Dilov's anxiety was about humanity's limits in its current stage—fearful humans couldn't treat anything, including machines, as equals. Toward the end of the book, a note discusses Icarians' game against nature, one that couldn't and shouldn't be won. Dilov's society was a place where folly kept the species back, its anxious conservativism preventing a true exploration of the Universe, with cyborgs and robots as ciphers for such failure.

Similar ideas permeate the humorous, bittersweet 1981 anthology *The Missed Chance: Stories from My Computer.* In this book, a bored author toys with his writing computer, feeding it narrative points and genres to see what stories it produces. Connected to the National Library and so to all world databases, it quickly produces coherent stories that the author can pass off as his own. Ranging from fairy tales to crime and science fiction, absurd twists abound as the author sets particular parameters: such as that the crime should have no real mystery—the killer, weapon, motive must be known from the start. Slowly Dilov—for he is the author—realizes it is only spitting out his own style back at him.[125] The final story, narrated by his idol Arkady Strugatsky, has him contending with his own double, sent in by the Soviet author. This is the computer's revenge after it is told it is to be replaced, and maybe scrapped, as it has taken too much of Dilov to be of use to others.[126] The computer mocks Dilov, reflecting his fear of the author's place in the world by amplifying them through its immense

6.11 Images of the covers for *The Road of Icarus* and *The Missed Chance.* (*Source:* Author's collection.)

power. Any fault of the machine is simply our own fault, as the computer can't truly produce but must be programmed. A faulty narrative is born out of human reasoning, not the machine's. Dilov's view is pessimistic about the regime's vested hopes in the computer.

Bulgarian robotic law production quotas continued, thanks to mathematician-cum-writer Nikila Kesarovski. A regular contributor to *Computer for You,* he wrote on the history of computing, its future, and information's meaning. A great popularizer of science, his programming guides helped children and adults, while essays demonstrated a great belief in information theory as being able to unlock human creativity by unifying disparate disciplines into a synthesis.[127] This belief was reflected in his short stories, three of them collected into 1983's *The Fifth Law.* The first, *A Crimson Drop of Blood,* shows a vision of the human body as cybernetic machine. Looking for alien consciousness, a scientist finds it in his own blood cells. Gradually deciphering messages received from an alien mind, he concludes that the society they are describing is his own body—a type of robot itself. Kesarovski's vision of nesting cybernetic machines—turtles all the way up and down—stems from his training as a specialist: He is more optimistic than Dilov. Yet a warning rings out in *The Fifth Law.* His most famous story had already reached a wide audience, especially among youth, as a strip in the *Duga* comic book series in 1980. The comics were massively popular, each issue published in the 120,000–180,000 range, a feature of most children's lives in the 1980s.[128] It had something for everyone—original children's characters for the youngest, detective and spy stories for teenagers, science fiction, historical comics based on Bulgarian and Roman history, games, and prizes—*The Fifth Law* would have been read by many.

A famous writer is killed by a man through a simple hug, baffling police. The investigation reveals that the fan is a robot that doesn't know he is one, violating both the First and Fourth Laws (Dilov was now canon). Hence the Fifth Law—"a Robot must know it is a Robot." The story delves into how the robot came to be, exploring corporate greed that created weaponized robots. A cyborg is created by melding a brilliant scientist with the machine, prompting a Terminator-like robot rebellion in a weapons facility in Guam. Holding the world hostage with the nuclear weapons there, they demand that the cyborg be delivered to them, as well as the best robot psychologists, so as to negotiate their terms with humanity. The story ends on an open but hopeful note amidst human threats to nuke the base as a preferable option to servitude to robots—while one of the psychologists recalls Christ's words to Peter and how he would betray him three times, and yet on that rock he can build his church.[129] For Kesarovski, the automated dream was dangerous but promising, if only humanity could understand itself as a cyborg organism and thus attain its next stage.

Finally, in 1989, Li͡ubomir Nikolov had had enough of Bulgarian propensity for robotic lawmaking. In his short story, "The Hundred and First Law of Robotics," a writer is working on the pleasant but limited "The Hundredth Law" short story, which states that a robot must never fall from a roof—in the nested story, it had just done so, killing a pedestrian. He himself is killed by a servant robot that just didn't want to learn any new laws, hence the final law: "Anyone who tries to teach a simple-minded robot a new law must immediately be punished by beatings on the head with Asimov's complete works (200 volumes)."[130]

6.12 The Fifth Law as comics. (*Source*: Author's collection.)

Bulgaria thus boasted the highest number of robotic laws per capita, born out of an industry even more successful than its IZOT factories—science fiction. The spread of such ideas is well illustrated by the literature, but also by the philosophy that started this chapter: Both indicate a widespread fascination and expectation tied to the regime's promises. Through education and popular engagement with the computer, a new generation grew up with a voracious appetite for the machine. The regime fostered this new socialist human, but at the same time was being called to increasingly answer for its failure to deliver the tools to create that individual—first by its own intellectuals, whom it had encouraged to think in terms of information-age socialism, then by journalists, and finally by the children and students of the 1980s.

The last word should belong to another writer—Velko Miloev—whose *Nanocomputer for Your Child* came out in 1988. A discovery is made—a brainwave that reacts to stupidity, and then computers that can detect the same. Quickly these computers contribute not to human knowledge but to fashion statements, or parading around your stupidity. In a search for those who would react properly to stupidity, the scientists turn to children—the perfect, honest, stupidity-detectors. The story ends with computer toys that subtly encourage children whenever they detect their reaction to surrounding stupidity, training them for a future where they will always do so. A jab at parents, at society, and a hope for the future of Bulgaria, Miloev's story was too little, too late. By then the adults in the room, many of them the intellectuals and technocrats trained by the regime for its dream, had become part of wider trans-national networks of thought and experience, with their own ideas about the regime's stupidity.

7

NETWORKED AND PLUGGED IN: COMPUTER PRIESTS AND THEIR PATHWAYS

The archives and interviews abound with the names of countries outside the socialist bloc where Bulgarian specialists and traders plied their trade or learned the latest technological tricks. Interviewees often speak of their experiences in West Germany, Austria, France, and Italy, while authors of books and memoirs are enamored with countries even farther away, especially Japan. The computer industry faced both inward and outward, and for it to compete, sell, steal, and implement, it needed a well-trained and well-connected class of specialists. From the very highest level of Ivan Popov and Ognīan Doĭnov down to the individual scientists and workers who were subject to the automated machines, hundreds of thousands of Bulgarians were entangled to various degrees in the machine of the future and the circuits it threaded throughout the world.

In August 1991, the Varna Shipbuilding Computer Centre became a private company, taking out large advertising spreads to show its capabilities in software for both shipping needs and other industrial tasks it had carried out throughout its existence.[1] It fulfilled its final socialist-era contracts around the same time, before suffering a decline in capital and orders as the Comecon and internal markets collapsed. By 1995, it had taken stock of its declining material position, and by 1997, it divided its shares among 1,356 current and former workers and owners, before 90 percent of the new company was put up for privatization in that year.[2]

It was a typical story—a high-technology enterprise that was privatized and limped on in the tumultuous 1990s, a story of deindustrialization and de-skilling. But it is also indicative that it trained cadres that did flourish beyond the end of the regime—the human capital, if not so much the computer center itself, remained.

This chapter is the story of some of the elements of this technocratic class—both engineers and their higher managers. It addresses the questions not only of how this class constituted itself, and how it was stratified, but also how it interacted with the nonsocialist world and thus became engaged in different intellectual projects than Marxism-Leninism. In his history of the US software industry, Ensmenger shows how computer specialists became not just links between the computer and society—as we have seen in previous chapters—but also business experts. His work shows how the computer intelligentsia were not engaged just in maintaining the tools of the new age but became ensnared in power relations, constituting new realms, too.[3] This is doubly important for a socialist state, where these experts, often Western-facing, as other chapters demonstrate, became one of the few groups that could be termed "business experts" in a command economy. This is particularly noteworthy at the highest echelons of the industry, which often overlapped with high ranks in the party.

However, the fact that these men (almost always men) were socialists, too, is important in that this story can't and doesn't end in 1989. Sociologists working on Central Europe have asked how we make capitalism without capitalists ever since the regimes fell. Eyal, convincingly, has shown how cultural capital was one of the few things that could be transformed—and this included technocratic knowledge and connection. In this view, it is only the individuals holding political power who lost, as the managerial class managed to constitute itself into a new class while creating the conditions of private property, too.[4] To a large extent, this does map onto the experience of this chapter's subjects—their own cultural, educational, and professional capital allowed many to either flourish within the country or outside it. However, Bulgarian specificities are important: This elite had tried to form itself within socialism, had been periodically defeated, and often had to struggle in the 1990s, as the Bulgarian economic and political transition took on a different trajectory than that of Central European states. Often this meant that Bulgarian technocrats' trajectories intertwined with the world of organized crime or the old political elites, who tried to transform their appearance.

But there is another part of the networked class that existed, beyond the engineers and their managers, which has often not been integrated into the story anywhere: children. The peculiarities of the Bulgarian case meant that the last socialist generation was also steeped in electronic dreams and classes, was told that this was the future, but due to economic shortages was often prevented from indulging in this future. Research on the origins of the internet or the Silicon Valley counterculture of the late 1960s has shown that the new technology enabled new dreams, "wizardry," creativity, and was a sort of countercultural innovation—often tracing a link between the hippie culture and the boom of the 1990s.[5] Bulgarian children were being told that they would be creative—if only they could get to a computer! Some who did, thus, came onto the world stage in conditions of anxiety and creativity, resulting in a virus explosion that put that generation on the map as socialism crumbled. Many of them then grew up to become the new IT entrepreneurs in the 1990s and 2000s, a success story that had its roots—at least in part—in this age. Socialism's unique conditions of utopia and failure thus combined to make a generation that had no power to also be global and "plugged-in."

This chapter thus blends the stories of those with the most power—the managers—with those with education and no capital—the engineers and scientists—and those still being educated—the students—into a narrative of the creation and stratification of a transnational, networked, and globalizing class. This was how socialism entered the information age—imperfectly and piecemeal. It is important to note *who* entered the information age, and *how*. Thus, Bulgarian socialism, often seen as gray and orthodox, was also a hotbed of numerous creative factions who tried to reform it, think with it, and eventually fight it. This chapter is the story of the people who worked in this industry and the transformation it brought about in their experiences, ideas, and post-1989 lives. It is also a story that shows that even the seemingly closed state was penetrated by outside ideas, and how it fostered technology that created uneven development within its own borders—not just geographical, but often intellectual. Simply speaking, increasing numbers of people engaged with electronics thought in ways quite different from those which the party held. Their tastes, knowledge, and contacts were new; and so were their ideas. As the chapter follows how they played a role in the fall of the regime, it also sketches out the contours of the post-socialist order, and some of the afterlives of the ambitious cybernetic project that Bulgaria undertook, calling for a reconceptualization of 1989 as a hard and fast endpoint for developments in the region.

INTERNATIONAL PROFESSIONALS

From the start, the Bulgarian electronics specialists were international in training and connections. As we have seen in the first two chapters, the first such specialists were trained in the best socialist institutes, especially in the USSR and the GDR. Specializations abroad remained a key part of education even after the country created its own dedicated courses. Stanulov, for example, defended his candidate of science dissertation in Leningrad rather than at the ITCR.[6] Needless to say, Bulgarian specialists were always part of a massive network of professionals stretching from Berlin to Vladivostok, a community that shared ideas and worked on joint projects.

But they were also facing West. Specializations in the capitalist world were almost impossible to get in some sectors but were—merit and political background permitting—available in the electronics area. Some, of course, were connected to STI activities, but this doesn't diminish the fact this connection also allowed specialists to become part of Western scientific collectives, forging connections across the Iron Curtain. Some thus acquired impressive resumes due to this travel. Naplatanov, as a director of a key institute with impeccable credentials, specialized in cybernetics at Stanford, and both California and Illinois Institutes of Technology in 1971–1972, even giving guest lectures at these institutions.[7] Afterwards, his bionic and theoretical output were at least partially influenced by his time among the leading departments in the field, helping turn his nine-volume textbook into an up-to-date manual for those Bulgarians who didn't have his travel opportunities.

Those who trained entirely within Bulgaria were not truly behind an intellectual Iron Curtain. It is worth repeating the obvious: Socialist professionals often crossed the barrier to conduct business or talks abroad. The international nature of science, as chapter 3 highlighted, was one of the ways that STI expected technology to be transferred.[8] Bulgarian computer specialists took numerous, in-depth trips abroad for conferences or specializations, as the regime needed them to be at the cutting edge so as to keep its Comecon advantage. The CSTP used bilateral cooperation agreements with the West to scout out what profiles they could best help with. So the French hosted Bulgarian programmers for six months per year in certain areas, with two-month courses in statistical and demographic programming centers.[9] The Ministry of Electronics received the largest share of specializations abroad and travel permits in the late 1970s—in 1977, alone 263 quotas for specializations were given, as were 155 for participation in international seminars, more than were received by the larger Machinebuilding Ministry.[10] Participation in UN programs was also pursued fully, such as in CNC machines in metallurgy.[11] In the late 1960s, Bulgaria had received aid through the UN Development Program that contributed $600,000 for computer equipment; 10 years later, it was a net contributor.[12]

The ITCR's archives are a testimony to the Western window. While in its first year—1964—it sent people out only to the East, its growth in the 1970s came alongside increasing Western contacts. In 1973, an engineer visited the Netherlands to study their PDP-9 and PDP-15 usage in Amsterdam and Utrecht hospitals. He noted how heart activity and maternity wards were automated, reducing clinical errors. He recommended similar developments for the PDP analog—IZOT 0310—which became the basis for the institute's medical electronization program, a huge success used in ASUs such as Botevgrad's.[13]

Naplatanov's bionic interests also sent people abroad, such as the 1976 five-month stint of Dimitŭr Mutafov in French biomedical labs. He studied the latest in neurology, physiology, EEG analyses, and visual cortex mapping—key for the ITCR's bionic program. He noted what types of oscillographs and computer software were used. Mutafov spent time in labs that worked on scanning and algorithms for digitizing book texts. Importantly, he forged friendships with a French professor over a shared interest in neural modeling, resulting in the Westerner passing on all information he had learned during a specialization in a laboratory at Canterbury University, UK, complete with electronics elements developed there. The professor also gave Mutafov access to the latest French, Japanese, Dutch, and German electron microscopes.[14] Such trips resulted not just in knowledge transfer, but also in long-term correspondence and debate across ideological lines, based on shared interests. Often these encounters were another channel for technology transfer that eluded STI, but on a personal level, they plugged Bulgarian scientists into the international knowledge economy. This information was shared within the domestic scientific community. Pencho Venkov, the man who had visited the Dutch labs, was acquainted with Mutafov's contacts and data, allowing Venkov to fully participate in the Biosigma'78 colloquium on biomedical information analysis in France. There he participated in US-British discussions on 3D modeling of microbiological structures, presented on ITCR's telemetry development, and he created further links within French science.[15] Such seemingly mundane academic experiences were important ways for internationalizing Bulgarian science.

The other topic learned abroad was business organization and management—scientific institutes were just as interested in how Western firms set up their offices and computer centers as in their science. In 1977, ITCR engineer Desho Mladenov visited a Danish company—RC—that serviced a Ford plant, coming away with organizational charts and flow diagrams and the company's future expansion plans. He noted their network automation of deliveries for the car repair shops, and he brought back catalogs and general know-how applicable to Bulgarian firms.[16] Another four specialists attended classes in the company's Copenhagen training center, evaluating their new software and hardware knowledge as top-notch.[17] Socialist engineers were just as interested in how their capitalist counterparts envisioned automation and practiced it as they were in pure research—and STI also copied Western business plans. Operational management was just as international a language as computing, and it was related to it.

IZOT and CICT were also globe trotters. Their archives are littered with reports that crisscross the world, both "ours" and "enemy." This included the twin hearts of the beast—the US and UK—where CICT scientists specialized in graphics and displays even as Reagan declaimed on the "evil empire" (the two events coincided precisely, in March 1983).[18] Others explored cooperation with Xerox in France or signed purchase agreements in London.[19] That same year, as STI was finding it ever harder to battle tighter CoCom controls, a CICT physicist—Valeria Gancheva—started a year-long specialization at Virginia Tech on the highly sensitive topic of magnetic memories based on new principles.[20] Others studied microprocessor programming in Dutch firms, contracts were signed in West Berlin, and joint networks were being planned with the Greeks.[21] NATO countries welcomed Bulgarian specialists in a sensitive industry obviously trying to catch up with the West well before Gorbachev's watershed efforts. CICT scientists were freely visiting IBM offices in Belgium to learn about the "office of the future," discussing fifth-generation computers with Western colleagues or learning about computer-aided design applicable at least in part to defense in Norway—all while the Second Cold War was ramping up.[22]

The banality is the point—the run-of-the-mill life of professionals was not impacted by superpower confrontations, allowing for scientific and industrial links to exist and even flourish. Bulgarian professionals were learning from their Western colleagues and making valuable contacts even as the war in Afghanistan raged. Many political avenues might have narrowed in the early 1980s, but as the country's industry grew increasingly sophisticated, IZOT and other organizations exploited all cooperation agreements to the fullest. What was being learned in the West was being applied to the socialist world too—not just billboards were going up, but also IZOT was acting more like a Western business. During the 1980 Moscow Olympics, it supplied gratis 12 IZOT 132D printers requested by the organizing committee only in exchange for being listed as an official sponsor and receiving free advertising space in official brochures.[23]

The internationalization of science was boosted under Popov's CSTP reign. His own background understandably made him a proponent of multinational initiatives, pushing the small state to be part of larger initiatives. He exploited the Soviet-American scientific rapprochement of the 1960s to engineer closer contacts with the US scientific world, at a time when Bulgaria was still one of Washington's most distrusted nations. He engineered ties with US institutes that used computer modeling in governance in 1971, paving the way for Bulgarian participation in a hallmark institution of transnational Cold War science: the International Institute for Applied Systems Analysis (IIASA), founded in Vienna in 1972.[24] As Rindzevičiūtė has wonderfully shown, the IIASA, a place where complex international questions in global governance were to be discussed and solved, was a surprisingly open place where both Western modernization and Eastern scientific-technical revolution theories met, a convergence of ideas but also of elite's responses to the challenges of the age.[25] Bulgarian science was thus part of this novel opportunity to approach policy through cybernetics and modeling from the very start. Rindzevičiūtė shows, that questions about demographics or pollution or even nuclear winter were paramount and were approached through models that Bulgarian scientists were extremely interested in—Popov managed to get these scientists in on the ground floor of this exciting collaboration.

Nacho Papazov's CSTP negotiated even lower fees for Bulgaria and other small states, and he pushed for more participation.[26] Economist Evgeny Mateev was one of the first to take advantage, finding many commonalities between Western colleagues' and his work on algorithms for automated resource distribution among production units in the economy.[27] By the early 1980s, Bulgaria was among the staunchest defenders of IIASA, at a time when the US reduced its membership dues, and lobbied for its expansion, such as an admittance of Mexico (another prong of Bulgaria's cultural diplomacy there). The CSTP deemed the IIASA as extremely important for its own projects in agriculture, energy sectors, risk theory, decisionmaking, and economic structural changes.[28] The country hosted numerous seminars—on demographics, children in the computing world, the human factor in innovation—ensuring that Vienna came to Bulgarian scientists who couldn't land coveted placements there.[29]

Another window on the world existed through the increasing coupling of Bulgarian science to global databases. The Central Institute for Scientific and Technical Information was the node through which to connect to national and international databanks. A key link was to the Moscow counterpart, the All-Union Institute for Scientific and Technical Information (VINITI), allowing Bulgarians access to the huge Soviet databases and its international subscriptions. By the late 1970s this link included video terminal links between the two institutions and regular mailings of magnetic discs. The national libraries in Sofia and Moscow were also linked, enabling thousands of requests to be fulfilled each year, a massive boon for the smaller state.[30] European and international scientific databases were accessed through Athens and Vienna, literally plugging the community into a global information exchange.[31] In 1983, the expansion of such systems of national scientific information was made a priority by the BCP, including widening the Vienna channel to offset the huge costs of the increasing academic avalanche—the country could afford to purchase just 15 percent of journals and 1 percent of monographs every year.[32] Wider access to databases that contained abstracts and summaries were desired so as to allow Bulgarian scientists to select the most relevant ones for purchase. Stoi͡an Markov pointed out that the country had terminal access even to the Library of Congress, however, so the shortcomings were mostly domestic and organizational, "because we have not yet created the network for data transfer in such a way that every user—in Plovdiv, Smolyan, Pazardzik—can receive this info through the transfer network that services the Bulgarian information system."[33] The international links were there, but information was still centered on Sofia domestically.

CSTP worked hard on developing algorithms and search criteria that would improve user access and navigation through the data deluge. Comecon, UNESCO, Soviet, and other databases were made easier to access, as were systems of scientific and dissertation abstracts or lists of foreign companies in particular sectors, each with up to hundreds of thousands of entries. Often Bulgaria had access to important caches, but until the early 1980s, it had been hard to oversee *what* was actually available domestically.[34] Becoming better organized and itemized in that decade, the system allowed scientists to access a wide-ranging treasure trove. What had started as a modest library of self-translated Western journals in the Voroshilov factory in 1950 had now transformed into an ocean of knowledge with millions of items, available to researchers through libraries and electronic terminals.

Interviewing members of the computer community reveals no ignorance of the world. Bulgarian researchers were fully aware of technological developments elsewhere, often through STI. The crisscrossing of international borders for specializations or conferences made them participants in transnational professional networks, and not just academic ones: they gained experience in how Western firms operated, too. Friendships forged in such circumstances lasted long after—Petŭr Petrov recalls visits by his German and Austrian colleagues long after the 1960s, invited for talks but also drinks: "that's when I developed a taste for decaf coffee."[35] Angel Angelov's wide-ranging career, including Japan, is testified

Development of robot is not only a mere development of mechanical unit called robot but also the development of human knowledge and abilities. It needs ceaseless efforts an research and endless co-operations of related persons.

In honor of the co-operation on robotics between Bulgaria and Japan,

H. Makino

14 June, 1984

H. Makino

Professor, Yamanashi University

7.1 Message to Angelov from a Japanese colleague. (*Source*: Petŭr Petrov personal collection.)

to in his 1980s jubilee book, where messages reveal the overlap between professional and friendly relations.

Sometimes the specialists had to defend national interests in the face of competition, especially within Comecon specializations: "you bring your suggestions twice a year to the working groups, in autumn and spring, and each nation says what its own interest is . . . then there are strict international tests, everyone makes notes, and only then can you produce—they too are aware of global technical developments." They recall many debates: "the Hugarians had their own problems and were trying to pull away, the Czechs, us Bulgarians . . . the Poles were the favoured child, they had this strong lobby in the West too and had prototypes from Control Data . . . but it was smaller than ours [capacity-wise]."[36]

Mostly, however, these meetings were cordial, and the excitement was palpable when talk turned to Western contacts. Stoi͡an Markov, the penultimate head of the CSTP, told me that apart from the USSR, "I will tell you what the five most important countries [were] for our technological development: West Germany, Austria, Japan, France, Norway."[37] They were the usual destinations for specializations in computing and provided licenses and training. There Bulgarian experts became international ones. Nikolaĭ Stanulov, for example, was one of two socialist cyberneticists published in the last volume edited by the discipline's father.[38] But the framework of these contacts for scientists was created by another set of people—party functionaries who were managers rather than intellectuals themselves. This technocratic elite also learned things in its dealings abroad and increasingly constituted itself as a vibrant, new group at odds with the older generation.

THE SOCIALIST BUSINESS CLASS

Only so much could be done with Soviet help or one-off deals with Japan if the party wanted to keep its high-profit industry ticking. Thus, ever pragmatic, the BCP undertook financial law reforms that eased contacts with Western firms, allowing for increased technology transfer, as well as codifying financial and industrial cooperation with foreign entities. The first reform, Directive 1196 in June 1974, relaxed rules on the kinds of contracts that enterprises could sign with nonsocialist companies. But the real framework came in March 1980 with Directive 535 on cooperation between Bulgarian and foreign judicial subjects, opening the door to wider transfer across borders. The 1981 Code of Economic Mechanisms supplemented it, attempting to square the circle of combining centralized control and economic self-regulation. These reforms were the legal face of the search for intensive growth. They were punctures in the Ministry of Foreign Trade's control over policy, giving more power to DSOs and enterprises. These could now undertake foreign trade at their own risk, with financial control increasingly passed to the sectorial ministries, which would be responsible for new debts. De facto, the economic organizations became judicially independent capital associations and islands of "bourgeois" law in the central economy: legal entities with their own capital and director councils. The tension, of course, remained, as the state was legally the owner of all means of production. As Ivo Khristov points out, these were the first mundane steps toward dismantling the economic system, driven by a state reflex in search of Western technological resources. It resulted in the first enclaves in the planned economy, parallel law codes, and the falling of masks—trade carried out by producers with legal protection for foreign partners to export the profit abroad or get tax breaks if they reinvested in Bulgaria.[39]

The first wave of foreign trade firms had passed in the 1960s with the end of the Texim empire (mentioned in chapter 4), when around 80 such firms popped up. During the 1970s, around 40 were created to increase industrial export. But the new laws created a huge wave by 1982, matched only by the last 1987–1989 surge of preustroistvo, Bulgaria's perestroika.[40] Khristov counts at least 450 firms abroad created during the whole socialist period, with at least $712 million in investment (not including wages, maintenance costs, etc).[41] As we've seen, they were often STI connected, and the waters are murky due to the 1990 archival culls. Khristov maintains that they failed to stimulate capitalist trade or expand markets, often bringing in a loss. He also critiques them for not transferring the latest technology, or for becoming conduits for market mechanisms—unfair critiques, for both plenty was transferred, and market revitalization was never their goal![42] Khristov is right in pointing out that they became conduits for state losses, as their profits were not transfered back into the country—a 1991 document from a commission to look into this cites at least $1 billion locked abroad, with only $115 million recovered.[43] This capital transfer, however, went hand in hand with the transfer of expertise—contrary to his conclusion, while the firms might not have been conduits for market ideas, the people involved often were shaped by these experiences into unorthodox socialists, to say the least.

The combination of legal reform and this policy of opening foreign firms was fertile ground for the emergence of new technocrats. Despite archival culls, traces remain and are often openly talked about in interviews. Immediately after the 1974 directives, the Electronics Ministry deepened contacts with Fujitsu, signing agreements for the transfer of MOS integrated circuits, leading to a qualitative jump in IZOT's element base. The firm was granted the status of most favored supplier, a deal worth 40 million levs.[44] It was also the year the Politburo discussed the possibilities that the world oil crisis of 1973 entailed—Western firms would look for new markets, and maybe even whole enterprises could be moved to Bulgaria! Smaller firms were to be particularly targeted due to their vulnerability, needing investment and markets—in exchange for their latest technology.[45]

Close cooperation with the Italian giant Olivetti was mooted in the later 1970s. It sold telephone exchanges throughout the Eastern Bloc, and the Bulgarians negotiated to replace the PDP-11s in them with their IZOT 0310 machines.[46] These were the first steps in wider cooperation that the Italians wanted to regain positions in the Bulgarian electronic typewriter market. Olivetti wanted to sell a license to ZOT-Silistra for the production of 18,000 machines worth $5.4 million for Western markets.[47] This deal was finalized by 1980, with Olivetti getting preferential treatment in Bulgaria in return for placing socialist goods in the Italian market. The deal offered help on the pressing problem of the "future office," with Olivetti specialists visiting Plovdiv factories and noting the great organization but technology that was older than the Italian standard.[48] Nevertheless, IZOT gained not just valuable markets but also information about the future of office automation, while specialists were impressed by Italian preparedness and criticized their organization for lacking concrete talking points for negotiations.[49]

Such learning experiences intensified after the 1980–1981 reform and even involved American giants. Control Data Corporation (CDC), through its Austrian auxiliary CTI, approached IZOT to see whether the Bulgarians could provide competitively priced parts.[50] Tours of the ZIT factory showed representatives 29 and 100 MB discs and the IBM 360 and 370 clones: the ES-1020 and ES-1035, respectively. The Austrians were interested in disc mechanical parts and circuit boards, raising the possibility of huge, long-term deals.[51] By now, the Bulgarians were wily enough to recognize that it was unlikely that their quality would satisfy CDC, so they planned to concentrate on producing just a couple of elements of good quality, to secure a smaller but profitable deal rather than letting the whole venture go to waste; alternatively, they would demand CDC licenses for whole devices.[52] The Bulgarians also passed on information about the elements produced by the crown jewel, Stara Zagora, hoping to convince CDC to transfer its soon-to-be-relocated Welsh disc package factory there—a factory that worked with elements made in Omaha.[53]

In 1980, Izotimpex met directly with the American trade director, where they proposed research cooperation as well as joint sales rather than purely Bulgarian deliveries of US items, all the while citing Directive 535.[54] CDC was uninterested in joint enterprises, but in Committee for the Unified System of Social Information (KESSI), which pushed them toward wider cooperation, especially after deeming the preliminary batch of disc parts to be satisfactory.[55] Of course, citing embargoes, CDC declined to commit to direct technology transfers. Internally, the IZOT stance was clear: "the aim is to shorten the times for development tasks of our analogous devices," referring to discs over 500 MB capacity, and certain CDC analogs of IBM products (including the star prize, a 2×635 MB disc package, which would catapult the Bulgarians to world levels).[56] Conversely, the Americans were deeply interested in the Bulgarian market, sending questionnaires on wage laws, tax policy, currency exchange rates, and joint enterprise directives.[57] By October an agreement was reached, with ZMD-Pazardjik (a magnetic disc factory) securing the production of 18 element types in runs up to 600,000 units. Trips were made to Wales to check the expected CDC qualities. The deal was worth 27 million levs, with potential for more. IZOT congratulated itself on "entering the system of sub-deliveries on a long-term basis with a respected Western firm," updating its technology, and cooperating with American and British engineers.[58] CDC expressed regards for the professionalism of Bulgarian organizations, KESSI in particular, touting their own systems as ES-compatible: a curious reversal. Importantly, they promised systems in configurations that could "fly under the embargo radar." The paper trail runs cold here, but not the cooperation: The factory implemented the technology, and contracts worth $13.3 million in parts were exported to CDC by 1983. The factory entered "the network of suppliers for CDC with a growing capitalist currency effect."[59]

This deal has been described at length to show the depth of cooperation possible after the reforms of 1974–1981, even with the enemy superpower. Bulgarian representatives increasingly defended their interests with success and maintained a professional demeanor. This increased confidence led to IZOT seeking more business abroad, beyond importing technology. A notorious example was "Busycom," an attempt to break into the American floppy drive market. Noting the huge US market—expected to top $78 billion in 1982—Izotimpex proposed the formation of a joint enterprise to get a cut of the profits. Contact was made with a British firm—Busycom, Inc.—to set up a joint business in San Francisco.[60] Its board of directors would include a representative from each country and would place disc drives, floppies, and microprocessor software in California. The Bulgarian goods would get a British branding, so as to be taxed at 4.5 percent rather than 80 percent. The British connection would also overcome "psychological barriers" associated with socialist goods—and, needless to say, circumvent CoCom.[61] The joint business planned to eventually incorporate two local firms—an LA information management company and a San Diego microcomputer services one—together with a Hong Kong enterprise, to create a multinational holding. This would expand the networks throughout California and into East Asia, exploiting favorable customs regimes through a British offshore holding. Half of the initial, modest, capital of $110,000 was provided by an IZOT willing to test the waters of how Bulgarian computing would fare in the superpower market.[62] The finances were approved, as the experts noted that the US floppy market constituted 70 percent of the world one. The company started with the aim of selling 50,000 units, hoping to raise this to 625,000 by the mid-1980s.[63]

The enterprise was a fiasco. By 1984, when there were ambitious dreams of 1 million drives being sold, the company had not yet been set up.[64] But the people involved were already in the US, principally Bisser Dimitrov, who enjoyed a meteoric personal rise despite the business failure. He had been in IZOT's trading department since 1970.[65] He had been the IBM representative to Bulgaria in the late 1970s, in the "technical services" wing.[66] He left the firm in 1981, using his skills and contacts in the country's attempts to export electronics in the West, such as through trips to Greece to place SM-4 machines and floppy drives.[67] His knack for the business made him Izotimpex's man in Busycom—he was already in California and registering the company a full 10 days before the discussion even came up in the Council of Ministers.[68] The registration was shambolic—the wrong documents were submitted, and US law sanctioned it for reselling embargoed goods to China. It ran up $230,000 in debt while Dimitrov drew a $54,000 salary, with the British and American investors complaining of general poor management. The company managed to place drives and parts worth $200,000 throughout its existence, never being the breakthrough envisioned. Yet, on a personal level, it allowed some experts to link themselves to the West, being de facto American company directors despite their socialist—and indeed, secret service—credentials. Dimitrov's career only improved—in 1986, his reward for failure was to be made the director of the Plovdiv branch of IZOT.[69]

Dimitrov's own narrative is that of a go-getter ready to break with staid communist behavior, a sort of electronics Iggy Pop. He talks of brushing off fears of hiring "enemies," such as American specialists:

Enemies or no enemies, you want the computers or you don't? . . . I have a product instantly. Maybe not the best, but a product which is by far superior [to] what is existing in Europe or the Eastern Bloc countries. So that's what's been the major difference in approach to what they [Izotimpex's other representatives] had before. An approach which I implemented.[70]

He prides himself on other deals, such as those with Japan. Failure does not enter the conversation, as he is a trendsetter who does things the "modern" way—which means the capitalist way. Innovative and entrepreneurial, he succeeds where Izotimpex fails. In Dimitrov's language, we see some of the kernels of the self-constitution of a transnational elite in the highest echelons of the industry.

Experiencing the nonsocialist world was the crux of how many apparatchiks diverged from the top of the party in at least some ideas. Japan was a crucial country for this divergence, capturing imaginations and setting key figures on different intellectual paths. This is most striking in Ogni͡an Doĭnov and Nacho Papazov. Doĭnov was the trade representative in Japan in the late 1960s when Papazov himself was ambassador, and as we have seen, both took leading posts once in Bulgaria—Doĭnov as Politburo member and economic tsar, Papazov as the longest-serving head of the CSTP. Both were replacements of Ivan Popov in different capacities, and Doĭnov was probably the next most important figure in industrial development. He was closer to the image of the transnational technocrat than his predecessor—and beloved of protégés such as Dimitrov, who saw him as "the number one guy in the high-tech business in whole [sic] Eastern Europe."[71]

As we have already seen in chapter 2, he was a deputy trade representative in Japan when he impressed Zhivkov on his visit there for EXPO'70. Those five Japanese years were a formative experience for him and his openness to world business developments. Memoirs should always be treated carefully as sources, especially as Doĭnov's came out when he was in exile and wanted by the post-socialist government. Yet his stances in Politburo meetings do testify to a more Western-influenced economic worldview, and his own words are useful—like Dimitrov's—in tracing how the nonsocialist world impacted the self-conceptualization of such agents. They dedicate many pages to his struggles with the poor quality of Bulgarian exports, teaching him the value of negotiation and compromise in order to offset shortcomings, turning into a lifelong obsession with

7.2 Ogni͡an Doĭnov (left) and Nacho Papazov (right, cutting the ribbon), as they appeared in the Indian press, symbols of the mobile socialist elite. (*Source*: Ministry of Foreign Affairs Archive, Sofia.)

raising the quality of domestic production.[72] The Japanese economic miracle impressed him hugely, and in those years, he forged links with major industrialists, such as the heads of Nippon Steel and Fujitsu-Fanuc.[73] It was a veritable school for cutting-edge management techniques:

there I could read many of the contemporary publications . . . for the first time I got to know the modern . . . approaches to solving economic problems—such as the critical path method, Program Evaluation and Review Technique, governance structures such as the matrix or group-task one, logistical tasks, optimisation models. I was especially grabbed by operational studies, by econometrics. The works of John Keynes, Milton Friedman were of interest to me, as was the macroeconomic textbook of Professor Samuelson. I read voraciously the books of economic futurists who predicted the global changes.[74]

This was an unthinkable knowledge base for 1960s Bulgaria, where such books were locked away in specialist party libraries, inaccessible to a mere trade representative. Japan allowed Doĭnov to create his alternative library:

the most valuable thing in my luggage as I returned . . . were books in English—over two hundred volumes on company management, optimisation models, systems analysis, micro and macroeconomic tasks and solutions. Many of them I used during my stay in Japan, some I brought back as a promise for my future. I gave some away to be read or translated, others I lost eventually during my emigration period.[75]

Japan was a window to another world. Doĭnov's future career and championing of robotics, automation, and rational management are more understandable in the light of these five formative years. He was a complete novelty for the Politburo, as a self-styled Western technocrat with a different language—not that of Popov, a man of a different time, and more orthodox than the young upstart. Zhivkov's new obsession with Japan as a model for a new leap also helped turn the spotlight onto him. Doĭnov recognizes this as an important part of his rise: "in these discussions the biggest interest was piqued by opinions . . . on why Japan managed to achieve such economic and scientific achievements in such a short amount of time."[76] In his position as economic advisor in the early 1970s, Doĭnov's outlook was often used as the "opposing" view in discussion, the devil's advocate against more orthodox economic plans. "His stay in Japan had a positive impact on him . . . Doĭnov transferred the Japanese experience of creating industries on the basis of high technology," states foreign correspondent Moncho Behar for the party's *Rabotnichesko Delo.*[77] As Doĭnov climbed the ranks, he championed production concentrations—a continuation of Popov, but in his case more influenced by Japanese agglomerating practices in economies of scale.[78]

By the 1980s, Doĭnov was a towering figure and a defender of the industrialist strata of the regime against the older elements. "Probably the most neglected leaders in socialist society were economic leaders"—and he saw himself as the leader of the planners and directors who had to make the economy tick.[79] This was a position he defended forcefully in Politburo meetings, as the archives corroborate: "we are allowing some cultural . . . workers to get 30–40 thousand levs for some unique creations, but there is no case of an economic actor—head engineer—who has solved problems in a talented way, to get more than three thousand in annual income."[80] Engineers and business leaders were "honest" and neglected, seen as a second-rate intelligentsia to that of the writers and humanists championed by Zhivkova—a "service" class to the "real" intellectuals, and with the remuneration to boot.[81] Disunited, they lacked the unions that creative workers possessed; at the same time, they were expected to deliver profit year after year—something impossible, as all economies go through boom and bust! His dream to unify their power got Zhivkov's approval and in April 1980, BISA was created—the Bulgarian Industrial Economic Association—with Doĭnov as acting head until 1984, and honorary head thereafter. BISA was explicitly based on the Japanese organizations of the Keidanren type (the Japanese federation of economic organizations, founded in 1946 to represent big business interests), as well as French and Swedish models. It lobbied for simpler mechanisms of investment, decentralization in financing, and it advised on trade with the West. BISA also trained enterprise directors in Western-style business management, negotiations, and basic market economics. Fans of Doĭnov contrast his language to functionaries such as Lukanov, plus his efforts to develop Bulgarian human capital.[82] In 1986, BISA secured a Japanese credit of $400 million, which included $3 million for a training center to raise production quality—the culmination of Doĭnov's dream since his Tokyo days.[83] Of course, this didn't mean that things necessarily improved on local levels everywhere—in 1985, workers in the Varna Shipbuilding Computer Centre complained that programmers were not being paid enough, and that their rise in productivity was still not being met with a corresponding wage rise.[84] Yet BISA was an ultimate expression of Doĭnov's stature as economic tsar and technocrat, a propagator of capitalist practices, and a patron of his colleagues, elevating them to a position more in line with their importance to national life.

Papazov had also become smitten by Japan. In 1989, after his tenure as both ambassador there and at the head of national science, he penned a memoir, *Japan: From the Samurai Sword to Artificial Intelligence*.[85] A wide-ranging book that was also a sort of primer for postwar history, and a perhaps Orientalist observation of Japanese traits, it shows a deep respect for workers' discipline and the wisdom of their leaders: the preconditions, in Papazov's eyes, for the miracle. He praised them for grasping modern managerial science without "Americanization," instead applying it to their society in novel ways—the only way any state could master modernity while protecting national interests.[86] He used cybernetic terms in explaining how Japan concentrated power in a few "steersmen," such as the ministry of trade and industry: the famed Ministry of International

Trade and Industry (MITI). He highlights how Japan understood the need for state intervention ever since the Meiji restoration, allowing it once again to take the best economic practices and wed them to tradition in the service of the nation rather than of commercialism.[87] The praise for a national-minded, state-led intervention in the economy can be said to have served him well as CSTP head earlier—he was the Bulgarian steersman in science and industry, too. His conclusion explicitly warns that if due to political conservativism, a country ignores that which is proven to work elsewhere, it becomes a slave to dogma. Japan's miracle had to be Bulgaria's model.[88] Tokyo was a shining star for influential Bulgarian leaders, not least Zhivkov himself. Much like Doĭnov, Papazov utilized contacts made as ambassador to benefit Bulgarian technology throughout the 1970s and 1980s, especially through a close connection to Fujitsu. It was indeed this network of contacts and knowledge of the Japanese mechanisms that most likely made him the choice to succeed Popov to the CSTP's directorship in 1971.

Other figures, too, were international in the models they envisioned. Stoi͡an Markov, the head of a reformed CSTP (renamed the State Committee for Research and Technology, or DKIT, in 1987)—in 1986–1987, and Politburo candidate-member in 1986–1988, was a man who owed much to Doĭnov's rise and became his right-hand man. Markov's scientific skills, honed in the USSR, made him one of the most educated members in the BCP's Everest—some involved with the IZOT-1014 matrix computer, Bulgaria's crown jewel in computing power, see him as the project's father.[89] His memories and indeed office are filled with mementos of meetings with major figures throughout the world, including Deng Xiaoping.[90] He is explicit: The electronics business was a ticket to the world. Similarly, Plamen Vachkov, director of the Pravetz factories, speaks of an industry that gave him a seemingly impossible head start: "In the early 1980s, I was in the Silicon Valley, meeting people like Bill Gates and Steve Jobs . . . I learned the word business there, while no one here was using it."[91] A 1986 article testifies to this new type of manager that fascinated many. The difference is obvious from the cover of the magazine (figure 7.3): Vachkov behind his desk that testifies to diligent output through many papers, besides a PC that is the mark of the scientific revolution, in a smart suit and significantly younger than the stereotypical director. In an interview, he attacks bureaucracy and states that "if we ignore 70–80% of documents [directives], there would be no consequence to the collective or production."[92] Management of personnel is more important than dry technocratic language, as real people are not an inert mass. He laments that he spends too much time in meetings, up to 30 percent, which harms his work performance, and he prefers delegation to subordinates in whom he has absolute trust. Throughout, he presents a business-minded figure rather than the orthodox director who shied away from such direct attacks on the bureaucracy. For Vachkov, results are paramount. Today he looks back nostalgically at what could have been and points to another country that served as his model back then—Singapore: "If you want a blueprint to developing a small country extremely successfully, read Lee Kuan Yew's *From the Third World to the First:* it's a 400-page textbook of how to make a state," he muses.[93]

The way that Vachkov's generation acted and even *looked* set them apart from the older functionaries often in pursuit of nineteenth-century pleasures, such as spas and hunting.[94] Like their travel itineraries, their tastes were modern. Moscow wasn't necessarily the universal center for them, and their knowledge of the West or East Asia often introduced a new style of desire. A key figure was Andrey Lukanov, the foreign trade doyen and Moscow's most trusted official in Bulgaria. Despite that, he and others like him were the conduit of a new gourmet culture in 1970s Sofia, with new exclusive restaurants opened, and a Westernization of the tastes of the new mandarins.[95]

Here is how an Australian journalist described Lukanov in a 1985 article for an Indian magazine:

> a typical member of the 'under 45 generation,' one of the young executives running the country's vital economic sectors. . . . We conversed while on a motor boat trip in the Black Sea. On the shore we looked at multi-storeyed hotels nesting among the green oak groves of the Golden Sands health resort, the sea sparkled, was iridescent with patches of sunlight.[96]

This could have been a Western CEO on the French Riviera. John Caswell, the American embassy's political economic section chief in the early 1980s, was impressed by both Doĭnov and Lukanov. For him they were examples of the "younger technocratic people particularly in the economic ministries [there] to try to make the economy a little bit more dynamic or responsive or whatever." Both were "of a younger generation, they were more, particularly Lukanov, cosmopolitan and aware of the outside world beyond the Soviet orbit." He obviously had more dealings with Lukanov, whom he called "the point guy . . . for dealing with Westerners," but was impressed by Doïnov, too: "he also seemed to be sharper, more intelligent than the average in terms of making the system work. So the communist leaders weren't all knuckle draggers."[97]

Electronics and its associated trade had thus helped create a managerial class connected to Western ideas and tastes, drawing on models from Tokyo or Singapore as much as from Moscow. Some started their careers in the industry's scientific institutes (Vachkov and Markov); others owed their rise to technological trade acumen or pretence (Doïnov and Dimitrov); and finally, there existed modern-minded apparatchiks such as Papazov who could act as channels and patrons. Unlike the lower-ranked engineers, these were men among the highest echelons of power and could utilize the language that arose in journals in the 1980s (see chapter 6) as well as their own views on economic organization to advance new visions for the regime at its end.

REFORM AND TRANSFORMATION

"This question is about the most valuable thing that our society has—the person as a creator of all material and spiritual goods in Bulgaria," said Zhivkov in a 1984 Central Committee discussion of an ambitious program aimed at raising the intellectual and creative abilities of Bulgarians.[98] From the mid-1980s, such discussions of what computers meant for society were becoming increasingly frequent at the party's heights, allowing the electronics managerial class to utilize both experience and discussions in the field to criticize the BCP through the prism of its Scientific Technical Revolution (STR) policy and its political and economic consequences.

It was during such debates that the Politburo once again took stock of its ambitious goals of scientific governance. Zhivkov's November 1984 report emphasized to the Central Committee that it was technology that determined who would win today's economic competition. Applying the huge scientific potential of the Eastern Bloc to contemporary problems was the only way to defeat capitalism, but that potential was still being

7.3 Plamen Vachkov as the modern manager. (*Source*: socbg.com.)

squandered.[99] Doĭnov was more optimistic, stating that Bulgaria was producing analogs of the most widespread computers, and could almost cover its own automation needs.[100] Yet the plenum made it clear—cybernitization had not yet happened.

An October 1984 discussion cast a shadow over the plenum. At that meeting, the leadership discussed the unwieldy-sounding "Complex Investigation of the Person and More Specifically His Brain with a View towards Raising the Intellectual Abilities of the Personality and the Development of its Creative Powers."[101] The plenum noted its debt to Zhivkova's vision of aesthetic education, noting that the issue was more pressing now, after her death, as artificial intelligence entered society, and humans' intellectual capabilities were at a new threshold.[102] Bulgaria faced a dual problem—insufficient technology to free workers from menial labor, and lack of education, meaning engineers had to simplify technology for users rather than utilizing devices' maximum capacities.[103] As quoted, Zhivkov highlighted "grey matter" as the resource, the creativity that was not yet fostered by the devices but was needed for the new economy.[104] The party called for 3 million levs of research into human-machine interfaces, into brain-AI parallels, and pedagogy to expedite "the more beneficial manifestation of the personality."[105]

In March 1987, Zhivkov discussed the "synthesis between science, information and computers" with leading party cadres involved in technology. It was an airing of grievances. Blagovest Sendov highlighted the lack of software and Bulgarian language interfaces, which would result in a generation raised on English and Russian. He also lamented the focus on artistic over technical intelligentsia—the "Gyaurovs of science" (named after the famous opera singer) needed access to their own "La Scalas," such as MIT.[106] Markov told Zhivkov about the communications revolution, "one of the most important preconditions for the democratization of human experience."[107] Bulgaria had the modems, networks, and specialists but needed to allow access to them—export orientation hampered this: BAS often lacked computers produced just a few kilometers away, which were sent directly to Soviet institutes.[108] Ex-IZOT director Ivan Tenev highlighted the potential dilution of scientific potential as new developments were too many, and Bulgaria had to pick and choose the appropriate focuses.[109] Zhivkov picked up on this, railing against a Comecon "border closure," the veritable diminution of cooperation in the late 1980s. Tenev countered that only the internationalization of production would allow them to achieve the next jump: after all, "borrowing" had been key to Bulgarian success so far.[110] CICT's Vladimir Lazarov criticized the economic minister, Ovcharov, for a value-neutral approach to electronics. On its own, technology was useless, as "there is a barrier to the use of this technology among our society right now," an argument straight out of the contemporaneous psychological studies. Work collectives had to be networked through small machines to allow the lateral spread of expert knowledge rather than routing them through centralized control.[111] Lydmil Dakovski, the rector of the technical university, returned to the Sendov stance at the close of the talks: Software and mathematical languages could and should protect Bulgarian nationhood and language.[112]

The array of criticism impacted Zhivkov. He agreed that an opening to the West was needed, the only way to make the next leap. He contrasted a more closed USSR while "we opened Bulgaria to the world and Western Europe, without allowing for crises to emerge inside the country"—curious for a country that was a pariah after the inauguration of its Revival Process.[113] Electronics had also fused with most production and social spheres, bringing to a close the era of discrete economic sectors.[114] Society now needed more choice in order to utilize the self-governance that computing allowed everywhere. The deficit was time—capitalism was drawing ahead, so it had to be involved in Bulgarians' work if the country were ever to catch up.[115] The computer was now firmly a tool to criticize and open up society.

Preustroistvo—the Bulgarian perestroika—kicked in after these plenums, and these electronics failures were further instrumentalized both by specialists and party functionaries who championed the industry. This came to a head in late 1988 in Central Committee debates on preustroistvo in the spiritual sphere, and at the December plenum. Ivan Stoi͡anov, hero of socialist labor and committee member, attacked the regime for never managing to turn out enough automation specialists: "in 30 years we turned people tied to the land into miners and power engineers, but we haven't yet managed to turn everyone into a real specialist, professional in their job."[116] Lyobomir Iliev, one of the towering figures of computing, lectured on the universal science of informatics and its ushering in of a new era. But sadly, in Bulgaria, it operated in a handicapped manner, for human culture is made up of science and art, creative and executive levels, and most importantly—state and society. The lack of one of each pair was crippling, and scientific workers knew that preustroistvo's power lay in addressing the latter imbalance.[117] The salvo continued with CICT's director, Petŭr Stanchev, calling for free and open information exchange with socialist allies and capitalist enemies alike. Scientists had to be given a bigger role, too, as they were the real movers of reform—all human history was the history of scientific achievement, and so science would now enter everything to create "a celebration of democracy and creativity."[118]

At the December plenum, Petŭr Mladenov, the man who would succeed Zhivkov in the palace coup of November 1989, took all these reasons and gave them further political clout. What the old regime had been quick to label "capitalist" or "bourgeois" was in fact the purview of all civilization, and this had made Bulgaria into a technological province of advanced countries. Without openness, reaching the desired Western level was impossible.[119] Plamen Vachkov, in turn, pointed out that further PC advancement was possible if the regime stepped back and gave the new socialist firms the leading role.[120]

This is not to overstate the importance of the technological elite—the coup of November 10, 1989, was led by foreign trade minister Lukanov. Ivan Chalakov convincingly argues that Lukanov's power was part of a prolonged struggle between the party and economic elites throughout socialism. At the end of the 1960s, the Texim empire of Georgi Naĭdenov was destroyed as a "capitalist" isle that drew Moscow's ire. Ivan Popov's figure, too, eventually threatened political elites due to his economic power, and he, too, was removed. Doĭnov was also removed in 1988 after a conflict with Lukanov—the solidification of the latter's power and best demonstration of the two elites: a "self-made" man (elevated by Zhivkov and Lukanov himself!), Western-oriented and with fresh ideas; or the old communist families trusted by Moscow, as was the case with Lukanov.[121] Doĭnov sought technological solutions to economic problems, while Soviet-centered Lukanov was master of trade and finances—these were the personifications of the industrialist and financial-political wings of the party. Doĭnov's fall a year before the regime's end was the third defeat of the socialist "business class" and a signal to allies. The economic elite described here had the skills and contacts to look to the West and identified strongly with the high-technology sector—Doĭnov's BISA was the best expression of their unified vision. The party nomenclatura, however, retained the most important levers of power—connections to Zhivkov and Moscow—and the party's huge financial reserves. This was the ultimate power that led the regime's last energy minister, Nikolaĭ Todoriev, to half-joke that if you wanted to know if you still had a job, you had to read the Monday edition of *Rabotnichesko Delo* to see what the old clique had decided during their weekend retreats.[122] Doĭnov himself felt this struggle with Lukanov personally, as he considered him a friend, and their families holidayed together. His memoirs paint the latter as a man who waited for Doĭnov to be away to criticize him and turn Zhivkov against him. He sees Lukanov as jealous, with an inferiority complex vis-à-vis Doĭnov himself, a Soviet stooge ready to use people and then discard them.[123] The bones he had to pick in his memoirs were obvious, written as they were in Vienna with a warrant on his head. Yet they do offer a glimpse into the personal experience of a real power struggle at the top of a party that historiography sometimes paints as monolithically gray—the traditional elite and the upstart one.

The struggle continued past 1989. What the economic elite couldn't do before, it could do at least in part after that year—gain political and other influence. It was the party elites that removed Zhivkov and transformed the BCP into the reformed, electable Bulgarian Socialist Party (BSP) in 1990, but the economic elites had connections outside the crumbling Comecon plus assets in the foreign trade firms, many connected to electronics. It is here that the historical record meshes with speculation—what some might call "journalism"—but the 1989–1991 liquidation of many firms, together with some personal trajectories, do strongly suggest the transfer of money into private hands. A 1991 investigation by the National Service for Constitutional Defence found that over $1 billion in profits in foreign firms and joint enterprises were missing.[124] Two of the most notorious cases were in the electronics sphere: the "Neva" and "Mont Blanc" projects, linked to the last heads of Bulgarian science and industry, Doĭnov and Markov themselves.

"Neva," started in 1987, envisioned the creation of a disc factory in Kostroma, USSR, to give the Soviets the capabilities that Bulgaria had monopolized. The Bulgarians insisted that the project was to be paid for in sorely needed dollars rather than in roubles. An STI front firm ("Insyst") was created, with three officers working as representatives.[125] Later that year, the Soviets signed an agreement with "Setron," based in Liechtenstein, for the import of know-how in Winchester-type drives. Setron itself signed a deal with Insyst to facilitate the transfer from the Stara Zagora factory, the bloc's most advanced such factory. Doĭnov talks of $200 million in Bulgarian investments, others of $250 million; Khristov calculates that if you add in Soviet investments, the project reached $600 million.[126] The labyrinthine financial operations involve Israeli and Panamanian companies, with numerous transfers, where at least $26 million went missing. The factory, obsolete by then, was completed in 1994, while the "Neva" money disappeared in numerous offshore companies. Bisser Dimitrov stated that "as a trading operation, Neva was spectacular; the technological effect—zero."[127]

"Mont Blanc" dates from the same year. Khristov, who has access to the 1990 "Case 4"—the investigation into the abuses of the socialist economy—calls it the "endgame" plan to extract money from the profitable disc industry. This was Neva's reverse—an expansion of exports to the West rather than to the USSR. The sum of $5 million was made available in 1988 to purchase a failing Northern Irish company, Data Magnetics Ltd. A key role was played by Atanas Atanasov, the head of the Stara Zagora factory and the new Disc Memory Devices (DZU) conglomerate that united disc production after 1987. Data Magnetics was in the floppy drive business and was a way to increase Bulgarian technological acumen but also to circumvent CoCom by placing Bulgarian discs abroad through a British brand: DZU would thus bring in scarce Western cash. By the end, around $11 million had been spent, to almost no effect. The bank investigation couldn't find where the money had gone, with no discs were ever placed abroad.[128]

The usually purged state security archives do contain a document with some messages from February–April 1990, the chaotic months of the first democratic elections. Sent through the Vienna residency, decoded unusually fast, they show urgency. Concerning "Mont Blanc," a Taiwanese firm seemed ready to cover the investment, and at least $6 million could be recovered immediately. The STI firm had to be liquidated immediately as "our experience up to now has shown that we can't maintain high-technology installations abroad . . . also, real sources of cash for paying for the activity are absent, even beyond the end of this year."[129] Concerning the bigger "Neva," a March message warned of an end to financing and "preparation for the destruction of the documents that can discredit our partners if the activity is revealed."[130] Finally, in April, a message talks of a meeting with partners in the project—probably the Hungarian Videoton that in the late 1990s would buy what remained of DZU:

> who insist on us to undertake steps in "cleaning up" and "destroying" the materials that would prove to "future governments [the involvement] of specific firms, people and forms of realisation." He was informed that around 600 people, representatives of the democratic forum and social liberal party, have undergone training in REDACTED West Germany and REDACTED USA and will become the backbone of the future Hungarian intelligence and counter-intelligence services. He feels that we underestimate the fact that a change has happened, which in the past we called a counter-revolution![131]

Where the money went, and to whom, is difficult to say. However, speculation is rife, and the likeliest candidates have to be those connected to both STI and the economic associations—as we have already seen, this was a symbiosis. It is precisely people such as Doĭnov who are figures that may have benefited, especially from "Neva." Doĭnov left Bulgaria in 1990 for Austria and was charged in absentia for "economic crimes," such as financial aid to developing countries, while doubts remain about more egregious crimes. Arrested in Vienna, he was never extradited, and he died in exile in 2000—but not before he worked with Robert Maxwell and various companies where he could utilize skills from his past life. He advised two notorious billionaires—Maxwell and Gregory Luchanskiy, the head of Nordex. Journalists cite his name in connection with "pumping" money out of the DZU conglomerate; the creation of Multigroup, one of the most powerful Bulgarian crime/business organizations of the 1990s; and a key figure in the first Bulgarian mobile operator. Whatever the truth, it is hard to doubt that Doĭnov's socialist-era skillset and his connections allowed him to live a luxurious lifestyle abroad.

Markov also left the country after a brief stint in the first elected government. He specialized in CERN in Geneva, but mostly lived in London, before coming back to Bulgaria in 2005. In 2009, the BSP government purchased an IBM Blue Gene supercomputer for 5.4 million levs, in a controversial deal but pushed through as a "research project" and utilized by a Markov-led team. By then he was working in BAS again—in the

Institute for Parallel Informational Processing, an institute that started its life in the 1980s under Sendov.[132] Markov had thus come full circle, back to where his career began.

The paths of others, such as Plamen Vachkov, shed light on the basic premise being made here: This was a social stratum whose members could make their fortunes much more easily than most Bulgarians after 1989. He recalls his facilitation of the outsourcing of a telecommunications firm from Ireland to Greece, benefiting from his contacts and knowledge of EU tax breaks. In the same breath as talking about the end of socialism, Vachkov recounts equipping Uzbekistan's school system with thousands of Pravetz PCs, ensuring profit for his enterprise in those difficult years, and one operating in the conditions of perestroika—a veritable window to try out the art of the deal in socialism. All this was neatly encapsulated at the end of our interview, as he remembered "I want to tell you that Pravetz created a corporation in Singapore which supplied us with components, and of which I was the sole owner as the general director of Microprocessor Systems Combine."[133]

In the 1990s, he was the general director in charge of electronics and communications in the notorious Multigroup conglomerate. He became head of the State Agency for Information Technologies and Communications in 2005, overseeing the country's IT market and ensuring fair competition.[134] His path through the murky post-1989 world of Bulgarian business and government is fairly open, and he is candid, without implicating himself in anything untoward. Whatever one may think of the allegations made against these figures, one thing is clear: These individuals had the abilities and connections to remain involved in business and politics well after the end of the regime, sometimes even increasing their power—especially when the reformed BCP assumed office. The Bulgarian computer thus straddles 1989, extending tentacles into the endless transition to democracy and the market. Another generation made this journey, too.

CYBORG AFTERLIVES

As the regime transformed, Bulgarian children became the vanguard of the new age: creative and entrepreneurial. Many were drawn into the Achilles heel of the industry—software—and banded together to solve its problems.

One of the first such enterprises, uniting university students, high schoolers, and teachers in Sofia, was "Avant-garde," praised in 1985's *Computer for You* as proving that "the intellectual revolution is in the hands of today's students," full of fantasy that was impossible to achieve for older scientific luminaries.[135] In that year alone, the enterprise produced 45 games and 10 programs, distributed through the software house "Eureka." It cooperated with Pravetz and IZOT, producing tools for the automated workplace, such as data processing and graphics editing software.[136] This was just the biggest of the software houses set up by young people who were drawn into software through the Technical and Scientific Creativity of Youth movement (TNTM). Another DKMS institution, its aim was to involve youth in science but also utilize their creative labor. It provided the material and educational base for invention, a counterpart in a way to the brigadier movement that made students into seasonal agricultural laborers—a way to plug gaps in the economy. Software creation was thus subsumed within TNTM and spread through the computer club network. The software boom encompassed provincial cities (such as Silistra, Haskovo, and Ruse, too), and by 1987, "Avant-garde" had grown into a national center to coordinate this nationwide activity.[137] Even in the official software houses, the young predominated—a software firm in Burgas, "Busoft," demonstrated its program on aviation administration through one of its programmers, a tenth grader.[138]

Many children were thus on board with the regime's dream. Yet computers were scarce, and there was never a guarantee of a job involving them once they had graduated. One of the manifestations of this frustration that combined with the right skills was the rise of the computer virus. In 1989, Veselin Bonchev ran the first detailed article on viruses in the magazine. He talked of viruses existing in Bulgaria, too, such as in a Sofia computer club, and that even Bulgarian ones existed—VT-88.[139] The magazine ran more exposés on viruses and offered free software to protect against them, but the explosion continued through the end of the regime. The first 1990 issue blamed users for the spread of viruses, admonishing users for pirating software (one of the few ways to get software in conditions of deficit): "the way you gladly sit down to a feast of someone else's bread."[140] By then, Bulgaria had its first celebrity electronics bandit—"Dark Avenger"—whose open letter the magazine published. He called the publication "Virus for You," belittling Bonchev as a supposed expert who didn't even know of their existence while the Avenger was creating them. He boasted of two viruses by mid-1989, both bearing Iron Maiden-themed names: "Eddie" and "Number of the Beast." Bonchev replied, calling on all to help "heal this ill with ambition brain."[141]

Bulgaria was already garnering a reputation as the world's "biggest creator and distributor of computer viruses," in the opinion of a German specialist.[142] "Eddie," "Yankee Doodle," and "Vaccine" all proudly bore the tag "Made in Bulgaria" and the creators' names, who saw them as points of pride. The magazine lamented that "the creation of viruses is far from a thing to be praised, and the halo of such bad fame, with its bag of viruses, will prevent us from easily entering the European home."[143] Too late—in December 1990, the *New York Times* ran a story stating that up to 90 of all 300 viruses written for IBM machines originated in Bulgaria. John McAfee stated that at least 10 percent of all calls to his company had to do with Bulgarian viruses, and 99 percent of those—with the Dark Avenger. The article suggests this was "a consequence of having developed a generation of young Bulgarians whose programming skills found few outlets." Bonchev, interviewed by the Americans, agreed: "These children quickly acquired software-writing skills, but had little or no chance to apply them constructively." A hacker points to simple revenge on a company as a reason.[144] By 1997, the big peak had passed, yet the magazine *Wired* delved into the story. David Bennahaum traveled to the country in search of the Dark Avenger. He describes a socialist digital culture born in the Eastern Bloc's most successful "computer country," where children did what they always do—play and explore. When *Computer for You* had run its first article on viruses, it planted the idea in everyone's brain, triggering the deluge: "Everyone was writing viruses," an interviewee states. Yet now two of the country's largest internet providers were run by people whose first exposure to the world was through hacking a Pravetz in high school. Bennahaum explores the community widely, speculating that Dark Avenger was motivated by malice and hatred toward Bonchev, as well as reclusiveness—he famously corresponded with an American, Sarah Gordon, whose computer his virus infected, and who requested a virus just for herself—to which he obliged; what followed was an e-mail exchange lasting years, ending when Sarah was to be married. The article clearly shows, however, that hacking was connected to both opportunity and playful creativity.[145] Bulgaria, praising the computer's virtue but giving only a limited chance to use it, was prime estate for this explosion. As Bonchev pointed out, there was another reason, too: The regime encouraged copying, theft, and reverse-engineering. The step to malicious software was but a small one.[146]

Dark Avenger's "code was the best," according to the *New York Times,* and many emulated him. The antivirus industry boomed in Eastern Europe, Bulgaria included, with some of the most widely used programs (such as Kaspersky, AVG, and Avast) coming out of it.[147] Bulgarian programmers, of course, didn't create just viruses. Among the first companies to be set up in Bulgaria the late 1980s, as reform allowed for such enterprises to exist, were in computing. Many didn't need much—the "Diana" software house started in an apartment's living room in 1989, enough to hold a few PCs. The average age of the firm employees was 29, all highly qualified and eager, and of a new mould: the company had created its own business codex of morals and practices, which focused on "respect for the customer and partner." As the founders stated, morals and business go hand in hand.[148] This was definitely not Dark Avenger's motto. Other firms were more ambitious from the get-go, utilizing their participation in world networks—literally—to help with investment and advertising Bulgarian enterprises as far away as New York. Using office space and the connection to world databases in the Sofia Trade Centre, "Interpred-Norma" was at the cutting edge of this brave new world, while Zhivkov was still in power. Irina Terziyska, this company's main computer engineer was fluent in English and French, and she sang the praise of the "future of networks" and the membership fee Bulgarian firms were paying the firm in order to participate in the world economy. She was placing advertisements from many major players in Bulgaria industry and was excited about the world opened up by her new job, unmoored from the command economy. A user joked that without this service, he wouldn't have ever known the cost of a champagne bottle on the world market was 55 cents.[149]

ИГРАТА ЗАГРУБЯ

За вируса Eddie и неговия злополучен автор

Без да сме имали ни най-малкото намерение, редакцията ни се оказа въвлечена в епистоларния жанр (литературни произведения, написани във форма на писмо към някое лице или във форма на преписка между двама или повече герои). Спомнете си историята с програмните продукти на битака. Сега се налага (напълно в духа на времето) да отпечатаме пълния текст на едно писмо, чийто автор е причинил хиляди часове неприятности на българските (а знае ли човек и на колко чужденци) потребители на програмни продукти.

Става дума за писмото на всеки читател ще има възможност да се запознае изцяло с гражданските и професионалните позиции на двамата автори и да си определи отношението към тях и към резултатите от техния труд.

Ние по принцип с анонимни писма не се занимаваме – нито ги разглеждаме, нито ги водим на отчет. От бюрото на главния редактор те попадат в кошчето за отпадъци. Но не е такъв случаят със злоумишления създател на Eddie. Редакцията смята, че колкото повече светлина има по този въпрос, толкова по-скоро ще бъде идентифициран злосторникът. Очевидно този човек има

7.4 Wanted: The Dark Avenger! (*Source*: Kompiûtŭr za Vas, National Library, Sofia.)

7.5 Busoft, one of the first Bulgarian private computer firms, was run from an apartment in Burgas, 1989. (*Source*: socbg.com.)

NETWORKS

According to Manuel Castells, the defining feature of the information age is networking: both of people and places, who were defined more by their network connection than by any local (often territorial) ones. Thus arose two modes of labor: the "talent," the ascending knowledge worker; and the "generic," the executants of menial production. Countries and even cities thus housed nodes that differed hugely in power and wealth, so London and New York areas were closer together than to their geographic neighbors—Wall Street and City of London had more in common than either one did with Harlem or Tower Hamlets.[150] Thus, socialism logically failed in such a paradigm, goes the usual theory: It didn't transition from industrialism to informationalism, where productivity arose from the capacity to optimize and combine factors of production. The socialist regimes isolated academics from industry, stifled innovation, and proved incapable of integrating the vaunted STR into industry, leaving them two options: total collapse or total restructuring—tantamount to the same thing.[151]

Castells' later work concentrates on the nature of such networks and the effect of how nodes and actors can be part of overlapping networks: both within a nation and internationally. These interconnected nodes, he holds, can amass their own power if they absorb relevant information from other networks that they are part of and process it efficiently. When nodes become unnecessary for the network's goals, the system reconfigures itself. The most important nodes thus are the "switches"—those that are at the intersection of local and global networks. The BCP was such a node, of course—but it was often connected to a Moscow-centered global network that was lagging. Other switches, connected to elsewhere, could import new programs to try to reconfigure a failing network.[152]

Seen in this light, the Bulgarian technical elite can be considered as one of the most connected nodes in the country's power network. These individuals traded, exchanged, talked, and stole across the Iron Curtain. They held important positions of power, defended their own interests, and introduced new business ideas. They were outmaneuvered by another node—the Soviet-connected elite—which facilitated Zhivkov's fall and the transformations of 1989–1990. Yet the technical elite used its connection to the world information economy to negotiate the transition period more successfully than most, while their criticisms of the mishandling of the STR by the party became a key part of the whole idea of restructuring the failing regime. Robert Castle argues that the delayed transition, with the reformed BCP winning frequent elections in the 1990s and into the 2000s and operating under a mostly statist economy until the 1997 crisis, was down to a lack of consensus: The BSP had no interest in discontinuing the power of the old party elite.[153]

Yet the highest electronics echelons used their connections to exert important roles in the economy and politics. By the 2000s, they were ready to gain positions of authority again. Meanwhile, tens of thousands of educated workers and scientists could make a start abroad or domestically, utilizing their own professional networks. There was, however, a fracture. It is easy to see when talking to Plamen Vachkov and his exploits in California in the 1980s, or when listening to Stoi͡an Markov's global hopping story. The CICT institute, a few kilometers from Sofia's center, was closer in discourse to similar places in Kiev (and also in the US or Japan) than it was to villages only a few dozen kilometers away. Bulgaria was building robots by the 1980s, but villages were still being electrified in 2008: Plochnik lies 35 kilometers from Plovdiv, the site of one of this story's biggest disc factories.[154] It might as well have been on the other side of the world in terms of the global technological narrative of these pages.

Most workers in the Bulgarian factories were net losers after 1989. Their skills were not so easily transferable or in demand. Once the protectionist wall of Comecon fell, Bulgarian workers putting together complex machines on conveyor belts faced competition from the Asian tigers. Some of the technological elite didn't see the Bulgarian workers as particularly educated either—a perfect encapsulation of Castells' division of labor. Vasil Sgurev talks of the poor discipline in factories, with the rooms "smelling of peppers," a reference to the rural origins and habits of the laborers.[155] Ivan Popov reprimanded workers for poor hygiene right from the start of the industry, as we have seen in chapter 1. There was a clear, self-conscious division between the worker who simply put together the machine and those who created it and thought about it. The hundreds of thousands of older workers who had to work with the computers delivered were often not well trained or contended with unreliable machines, especially those workers in the huge agricultural-industrial complexes, as "they received the poorest produce," as Botev and T͡Sonev recall.[156] The last socialist generation, receiving computer classes, was better placed to become part of the new information age than the older generations who were left behind.

The global electronics industry that Bulgaria created had also fostered an internationally connected cadre. Some amassed enough power and clout to challenge the state in terms of economic reform, contributing to 1989. The next decades were also easier for them, as the computer industry had an afterlife in a vibrant software and IT sector, often staffed or created by those who worked in the industry or were students at the end of the regime. Some of the higher-placed individuals continued to have a say and position in politics well beyond 1989; most who worked in the factories, however, became part of the wider deindustrialization story of Eastern Europe. Networks of unequal power persisted, bridging the transition between the old regime and the new, enabled by the information age in which these actors operated and still operate, but also won and lost.

CONCLUSION: THE UNEVEN FUTURE

In 1989, the electronic association at the Ministry of Electronics reported on the financial and business state of the industry. Convertible currency—dollars or deutschmarks or pounds—was always sought by the regime, and the electronics industry was consistently failing to get enough of a market share in the West in those terms. The report was clear: Even keeping the markets won in the USSR required "production at a high technological level, which at the current level is impossible without the usage of equipment from the 2nd line, even if minimal."[1] Using dollars to create the PCs and systems for the East made little sense, as it only got back rubles—ultimately, the late 1950s debt crisis that the electronic industry was created at least partly in response to, came back with a vengeance, driven in part due to that same industry's demand for cash. The market in Bulgaria itself was not sufficiently "saturated" with the technology, the reports argued. What was the ultimate result of all this effort? The next stage could only come about through close cooperation with Western firms and Bulgarian enterprises learning to self-finance, market, and truly reinvent themselves as capitalists: "democratization and decentralization [have to] play a key role in the more dynamic development of socialist countries."[2] Good developments in parallel computing or teleprocessing systems, key to Bulgarian positions in the Soviet, East German, or Czechoslovak markets, were in jeopardy by continued inefficiencies in some factories, which prevented the ingenious systems from actually performing at the level required, making them comparable to other socialist competitors' machines.[3] In essence, the outlook in 1989 was bleak: The industry was facing a qualitative jump if it was to keep up its positions in the East, but that required both more and more Western cash, scarce as always, and Western cooperation and styles of operation. Financially, at least, socialist computing was becoming a potential liability.

Such a report seems to answer the larger question that hangs over this study, which is at the intersection of the history of technology and the history of the socialist regime in Bulgaria and the Second World more broadly: Did communism end because it ultimately failed to enter the information age? This view has been with us from the very start of this book, the conventional narrative that has been told and often advanced—not so much by specialists but by the broader overviews of the information age or the events of 1989–1991. This book has shifted the lens to allow what *did* happen in socialist Bulgaria's technology history to be considered on its own terms. However, the question is difficult to escape and is worth returning to, even briefly. This book starts at the watershed moment of socialist modernization—the shift between two types of industrial revolutions. The advantages of economic backwardness are by now well known from the literature, and socialist modernizers proved excellent at exploiting them to bring an accelerated modernity to Bulgaria after 1944.[4] They were innovators, as they didn't continue following the economic priorities and even possibilities they found in this agricultural country; instead, they concentrated resources in order to restructure and create whole sectors of the economy. In that, we may even call them entrepreneurs in their vision—despite their inherent brutality. They proved, at that first stage, to be capable of integrating innovations much quicker than their capitalist competitors, as they started from a low base and could utilize the best technology available to them. This is the standard story, well supported by the facts that places like Bulgaria were among the fastest growing economies in the 1950s and 1960s, and part of the arguments of scholars such as Janos Kornai: The first phase of socialist acceleration was successful. But as Ivan Chalakov notes and argues closely, seeing "socialist modernizers" as an undifferentiated and homogenous group is false.[5] Broadly speaking, there exist an economic-technological nomenclature and a political one, increasingly at odds with each other. This book has followed this argument to see how the second period of socialism perceived these struggles emerge and play out. In the first period, the economic strata found its interests aligned with the political one, and in fact, the former needed the latter's political and repressive apparatus to innovate. Through nationalization but also the mass mobilization of labor, mass education, and the creation of an informational and research infrastructure (from universities to libraries), the party created the conditions for the takeoff that the previous economy had not. The watershed that comes in the 1960s, as this phase draws to an end, is also a question of where the intensification of the economy will come. The economic-technological class felt it had an answer, but increasingly, it found that to put that answer into practice, it had to fight the interests of the political strata. The former was now operating in the world of fast-changing configurations and the possibility of risk when innovating; the latter continued to operate conservatively in the world of tried and tested methods—someone had already done the innovations elsewhere, so the politicals could apply them, risk-free, to the economy. This had worked for the steel mills, and initially it worked for the computers, too. But this approach did less and less well.

The socialist plan proved very capable in modernizing when it knew where technological development was going. By the 1960s, it listened to well-placed technological party elites, above all Ivan Popov, and trusted in their advice on where development was going. But as this elite gathered power due to their fostering of a successful high-technology industry, which captured large swathes of the international socialist market, it also became a threat due to its increased political confidence. Moreover, the watershed that was the creation of the computer industry also had an inbuilt flaw from the very start: accelerate and overtake. But unlike the usual Cold War story of socialism aiming to accelerate and overtake capitalism in the grand struggle for the soul of humankind, the Bulgarian computer was to accelerate and overtake its *socialist* allies. The competition was with the GDR or Poland, the prize was their own markets and above all the vast Soviet one. To overtake, then, one had to prove not that one had the best technology, but the economic organization to create that technology the fastest. This is why the intelligence services became a key part of this story, and why licensing, copying, and reverse-engineering take over fundamental research. As we have seen, such decisions were made by luminaries such as Popov rather than Zhivkov—it was the economic-technological elite that bet the house, right from the start, on overtaking its Comecon allies by copying: not necessarily because that strategy was cheaper (although that was a boon), but because it was *faster*. We cannot understand the fall of the industry if we do not situate it at those key years in the mid-1960s, when the battle was started over Comecon specializations. The kernel of increased costs over time, especially in the 1980s, as well as playing catch-up to the West (but not the East) took root then and there. Much like Charles Maier's argument about the GDR, by the 1980s, Bulgarian electronics manufacturers were becoming voracious consumers of Western currency in their attempt to keep Eastern markets. But the ultimate collapse of the industry is tied to the collapse of that Eastern market itself, its international economic order, and its overall relation to and comparison with the West.

Fundamental research and innovative thinking were of course present. This book has traced paths that criss-crossed the best institutes and universities in the socialist bloc and the Western world, following the footsteps of members of the economic-technological class—both the party nomenklatura (the Eastern Bloc class of high-level administrators whose positions depended on the party's blessing) and the thousands of scientific workers who made up its body. As a professional class, they had the knowledge and ability to innovate—when they didn't, it was because of their position as a political class within the framework of the late socialist regime. There was no law of biology that made Bulgarians less capable of computer design than the Japanese, and no law of physics that made the Bulgarian hard drive fundamentally worse than an American one. But there were laws of politics, with the knock-on effects on industrial production; shortages; quality control; and ultimately, the conservative instinct to let someone else take the risk and weather its costs. Japan started off with similar logic of reverse-engineering and licensing, which didn't prevent it from also becoming a country of innovation in the sector; China seems to be on a very similar road, too. By using the prism of the computer, this book has shown that the differentiated nature of the elites of socialism also allow us to answer the question differently: Socialism had indeed entered the information age, and part of the story of its fall is precisely that it did. The failure to deliver the full gamut of the promise that the BCP made in the 1960s was instrumentalized by those pushing for reform, but it was also a powerful tool precisely because the information age existed in Bulgaria, imperfectly: The party was doing a bad job of something, rather than no job at all. The book has asked: Who entered the information age, and what effects did it have? Traders, spies, computer specialists, automation experts, children—many had entered the information age, and they wanted it fully.

The sign of acceleration hangs over this book, too. The computer was a tool to manage the accelerated flow of information, and it seemed to bring the future to the cusp of realization. It accelerated some of its precepts into career trajectories that served some Bulgarians well, even after 1989, or on jet airliners to San Francisco, Tokyo, or New Delhi, when most Bulgarians had been waiting for over 10 years just to get a car. The computer accelerated the dreams of the party, and it accelerated the dreams of intellectuals—the command economy could be reformed in just a few short years through automation and perfect networking, or the new creative power of humanity could be achieved tomorrow, if only we found the right way to integrate it with the calculating power of the machine. It even accelerated workers' anxieties, intensifying them; and children's participation, making them workers before their time. The time had arrived, but as William Gibson said, it had not arrived everywhere. This book has sketched briefly some of the groups and sites where these accelerations, intensifications, and anxieties manifested—from workplaces and journals, to children and workers. These sketches are incomplete, and the author would call for more work in this direction, especially on the questions of gender in socialist computing (which has regretfully only been hinted at) or the nature of hacking communities in the Eastern Bloc. Alas, time accelerated for this book, too.

When communism fell, the computer was part of the story, but not the whole story. Political decisions opened up the Berlin Wall and ended Soviet power in Eastern Europe. They also ended the economic links that had made the whole network of exchange and the logic of the industry possible. Different political decisions would have resulted in different stories. Bulgaria could have limped on as a computer power within the framework of a Comecon that was increasingly overshadowed by the West and a rising Asia in terms of economic power; its industry could have continued outperforming its allies while falling further behind the rest of the world. Different political choices made after 1989 could also have resulted in an amended story of not just human capital afterlives, but of factories that changed their blueprints, their business practices, and eventually their markets. But once again, the reasons for why this didn't happen are not to be found in the nature of the commodity being produced, its relative merits or successes, or the lack of expertise in the country. Instead, they are found in the political and economic framework and logic that the industry operated within—both before and after 1989.

The afterlives of this industry continue to be felt in politics, economics, and popular memory. For some, the image of a socialist "Silicon Valley" is folded into a nostalgia for a Bulgaria that was growing and industrializing, its goods exported around the world, and life was secure and stable. For others, it can just as easily be harnessed to a vision of backwardness, where socialism held back Bulgarian innovation and the "natural development of the country," relegating the country to perpetual backwardness, because every computer design was stolen from the West. For many, the Pravetz and Atanasoff are symbols of Bulgarian technical cheating or genius; either a memory of when the country was at the cutting edge of the new age, or a reminder of an innate ability of its people to contribute to humankind, if only the BCP had not locked the country into the Soviet sphere. Today, vastly more Bulgarians have computers than they did during socialism, yet this information age has a prehistory in the country, a local configuration of global developments with its own reasons for existing the way it did and with its own cultural and intellectual offshoots. This book has shown how different an information age can look, and that computers do not thrive and develop just under the conditions of democracies, open-access contacts, freedom, and creative commons. They can be tools to foster a creative worker, to achieve utopia by eliminating menial labor, and to discipline labor through surveillance and recording. Even for Georgi Konstantinov, the anarchist we met in chapter 6, his politics and his computer experience combined during his exile to France in the 1970s, where he started publishing pieces on the forthcoming "robotronic" revolution. Even today, his works on anarchist philosophy and praxis are concerned with the melding of robots and information, which have made the possibility of a post-capitalist world thinkable—in a way, he is an inheritor of the party's dream despite his anti-state socialist thinking. The computer can thus also serve particular interests, as the party found out. Just because it is input into a computer did not make data objective fact.

The journeys the Bulgarian computer, its users, and experts on were spatially and intellectually stimulating. They forged contacts that are often obscured by the Cold War logic that supposedly informed most decisions during this period. They gave power to particular interests, showing the imperfect penetration of any modernization into our world. The global information economy created common terms and concrete links between sites thousands of kilometers from each other, leading to the possibility of severing links to the immediate hinterland. The computer was vested with meaning and importance by its users. It created a common horizon for Bulgarians to converse with and trade with Soviets, Japanese, and Indians. It also created tools to discipline, modernize, automate, and sell. Ultimately, it allowed for applications akin to the famed "black box" of computer science, its internal workings hidden and not readily understood. The input of computers into society did not always produce the same output everywhere, or in the way it was intended. It was in this "black box" of the intermediate stage, where actors negotiated between the universal and transnational language and particular local and personal interests, where history was made. But the story of the Bulgarian computer did not stop at the "end of history" in 1989. In fact, the trajectories of both politics and lives after that date have their origins in the story told here.

In 2014, the Bulgarian firm "Telerik" was sold to a US company in a deal worth $262 million. Founded in 2002 by four graduates of Sofia's ex-VMEI and the American University in Bulgaria, it produced tools for web development.[6] Svetozar Georgiev, one of the founders, recalls that his start came on a Pravetz-16 brought home by his dad, teaching himself to program on it.[7] "Telerik" was probably the most successful home-grown business of any kind in post-1989 Bulgaria. By 2019, the share of the Information and Communications Technology (ICT) sector in the country's GDP was over 6.6 percent, the second highest in the whole of the EU (behind Malta)—the vast majority in services, rather than hardware, as it had been before 1989.[8] Bulgaria is still a favored place to outsource technical support centers by firms such as HP or Cisco. The ICT sector is among the most sought after, commanding some of the highest salaries for graduates, and some individuals involved with it pay tribute to the longer history: "it's a strength that we've already had two generations of engineers with a strong R&D background."[9]

Thirty years after communism ended, its factories are shuttered, but the human capital survives. A generation of young people educated just before 1989 put their skills to use—for both malicious and productive ends. Thousands of others, including engineers and technicians, also fared well during the difficult transitio—either starting companies or migrating to the capitalist Silicon Valley. Krasimir Markov recalls how IZOT skills served his colleagues well in the 1990s. He met a colleague in 1993 who told him:

> I made such good money the last three years, he says. I ask him—how, Stoycho, the industry collapsed! But he says—yeah, but the Americans found me and sent me to service their IBMs in the old Soviet Union for $5000 a month! Do you think an American wants to go to Siberia to service machines? But for us that was a monstrous salary.

Many engineers jumped at the opportunity—having worked on IBM-compatible ES machines, often with previous visits to the USSR, with no qualms about going beyond the Urals for unheard-of sums.[10] The Iron Curtain had not stopped Bulgarian specialists from cultivating world-level skills before 1989, while a new generation was ready to tackle the software that had replaced hardware as the country's primary product. Although not such a defining sector as it had been under communism, nevertheless, computing remains one of the few bright spots in the post-socialist economy in terms of wages and revenue. Socialism's factories failed—but the people remained.

Atanasoff, with whom we started this story, is another perfect representation of the crossings between West and East that this book has told: the American professor who drove through the Iron Curtain in his hired German car to meet Bulgarian computing science in the persons of Sendov and others. He, like the industry, collapsed the Iron Curtain at will, and his global crossings met local circumstances and needs—in his case as a prestigious symbol. The Cold War could not keep Bulgaria locked out of the "thickening" connections of exchange, meetings, and purchases.[11] Bulgaria was an active participant in this global interchange, and not just as a recipient of Western knowledge. Its institutes and factories churned out goods for the East and South, and its engineers and thinkers harnessed them to projects of socialism, economic reform, and cultural elevation. The computer opened a window to the world geographically and intellectually, helping its practitioners to become globetrotters, sharing a common language in a cybernetic key, and interacting in the literal and metaphorical trading zones of science and business—conferences, fairs, personal correspondence, technical arguments, and shady deals.

A global narrative can elevate motion over place, treading the paths that circled the globe, while ignoring the local conditions in which these roads originated. Bulgarian interests shaped what the computer was to do. First, it was there to make cash for the regime. It was also there to modernize the economy by eliminating wastage, bottlenecks, and maybe even the unreliable worker. Bulgarian experts received the relative freedom to trot around the West and Global South not because the regime was interested in their intellectual dreams and widening horizons, but to hone their skills so they could build up the industry at home. And while the Iron Curtain was intellectually and technologically porous, it served to create economic "closed worlds" that both made the Bulgarian industry possible and would also eventually make it prohibitively expensive. Plamen Vachkov recalls a roundtable in Hannover in the mid-1980s, which he attended in his capacity as director of the Pravetz factories. A journalist asked him how he would respond to the accusation that his Pravetz computers were reverse-engineered Apple clones, while their operating system was a direct copy of MS-DOS. His reply was "if you tell me how much [licensing] fees I owed, I would write the check there and then," waving around a check book.[12] The answer was met with much laughter, but it is doubtful whether everyone understood the core of Vachkov's joke—the check would be in the nonconvertible roubles that would do neither Apple nor Microsoft much good. Comecon and its logic were sealed off from the West in terms of competition, creating the financial and market mechanisms that allowed IZOT and Pravetz to arise and flourish. The local mattered. The Second World constituted itself as an alternative modernity, where socialist states could in some ways start from scratch. But the industrial horizons often remained Western, and the dream of convertible currency was tied to a desire to open up the other "closed world."

The cracks through which the actions of global trends passed into Bulgaria were controlled by the state, which carefully chose which actors could serve its interests as conduits. Who negotiated and when they negotiated with Western companies, or sold to Indian ones, were political choices. Situating this history in the specific commodity and the experts it enchanted has helped this book to zoom between transnational and local levels, keeping the global connections hitched to the realities of power. How Bulgaria interacted with the world was determined by who needed the computer, and for what ends, and this anchor in local interests has lessened the risk of losing the grand picture of why and how this industry developed, even when chasing the threads that emanated from Sofia out to Moscow, Tokyo, New Delhi, or Silicon Valley.

Having already quoted science fiction writers, I will indulge in one further quote in lieu of historians or philosophers: The future is already here, it is just not very evenly distributed. William Gibson's phrase contains a whole view of history. Certain places in Sofia and other places—the huge factory in Stara Zagora or the specialized space of the Varna Shipbuilding Computer Centre, or even a provincial high school that had a few Pravetz PCs—were nodes in much larger networks. It is clearest at the highest echelons of the party and managerial class, where West and East often met. But overall, the computer meant that parts of Bulgarian society became much closer in terms of language, expectations, and skills to other nodes of the future. Bulgarian institutes could converse with their American and German counterparts, while Indian and Japanese scientists visiting Bulgaria talked in the same computer jargon but also increasingly the same financial language. Danish and Bulgarian engineers could swap stories about problems they had run into while working on the same type of machine. But everyone was also embedded in specific local conditions of power and access. They were part of the uneven and lumpy road of Bulgarian modernization. Not only did villages nestled in the mountains near places like Pravetz and Botevgrad exist in a different world, but often even the workers within the electronics factories themselves. They "smelled of peppers," after all, a generation removed from the land. The discourses of the managers and engineers were conceptually, physically, and economically far from many Bulgarians. The future and the past coexisted within meters of each other, even in the same conversation.

To be part of the computing world, you didn't need to understand its intricacies. The computer enabled the rise of different specialist groups, with different roles and thus different access to power, which also meant they had divergent trajectories and interests in what to use the computer industry for. The self-confident strata of technocratic managers, who conversed with the West and used its business language, who talked of industrial organizations based on the Japanese model, who *looked* different to the old party, were politically connected and championed the industry but also could use it in their own intraparty struggle for recognition and position. Other party members were part of networks that stretched to Moscow; they held different interests—the computer was interesting to them mostly as a source of revenue rather than as a tool of modernization—and their positions increasingly differed from the younger managers. They won in the short term, overthrowing Zhivkov, but as this book has shown, it was the other managers who could utilize their expertise and connections within their own networks to navigate the transition and turn their cultural capital into real capital. It is beyond the scope of this book to break down the divisions among the elites during the formation of new political and economic power clusters after 1989. But it is worth emphasizing what these chapters have shown: Expertise could be turned into money as multiple Bulgarian managers (and also engineers) operated at a truly modern level of professionalism, despite the relative backwardness of the Bulgarian industry. But what could be turned into the most money was being placed at the crossing points between the closed worlds—being a trade representative, a middleman in the technology transfer that this book is concerned with, put you in a privileged position in terms of both the global story (access to the "other" world) and the local story (access to the local world's political power). As chapter 3 has shown, trade and exchange were closely linked to a particular position in the structure of socialist power: the party's intelligence services. Even highly placed figures such as Ivan Popov and Ogni͡an Doĭnov could champion the industry, push reform, defend its interests, and then be removed for political reasons; but their industry's trade representatives were much more likely, despite their lower positions, to be better placed to turn their connections into capital after 1989.

The computer expanded the horizons for hundreds of thousands of Bulgarians. For some, it allowed them to gallop ahead into the rising information economy as participants in its transnational business networks. Others immersed themselves in the information age as a techno-intellectual pursuit, as engineers and scientists. Some wanted to build socialism with it, or free humans from labor; some saw it as a means to reform to a more open society. The regime promised a lot, but it failed to deliver, leaving vast swaths of society outside the promised cybernetic age. Integration and disintegration went hand in hand, as elements of the financial and political elite became part of the global network of the postindustrial class, while many others were left behind, very much industrial. Success and failure were joined at the hip, and the question is akin to asking: "Is what is good for Apple also good for the US?" The future was there for some, but it couldn't trickle down to the rest of the economy and society.

What is evident is that what was good for IZOT was also good for tens of thousands of technical intellectuals who became part of the knowledge economy, which allowed them to traverse the 1990s and 2000s in different ways. The last generation of socialism's children were also beneficiaries, poised on the cusp of the true information age that arrived when the global flooded the local through the internet—the first e-mail in the country was sent from a Pravetz in VMEI-Varna in 1987 to a Scottish recipient. The same team, led by the professor and graduate student who made it possible, then set up one of the first firms in the country aimed at networking, and from 1991, it was allowed to register sites with the national .bg domain.[13] Ends and starts blur, and 1989 is a watershed inasmuch as it allowed freer communication and integration, which had already been taking place—as well as the collapse of an Eastern market that had made it all possible.

Youth was part of this world in other ways, which continue down to today. Sendov, the man who invited Atanasoff to Bulgaria, also has a hand to play here. In 1989, Pravetz hosted the world's first International Olympiad of Informatics, which became the second largest Olympiad after the mathematics one. For three days in May, children from 13 countries as far away as Cuba, Vietnam, and Zimbabwe, butted heads against a series of algorithmic problems, armed with Bulgarian PCs. In the end, two Bulgarian students plus one student each from the USSR, Czechoslovakia, Hungary, and West Germany, received gold prizes.[14] This was the culmination of a Sendov proposal to UNESCO dating back to 1987, the natural progression to his new educational method tried out in Bulgarian schools and his international conference on children and computing in 1985—in fact, he had already organized two competitions as part of his conferences on "Children in the Information Age," in 1987 and 1988. The International Olympiad of Informatics would become an annual event, with 87 countries represented in 2020. Bulgaria might not have originated the computer, but it did birth an international "trading zone," where the next generation of computer experts would labor and create using the computer paradigm that is even more a part of our life today.

In 2008, the Voroshilov factory, which had been an early incubator of engineering knowledge, was dynamited to make way for new office buildings. The huge plant in Stara Zagora has not been producing computers for many years, instead turning to small-scale consumer appliances or putting together parts shipped from Asia. IZOT's central home, consisting of the institute and the ZIT factory, now houses over 200 small firms that continue their activity. Almost all are created by ex-IZOT engineers and scientists, working within the literal framework of the old world. Koĭcho Dragostinov took me on a tour of his firm, which manufactures devices for a French company, utilizing Chinese parts. Dragostinov's expertise had taken him, during socialism, to positions as far as Nigeria, an example of the story this book has been telling. IZOT's aging buildings are still full of the friendships and collegiate links he had built during that time, which he kindly shared with me, a space full of both the old starts of this industry and its present state, which exists because of those beginnings.[15]

Others took the path West, such as my father. Nestled in the June 1987 issue of the *Korabostroitel* ("Shipbuilder") newspaper, the publication of the industry in Varna, is his picture as a young scientist in the Shipbuilding Institute's Electronic section. He had two dreams, he says: the installation of a Bulgarian PC on a domestically built ship, and a programmable controller. He praises the institute for being the right place to be if you had ideas, as it provided the conditions for his creative work. Soon, however, the framework would collapse and by the late 1990s,

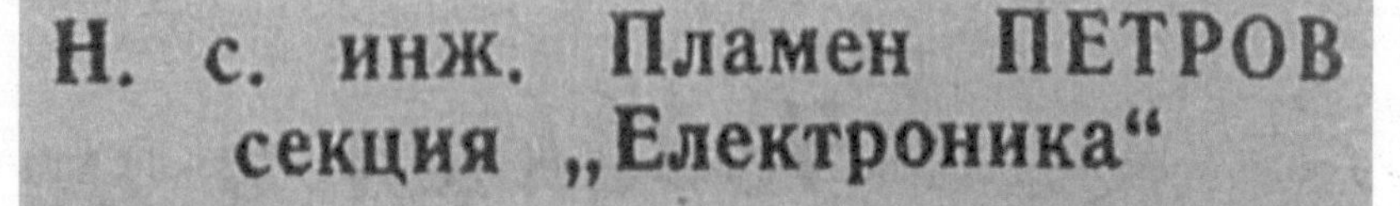

Н. с. инж. Пламен ПЕТРОВ
секция „Електроника“

„В Института работя от две години. Пряката ми дейност е свързана с темите „Уредба за

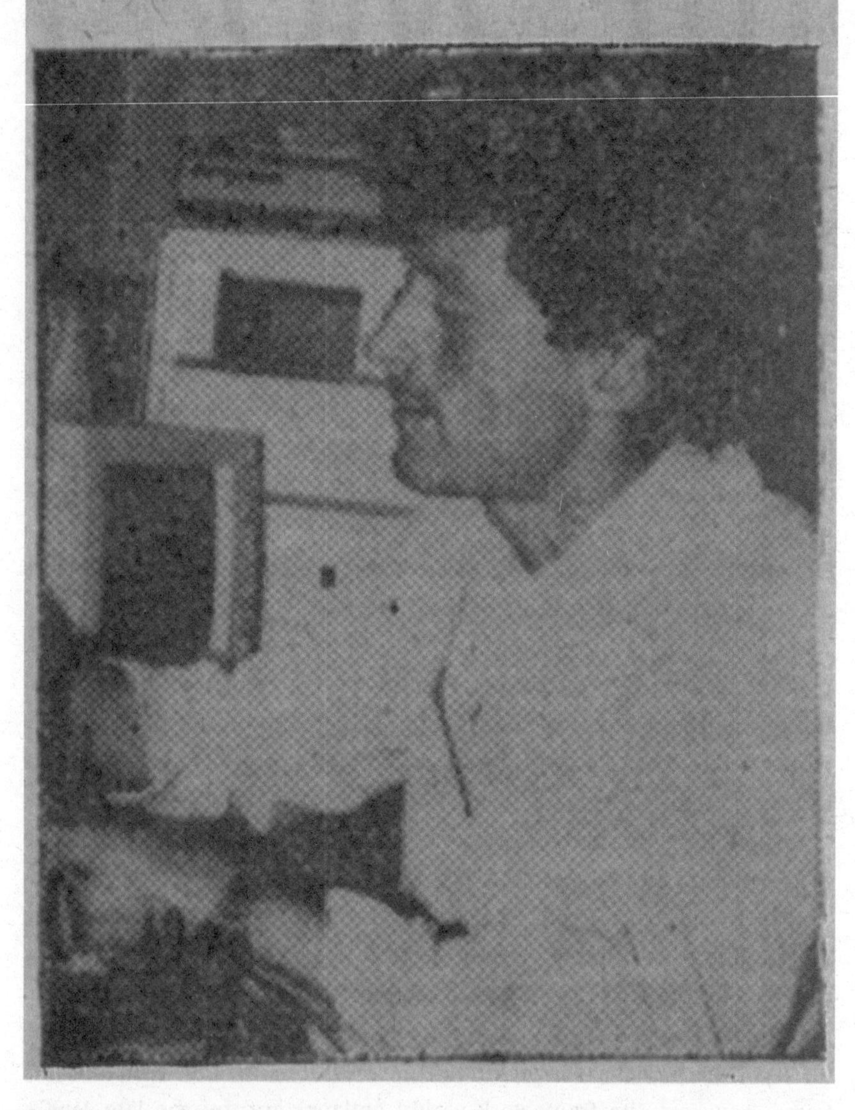

C.1 The author's father as a young electronics expert. (*Source*: Korabostroitel, Regional Library, Varna.)

he—and my whole family—ended up in Great Britain, and my dad's electronics dreams transferred from ships to software in the banking sector. In between, his electronics expertise allowed him to criss-cross Germany, the UK, and Japan in conferences and lecture courses.

* * *

In the hills just outside Varna stands a small house built by my grandfather. Its aerial is a strange circular object, rusted to a reddish-brown by the elements, but it still allows the old TV inside to pick up the Sunday sports. It is made from an IZOT hard drive, cribbed from somewhere by my father. Such makeshift antennas can still be found on many village houses if you take a drive through rural Bulgaria. In that house, time can stretch in a languid fashion if you decide to take a walk through the small orchards and vineyards surrounding it, even though if you turn on your phone's network, you can be plugged right back into the accelerated flow of the information age. Which one you decide to do is, ultimately, your choice.

APPENDIX A: SNAPSHOT OF AUTOMATION IN 1989

This appendix is a snapshot at the end of the regime of its achievements in its automation dream—the number of computers and automation systems in industry and services. The details are taken from the Central Statistical Authority's 1989 annual report (pages 137–138), and numbers reflect the situation at mid-1989. The preceding years are included to show the quick growth once Pravetz PCs went into production.

Computers in Use

	1985	1986	1987	1988	1989
Total	8,748	17,157	22,792	33,415	39,477
Large Computers (ES-type and equivalents)	No data	337	310	282	255
Mini-computers (SM-type and equivalents)	463	533	580	580	702
PCs	No data	16,287	21,902	32,553	38,520

(continued)

Automation Systems in Use (Selected), 1989

Type	Number in Use
CNC Machines in Metalworks	5,342
Industrial Robots	248 (163 currently installed)
Automatic and Technical Modules	1,259
Automated Technical Lines in Machine Building	284
Complex Automated Lines in Non-Machine Building	1,356
Automated Lines for Printed Circuit Board Montage	21
Complex Mechanized Technical Lines	3,639
Flexible Automated Production Systems in Machine Building	43
Automated Systems of Design and Preparation of Production on 32-bit Basis and 16-bit Basis	176 & 204
Microprocessor Systems for Automation of Design	1,297
Automated Dispatch Systems in Production	504
Automated Information Processing Systems	1,512
Automated Systems of Economic Governance on Basis of Multi-User Computers	277
Microprocessor Automated Workplaces in Economic Governance	1,916
Expert Systems	69
Automated Warehousing	124
Microprocessor Automated Workplaces in Administrative Services for Citizens	1,242

Source: *Statisticheski Godishnik na Narodna Republika Bŭlgarii͡a1989* (Sofia: Tzentralno Statistichesko Upravlenie, 1990).

APPENDIX B: TYPES OF MACHINES PRODUCED

Below is a nonexhaustive list of the computers and hard discs that equipped Bulgarian workplaces and won the country its markets. Tape memories and peripherals (such as printers, plotters, monitors, teleprocessing systems, and modems) were also produced but are not shown here. The following machines are chosen as important to the automation of work and as the main cash cows for the industry. Western equivalents are given as a guide for the reader.

Large Computers

Name	Entry into Production	Operations per Second (Gibson scale)	RAM	Western Equivalent
ZIT-151	1967	—	—	FACOM 230–30
ES-1020	1972	10,000	256 KB	IBM 360/40
ES-1022	1975	80,000	Up to 512 KB	IBM 360 series
ES-1035	1978	200,000	1 MB	IBM 370
IZOT 1014/ ES-1037	1987	Up to 2 million (MIPS) for each processor	2–16 MB	IBM 4331/4341

(continued)

Minicomputers

Name	Entry into Production	Western Equivalent
IZOT 0310	1974	PDP 8L
IZOT 1016/SM-4	1979	PDP 11
IZOT 1054S/SM-1426	1987	VAX 11/730
IZOT-1055S/SM-1706	1987	VAX 11/750
IZOT-1056/SM-1504	1988	VAX 11/780

Personal Computers

Name	Entry into Production	Specifications (CPU & RAM)	Western Equivalent
IMKO-1	1980 (prototype), 1981 (limited production)	Intel 8080A/2 MHz & 16 KB	Apple-II+
IMKO-2/Pravetz-82	1982	Synertek 6502/1.018 MHz & 48–64 KB	Apple-II+
IZOT-1031	1984	U 880D/2 MHz & 64 KB	Z80-chipset
Pravetz 16	1984	Intel 8088/ 4.77 MHz & 256–640 KB	IBM PC/XT
Pravetz 8E	Imported from Taiwan, 1985—no domestic production	6502/1.018 MHz & 64–1080 KB	Apple-IIe
Pravetz 8M	1985 (civilian); 1986 (military version)	6502/1.018 MHz & 64-128 KB & Z80/4 MHz & 64–128 KB	–
Pravetz 8D	1985	6502/1.018 MHz & 48KB	Oric Atmos
Pravetz 8A	1986	SM 630/1 MHz (Bulgarian analog to 6502) & 64–1080 KB	Apple-IIe

Name	Entry into Production	Specifications (CPU & RAM)	Western Equivalent
Pravetz 16 E/ES	1988	NEC V20/8 MHz & 640 KB	–
Pravetz 286	1988	Intel 80286/8–12 MHz & 64 KB–3 MB	IBM PC/AT
Pravetz 8C	1989	SM 630/1 MHz & 128 KB	Apple-IIc
Pravetz 386	1989	Intel 80386SX/12 MHz & 1–10 MB	Compaq DeskPro 360
Pŭldin-601	1989	SM601/1 MHz & 64 KB	None—clean patent
Pravetz 8S	1990	SM 630/1 MHz & 128–1024 KB	Apple-IIc

Hard Discs

Name	Entry into Production	Capacity (MB)
ES 5052	1971	7.25
ES 5061	1973	29
ES 5066.01	1977	100
ES 5067	1977	200
ES 5063	1982	317[1]
ES 5063.01	1986	625
SM 5508	1987	10
SM 5510	1990	160

[1]The hard drives were of the Winchester type from 1982 onward.

NOTES

INTRODUCTION

1. The story's details are taken from Blagovest Sendov, *John Atanasov: Elektronniyat Prometey* (Sofia: UI Sv Kliment Ohridski, 2003). That is just the latest book to appear in Bulgaria regarding Atanasoff, who spawned a number of texts dating back to the socialist period.

2. TsDA f. 1B op. 68 a.e. 1836 l. 201. Archive abbreciations can be found in the frontmatter.

3. *Statisticheski Godishnik na Narodna Republika Bŭlgariia 1988* (Sofia: Tzentralno Statistichesko Upravlenie, 1988), p. 165.

4. Kiril Boianov, "Kratki Svedeniya za Razvitieto na Izchislitelnata Tehnika v Bŭlgariia," in *Godishnik na Sektsiya Informatika*, vol. 7 (2014), p. 16.

5. Milena Dimitrova, *Zlatnite Desitiletiia na Balgarskata Elektronika* (Sofia: IK Trud, 2008), p. 62.

6. By comparison, Singapore—a leading Asian Tiger, especially in electronics—currently sees 17 percent of its manufacturing workforce engaged in the sector (https://www.edb.gov.sg/en/our-industries/electronics.html; last accessed July 18, 2022).

7. Interview with Horace G. Torbert by Charles Stuart Kennedy, US Ambassador to Sofia 1970–1973, in "Bulgaria Country Reader" at *Association for Diplomatic Studies & Training Oral History Collection* (https://adst.org/Readers/Bulgaria.pdf ; last accessed July 18, 2022).

8. The history of commodities has experienced a notable boom and has reached a wider popular audience, too—see Mark Kurlansky's books on fish, salt, and milk: *Salt: A World History* (New York: Walker and Co., 2002); *Cod: A Biography of the Fish That Changed the World* (New York: Walker and Co., 1997); and *Milk! A 10,000-Year*

Food Fracas (London: Bloomsbury, 2018). Sven Becker's *Empire of Cotton: A Global History* (London: Penguin Books, 2015) is only one of many widely acclaimed books that came out of mainstream academia and tell global stories through commodities. But here I am mostly referring to the country-specific, excellent study by Mary Neuburger: *Balkan Smoke: Tobacco and the Making of Modern Bulgaria* (Ithaca, NY: Cornell University Press, 2012), and her latest, food-centered, *Ingredients of Change: The History and Culture of Food in Modern Bulgaria* (Ithaca, NY: Cornell University Press, 2021). Not least, my own book is in some ways an homage to the title of *Balkan Smoke.*

9. This idea draws on Peter Galison's influential study *Image and Logic: A Material Culture of Microphysics* (Chicago: University of Chicago Press, 1997), where he posits the computer as an extremely important trading zone in the modern age.

10. Vladislav Zubok and Constantine Pleshkov, *Inside the Kremlin's Cold War: From Stalin to Khrushchev* (Cambridge, MA: Harvard University Press, 1997).

11. Alexei Yurchak, *Everything Was Forever Until It Was No More: The Last Soviet Generation* (Princeton, NJ: Princeton University Press, 2005).

12. Francis Fukuyama, *The End of History and the Last Man* (New York: The Free Press, 1992), p. 93.

13. The argument first appeared in essay form in Manuel Castells and Emma Kiselyova, *The Collapse of Soviet Communism: A View from the Information Society* (Berkeley: International and Area Studies, University of California, 1995). It was then integrated in Castells' *The Information Age: Economy, Society and Culture*, consisting of three volumes: *The Rise of the Network Society* (Cambridge, MA: Blackwell, 1996), *The Power of Identity* (Cambridge, MA: Blackwell, 1997), and *End of Millennium* (Cambridge, MA: Blackwell, 1998

14. See Charles Maier, *Dissolution: The Crisis of Communism and the End of East Germany* (Princeton, NJ: Princeton University Press, 1997), p. 73; the whole of chapter 2, on economic collapse, is worth citing as an influential example of this historiography.

15. For one example, see Gale Stokes, *The Walls Came Tumbling Down: The Collapse of Communism in Eastern Europe* (Oxford: Oxford University Press, 1993). While older now, the book's description of the regime has not substantially changed in general overviews of Eastern European socialism, which often pass over Bulgaria in a couple of sentences. There is a wide range of literature on Yugoslavia as different. For example, Yugoslav socialism's difference has been highlighted through its consumer culture in Patrick Patterson, *Bought and Sold: Living and Losing the Good Life in Socialist Yugoslavia* (Ithaca, NY: Cornell University Press, 2011) or through its engagement with the nonaligned world, a great example of which is Patterson's more recent *Nonaligned Modernism: Socialist Postcolonial Aesthetics in Yugoslavia 1945–1985* (Montreal: McGill-Queen's University Press, 2019). For Romania, see Vladimir Tismaneanu, *Stalinism for All Seasons: A Political History of Romanian Communism* (Berkeley: University of California Press, 2003). For Albania, see Elidor Mëhilli, *From Stalin to Mao: Albania and the Socialist World* (Ithaca, NY: Cornell University Press, 2017).

16. Interview with Donald C. Tice, Political/Economic Officer 1964–7, in Bulgaria Country Reader" at *Association for Diplomatic Studies & Training Oral History Collection* (last accessed: January 20, 2021)

17. In this I draw on the forceful point made by Mëhilli, *From Stalin to Mao*: We need to treat the Second World as a real project of modernity and a real space where ideas and materials circulated. For a longer-term and thought-provoking view of this, see Stephen Kotkin, "Mongol Commonwealth? Exchange and Governance across the Post-Mongol Space," *Kritika: Explorations in Russian and Eurasian History*, vol. 8, no. 3 (Summer 2007), pp. 487–531. See also Łukasz Stanek, *Architecture in Global Socialism: Eastern Europe, West Africa, and the Middle East in the Cold War* (Princeton, NJ: Princeton University Press, 2020).

18. György Péteri, "Nylon Curtain—Transnational and Transsystemic Tendencies in the Cultural Life of State-Socialist Russia and East-Central Europe," *Slavonica* vol. 10, no. 2 (November 2004), pp. 113–123; Michael David-Fox, "The Iron Curtain as Semi-permeable Membrane: Origins and Demise of the Stalinist Superiority Complex," in Patryk Babiracki and Kenyon Zimmer (eds.), *Cold War Crossings: International Travel and Exchange across the Soviet Bloc, 1940s–1960s* (College Station, TX: Texas A&M University Press, 2014), pp. 14–39.

19. The history of the Bulgarian security services has been told, understandably, mainly through its repressive apparatus and its alliance with the KGB—Jordan Baev, *KGB v Bŭlgari͡a* (Sofia: Voenno Izdatelstvo, 2009); and Momchil Metodiev, *Mashina za Legitimnost: Rolyata na Dŭrzhavna Sigurnost v Komunisticheskata Durzhava* (Sofia: Ciela, 2008). Journalist Khristo Khristov has however delved more into both the ring of companies that the Security Services created and some of their scientific-technological work in his book, Khristo Khristov, *Imperiyata na Zadgranichnite Firmi: Suzdavane, Deynost I Iztochvane na Druzhestvata s Bulgarsko Uchastie zad Granitsa 1961–2007* (Sofia: Ciela, 2009)—and in a series of articles in the newspaper *Kapital*. In the fields of electronics, cooperation, and Polish efforts, Miroslaw Sikora has done groundbreaking work: "Cooperating with Moscow, Stealing in California: Poland's Legal and Illicit Acquisition of Microelectronic Knowhow from 1960 to 1990," in C. Leslie and M. Schmitt (eds.), *Histories of Computing in Eastern Europe. HC 2018. IFIP Advances in Information and Communication Technology*, vol. 549 (Cham, Switzerland: Springer, 2019).

20. Nick Cullather, *The Hungry World: America's Cold War Battle against Poverty in Asia* (Cambridge, MA: Harvard University Press, 2013). In his introduction, Cullather explicitly calls out how both the American and Soviet models were predicated on being able to yield statistical proof of their superiority.

21. Tony Smith, "New Bottles for New Wine: A Pericentric Framework for the Study of the Cold War," *Diplomatic History* vol. 24, no. 4 (2000), pp. 567–591.

22. Theodora Dragostinova, *The Cold War from the Margins: A Small Socialist State on the Global Cultural Scene* (Ithaca, NY: Cornell University Press, 2021).

23. There is a rich literature on both Big Science and state/military support for the rise of computing. See Christophe Lecuyer, *Making Silicon Valley: Innovation and the Growth of High Tech, 1930–1970* (Cambridge, MA: The MIT Press, 2007); Peter Galison and Bruce Hevly (eds.), *Big Science: The Growth of Large-Scale Research* (Stanford, CA: Stanford University Press, 1992). Slightly tangential to this point, but key to the discussion of how expertise is formed, is John Agar's *The Government Machine: A Revolutionary History of the Computer* (Cambridge, MA: The MIT Press, 2003).

24. See Paul Edwards' seminal *The Closed World: Computers and the Politics of Discourse in Cold War America* (Cambridge, MA: The MIT Press, 1997).

25. See both Suvi Kinsikas, *Socialist Countries Face the European Community: Soviet-Bloc Controversies over East-West Trade* (Frankfurt am Mein: Peter Lang, 2014), and Angela Romano and Federico Romero (eds.), *European Socialist Regimes' Fateful Engagement with the West: National Strategies in the Long 1970s* (New York: Routledge, 2021), the result of the PanEuro1970s project at the European University Institute, Florence.

26. Mëhilli, *From Stalin to Mao*, pp. 9–12

27. Oscar Sanchez-Sibony, *Red Globalization: The Political Economy of the Soviet Cold War from Stalin to Khrushchev* (Cambridge: Cambridge University Press, 2014) remains a defining work. See also: James Mark, Artemy Kalinovsky, and Steffi Marung (eds.), *Alternative Globalizations: Eastern Europe and the Postcolonial World* (Bloomington, IN: Indiana University Press, 2020); Besnik Pula, *Globalization under and after Socialism: The Evolution of Transnational Capital in Central and Eastern Europe* (Stanford, CA: Stanford University Press, 2018); Philip Muehlenbeck and Natalia Telepneva (eds.), *Warsaw Pact Interventions in the Third World: Aid and Influence in the Cold War* (London: IB Tauris, 2018). An influential call and contribution has been made by the "Beyond the Iron Curtain: Eastern Europe and the Global Cold War" special issue in *Slavic Review*, vol. 77, no. 3 (2018), pp. 577–684.

28. Łukasz Stanek, *Architecture in Global Socialism: Eastern Europe, West Africa, and the Middle East in the Cold War* (Princeton, NJ: Princeton University Press, 2020) is an astounding contribution. See also Kristen Ghodsee, *Second World, Second Sex: Socialist Women's Activism and Global Solidarity during the Cold War* (Durham, NC: Duke University Press, 2019) and Theodora Dragostinova's work *The Cold War from the Margins*.

29. See Odd Arne Westad's chapter 16, "The Cold War in India," in his *The Cold War: A World History* (London: Penguin, 2018), pp. 423–449.

30. The literature on modernization, development, and Second–Third World contacts is growing. See David C. Engerman and Corinna R. Unger, "Towards a Global History of Modernization," *Diplomatic History*, vol. 33, no. 3 (June 2009), pp. 375–385, as well as the whole special issue; David C. Engerman, "The Second World's Third World," *Kritika*, vol. 12, no. 1 (Winter 2011), pp. 183–211. Also see Engerman's "Learning from the East: Soviet Experts and India in the Era of Competitive Coexistence," *Comparative Studies of South Asia, Africa and the Middle East*, vol. 33, no. 2 (August 28, 2013), pp. 227–238. For an example of another socialist state's efforts, this time in the Middle East, see M. Trentin, *Engineers of Modern Development: East German Experts in Ba'thist Syria, 1965–1972* (Padua, 2010).

31. Robert Kline, *The Cybernetics Moment: Or Why We Call Our Age the Information Age* (Baltimore: Johns Hopkins University Press, 2017). For more on cybernetics, its history, and various contexts, see Flo Conway and Jim Siegelman, *Dark Hero of the Information Age: In Search of Norbert Wiener, the Father of Cybernetics* (New York: Basic Books, 2005); Andrew Pickering, *The Cybernetic Brain: Sketches of Another Future* (Chicago: University of Chicago Press, 2010); David A. Mindell, Jérôme Segal, and Slava Gerovitch, "Cybernetics and Information Theory in the United States, France, and

the Soviet Union," in Mark Walker (ed.), *Science and Ideology: A Comparative History* (New York: Routledge, 2003), pp. 66–97.

32. Slava Gerovitch, *From Newspeak to Cyberspeak: A History of Soviet Cybernetics* (Cambridge, MA: The MIT Press, 2002); studies of Bulgarian computing in the language do a great job of tracing the reasons for its creation and its industrial development, but they do not touch on the politico-economic impacts it had on official policy in terms of the party's governance. See Kiril Boi͡anov, *Shtrihi ot Razvitieto na Izcheslitelnata Tehnika v Bŭlgarii͡a* (Sofia: AI Prof Marin Drinov, 2010) and his memoir *Istinata e Kladenets: Zhivotŭt Mi v Kompyutŭrnata Era* (Sofia: AI Akad Marin Drinov, 2018); Dimitŭr Shishkov, *Zvezdnite Migove na Bulgarskata Kompi͡utŭrna Tehnika I Kompi͡utŭrna Informatika 1956–1966* (Sofia: IK Tangra, 2002); Milena Dimitrova, *Zlatnite Desitileti͡a na Bulgarskata Elektronika* (Sofia: IK Trud, 2008); Evgeniĭ Kandilarov, "Elektronikata v Ikonomicheskata Politika na Bŭlgarii͡a prez 60te–80te Godini na XX Vek," *GSU-IF*, vol. 96/7 (2003/4), pp. 431–503. I must also mention other works by actors in the industry, such as Yordan Mladenov and Ognemir Genchev, *Panorama na Elektronnat Promishlenost na Bŭlgarii͡a* (Published online, 2003). Yordan Trenkov's 4-volume *Entsiklopediya na Elektronikata* (Sofia: IK Tehnika, 2010) is a technical reference encyclopedia on electronics, but due to its Bulgarian authorship, it also includes information about various domestic electronics developments from the period, proving invaluable when chasing down obscure disc drives, etc.

33. Benjamin Peters, *How Not to Network a Nation: An Uneasy History of the Soviet Internet* (Cambridge, MA: The MIT Press, 2016).

34. Eden Medina, *Cybernetic Revolutionaries: Technology and Politics in Allende's Chile* (Cambridge, MA: The MIT Press, 2012) remains a guiding light. Work on Latin American science has been very illuminating—see the volume edited by Eden Medina, Ivan da Costa Marques, and Christina Holmes, *Beyond Imported Magic: Essays on Science, Technology, and Society in Latin America* (Cambridge, MA: The MIT Press, 2014). For the Cold War angle, see Gabrielle Hecht (ed.), *Entangled Geographies: Empire and Technopolitics in the Global Cold War* (Cambridge, MA: The MIT Press, 2011). For a Soviet example of a technological community that uses the computer to advance its own claims, build its own project, and exchange information across the Cold War divide, see Ksenia Tatarchenko, *A House with the Window to the West: The Akademgorodok Computer Center (1958–1993)* (PhD dissertation, Princeton University, Princeton, NJ, 2013).

35. I borrow the concept of "cyborg science" from Phillip Mirowski, *Machine Dreams: Economics Becomes a Cyborg Science* (Cambridge: Cambridge University Press, 2002). For the computer specialists as "high priests" of this new universal science in an American context, see Nathan Ensmenger, *The Computer Boys Take Over: Computers, Programmers, and the Politics of Technical Expertise* (Cambridge, MA: The MIT Press, 2010). I am also reacting more broadly to Jeffrey Herf's work, which shows how the technical intelligentsia is a creative class that needs to be taken seriously as a political, cultural, and social force, too; see Jeffrey Herf, *Reactionary Modernism: Technology, Culture, and Politics in Weimar and the Third Reich* (Cambridge: Cambridge University Press, 1984). For the role of women in computing, see Mar Hicks, *Programmed Inequality: How Britain*

Discarded Women Technologists and Lost Its Edge in Computing (Cambridge, MA: The MIT Press, 2017). Joy Lisi Rankin's *A People's History of Computing in the United States* (Cambridge, MA: Harvard University Press, 2018) has also been inspirational in showing the myriad ways users and those usually excluded from computing history contributed.

CHAPTER 1

1. See Angela Romano and Federico Romero (eds.), *European Socialist Regimes' Fateful Engagement with The West: National Strategies in the Long 1970s* (London: Routledge, 2021).

2. These ideas can be found in Rosenstein-Rodan's article "Problems of Industrialization of Eastern and South-Eastern Europe," *Economic Journal,* vol. 53, no. 210/211 (1943), pp. 202–211; Kurt Mandelbaum's are in his short but influential *The Industrialization of Backward Areas* (Oxford: Blackwell, 1945).

3. See Rumen Avramov's magisterial three-volume *Komunalnii͡aat Kapitalizŭm* (Sofia: Bulgarska Nauka I Izkustvo, 2007) for a discussion of Bulgaria's pre-BCP economic history.

4. Ilii͡ana Marcheva, *Politikata za Stopanska Modernizatsiya v Bŭlgarii͡a po Vreme na Studenata Voina* (Sofia: Letera, 2016), p. 39.

5. Michael R. Palairet, *The Balkan Economies c. 1800–1914: Evolution without Development* (Cambridge: Cambridge University Press, 2003).

6. Marcheva, *Politika za Stopanska Modernizatsiya,* p. 40.

7. On Bulgarian tobacco, see Mary Neuburger's *Balkan Smoke: Tobacco and the Making of Modern Bulgaria* (Ithaca, NY: Cornell University Press, 2012).

8. Marcheva, *Politikata za Stopanska Modernizatsiya,* p. 35.

9. L. Berov et al., *Razvitie na Industriyata v Bŭlgarii͡a 1834-1947-1989* (Sofia: Nauka I Izkustvo, 1990), pp. 139–140.

10. Ilii͡ana Marcheva, "Problemi na Modernizatsiyata pri Sotsializma: Industrializatsiyata v Bŭlgarii͡a," in E. Kandilarov and T. Turlakova (eds.), *Izsledvanii͡a po Istorii͡a na Sotsializma v Bŭlgarii͡a 1944–1989* (Sofia: Grafimaks, 2010), pp. 179–180.

11. Berov et al., *Razvitie na Industrii͡ata,* p. 149.

12. Marcheva, "Problemi na Modernizatsiyata," p. 182.

13. Jan T. Gross, "Social Consequences of War: Preliminaries to the Study of Imposition of Communist Regimes in East Central Europe," in *East European Politics and Societies,* vol. 3, no. 2 (March 1989), pp. 201–202.

14. Quoted in Marcheva, "Problemi na Modernizatsiyata," p. 184.

15. Berov et al., *Razvitie na Industrii͡aata,* p. 272; Marcheva, "Problemi na Modernizatsii͡aata," pp. 194–195.

16. It is worth mentioning that early twentieth-century Bulgaria saw the only majority agrarian government in Europe and was a key part of the "Green International."

See Alex Toshkov, "The Rise and Fall of the Green International: Stamboliiski and His Legacy in East European Agrarianism 1919–39," PhD thesis, Columbia University, 2014.

17. Marcheva, "Problemi na Modernizatsii͡aata," p. 204.

18. All figures are from Ulf Brunnbauer, *'Sotsialisticheskii͡aat Nachin na Zhivot': Ideologii͡a, Obshestvo, Semeistvo I Politika v Bŭlgarii͡a (1944–1989)* (Ruse, Bulgaria: MD Elias Kaneti, 2010), pp. 188–189.

19. All figures are from Brunnbauer, *Sotsialisticheskii͡at Nachin na Zhivot,* pp. 208–209.

20. Taken from recordings of the speech, accessible online at the Bulgarian National Radio: http://bnr.bg/radiobulgaria/post/100483520/1958-sedmiat-kongres-na-bkp-obavava-pobedata-na-socializma (last accessed: July 19, 2022).

21. Todor Zhivkov, *Izbrani Suchenenii͡a,* vol. 4 (Sofia: Partizdat, 1975), p. 52.

22. Collectivization was also at the root of the "Goryani" armed resistance movement of dissatisfied peasants and military officers that spread throughout the early 1950s. Most armed resistance was put down by 1956, but the movement's illegal radio station broadcast until 1962. In some ways, the Congress also celebrated a real victory against interior enemies. For more on the Goryani, see the two volumes of documents by the State Archive Agency, *Goryanite* vol. 1 (2001) and vol. 2 (2010).

23. For a good discussion of the Congress, see Evgeniĭ Kandilarov's "Ot 'Realen' kum 'Demokratichen' Sotsializum: Iz Zig-Zagite na Ideynoto I Programnoto Razvitie na BKP sled Vtorata Svetovna Voina" E. Kandilarov and T. Turlakova (eds.), *Izsledvaniya po Istorii͡a na Socializma v Bŭlgarii͡a 1944–1989* (Sofia: Grafimaks, 2010), pp. 97–99.

24. Marcheva, "Problemi na Modernizatsiyata," p. 200.

25. On the little-researched topic of Chinese influence on late-1950s Bulgarian development, see Jan Zofka, "China as a Role Model? The 'Economic Leap' campaign in Bulgaria (1958–1960)," *Cold War History,* vol. 18, no. 3 (2018), pp. 325–342.

26. Brunnbauer, *Sotsialisticheskiyat Nachin na Zhivot,* pp. 139–140.

27. Marcheva, "Problemi na Modernizatsiyata," p. 201

28. Ivailo Znepolski (ed.), *NRB: Ot Nachaloto do Krai͡a* (Sofia: Ciela, 2011), pp. 292–293.

29. Znepolski (ed.), *NRB,* p. 297.

30. TsDA f. 132 op. 1 a.e. 19 l. 79–88.

31. Janos Kornai, *The Socialist System: The Political Economy of Communism* (Princeton, NJ: Princeton University Press, 1992), p. 346.

32. Martin Ivanov and Daniel Vachkov, *Istorii͡a na Vŭnshnii͡a Dŭrzhaven Dŭlg na Bŭlgarii͡a 1878–1990,* Vol. 3 (Sofia: Pechatni Izdaniya na BNB, 2007), p. 114; chapter 12 in the book, on which much of this section is based, is a great overview of the crisis.

33. Ivanov and Vachkov, *Istorii͡a,* Vol. 3, p. 115.

34. For more on the "grand tour," see Ivanov and Vachkov, *Istorii͡a,* Vol. 3, pp. 116–118.

35. Khristo Khristov, *Tainite Failiti na Komunizma: Istinata za Kraha na Bŭlgarskii͡a Sotsializŭm v Sekretnite Arkhivi na Delo N 4/1990 za Ikonomicheskata Katastrofa* (Sofia: Ciela, 2007), p. 54.

36. Khristov, *Tainite Failiti.*

37. Ivanov and Vachkov, *Istorii͡a,* Vol. 3, p. 129.

38. Khristov, *Tainite Faliti,* p. 54.

39. Todor Zhivkov, *Memoari* (Sofia: IK Trud I Pravo 2006), p. 213.

40. TsDA f. 132 op. 5 a.e. 1 l. 23.

41. Daniel Vachkov and Martin Ivanov, *Bŭlgarskii͡at Vŭnshen Dŭlg 1944–1989: Bankruptŭt na Komunisticheskata Ikonomika* (Sofia: Ciela, 2008), p. 247. This book is the best in-depth account of the regime's indebtedness; for a more outright critical view, based on the 1990 judicial case against Zhivkov, see Khristov's *Tainite Failiti.*

42. Andre Steiner, "The Council of Mutual Economic Assistance—An Example of Failed Economic Integration?" *Geschichte und Gesellschaft,* vol. 39, no. 2 (April–June 2013), p. 241.

43. Steiner, "The Council of Mutual Economic Assistance," p. 242.

44. See Romano and Romero, *European Socialist Regimes' Fateful Engagement with The West.*

45. Steiner, "The Council of Mutual Economic Assistance," p. 243.

46. Ilii͡ana Marcheva, *Todor Zhivkov—Puti͡at kŭm Vlastta. Politika I Ikonomika v Bŭlgarii͡a 1953–1964* (Sofia: IK Kota 2000), p. 116.

47. Marcheva, "Problemi na Modernizatsii͡ata," p. 200.

48. Marcheva, "Problemi na Modernizatsii͡ata," p. 197.

49. Vachkov and Ivanov, *Bŭlgarskii͡at,* p. 127.

50. Vachkov and Ivanov, *Bŭlgarskii͡at,* p. 131.

51. Marcheva, "Problemi na Modernizatsiyata," p. 202.

52. Steiner, "The Council of Mutual Economic Assistance," p. 244.

53. Maria Mursean, "Romania's Integration into Comecon. The Analysis of a Failure," *The Romanian Economic Journal,* vol. 11 (2008), p. 46.

54. Steiner, "The Council of Mutual Economic Assistance," p. 244.

55. Johanna Bockman and Gil Eyal, "Eastern Europe as a Laboratory for Economic Knowledge: The Transnational Roots of Neoliberalism," *American Journal of Sociology* vol. 108, no. 2 (September 2002), pp. 324–328.

56. Randall W. Stone, *Satellites and Commissars: Strategy and Conflict in the Politics of Soviet-Bloc Trade* (Princeton, NJ: Princeton University Press, 1996), p. 5.

57. Dina Rome Spechler and Martin C Spechler, "A Reassessment of the Burden of Eastern Europe on the USSR," *Europe-Asia Studies* vol. 61, no. 9 (November 2009), pp. 1645–1657.

58. Dennis Deletant and Mihail Ionesci, "Romania and the Warsaw Pact 1955–1989" (Cold War International History Project, Working Paper no. 43; April 2004), p. 27.

59. Jordan Baev, *Sistemata za Evropeiska Sigurnost I Balkanite v Godinite na Studenata Voina* (Sofia: Damyan Yakov, 2010), p. 99.

60. Baev, *Sistemata za Evropeiska Sigurnost,* p. 117.

61. Milena Dimitrova, *Zlatnite Desitiletiia na Bulgarskata Elektronika* (Sofia: IK Trud, 2008), p. 112.

62. Interview with Stoian Markov, July 28, 2015.

63. Ivan Popov, who died in 2000, did not leave a publicly accessible personal archive, or memoirs, to the detriment of history. However, the contours of his life are well known due to his biographical sketches as a party member, while all interviewees, memoirs, or narratives of the industry talk about him at length, leaving us with a plethora of impressions by subordinates and colleagues. Unless otherwise stated, the biographical sketch here draws extensively from Appendix 3 of Kiril Boianov's *Shtrihi ot Razvitieto na Izchislitelnata Tehnika v Bŭlgariia* (Sofia: Akademichno Izdatelstvo Prof. Marin Drinov, 2010), pp. 178–194; the chapter "Ivan Popov" in Dimitrova, *Zlatnite Desitiletiia,* pp. 95–112; and Jouko Nikula and Ivan Tchalakov, *Innovations and Entrepreneurs in Socialist and Post-Socialist Societies* (Newcastle: Cambridge Scholars Publishing, 2013), pp. 78–112.

64. Interview with Petŭr Petrov, June 10, 2016.

65. "Kŭsi Sŭedeniniia," *Rabotnichesko Delo,* no. 46 (February 15, 1952).

66. Dimitrova, *Zlatnite Desitiletiia,* p. 101.

67. Dimitrova, *Zlatnite Desitiletiia,* p. 25.

68. A feeling familiar to many doctoral students across time and space.

69. *Entsiklopediia Bŭlgariia,* vol. 3 (Sofia: Izdatelstvo na BAN, 1982).

70. Peter Totev interview in Dimitrova, *Zlatnite Desitiletiia,* pp. 37–38.

71. Peter Totev interview in Dimitrova, *Zlatnite Desitiletiia,* p. 40.

72. Stoian Dzhamiĭkov, *Zapiski na Konstruktora* (Sofia: Avangard Prima, 2015), pp. 96–97.

73. Dimitrova, *Zlatnite Desitiletiia,* p. 43.

74. Liubomir Antonov, *Kakvi Sŭm Gi Vŭrshil* (unpublished memoir, available at http://bbaeii.webnode.com/bylg-electronica-i-inormatika/; last accessed July 19, 2022), pp. 66–67.

75. Antonov, *Kakvi Sŭm Gi Vŭrshil,* p. 68.

76. Antonov, *Kakvi Sŭm Gi Vŭrshil,* pp. 74–75.

77. For the turbulent history of Soviet cybernetics, see Slava Gerovitch, *From Newspeak to Cyberspeak: A History of Soviet Cybernetics* (Cambridge, MA: MIT Press, 2002).

78. Kiril Boianov, "Purviiat Izchislitelen Tsentŭr v Bŭlgariia—Nachalo na Informatsionoto Obshetsvto u Nas," in *Bulgarska Nauka,* vol. 6, no. 4 (2011), pp. 53–58.

79. E-mail exchange with his daughter Yana Hashamova, February 28, 2015.

80. Boianov, "Pŭrviiat Izchislitelen Tsentŭr."

81. Antonov, *Kakvi,* p. 77.

82. Interview with Petŭr Petrov, March 19, 2015.

83. Petŭr Petrov, "Angel Simeonov Angelov na 85 Godini" (copy of an article celebrating Angelov's 85th birthday, kindly presented to me by Petrov).

84. Evgeniĭ Kandilarov, "Elektronikata v Ikonomicheskata Politika na Bŭlgariia͡ prez 60te-80te Godini na XX Vek," in *GSU-IF*, vol. 96/7 (2003/2004), p. 440.

85. Kandilarov, "Elektronikata v Ikonomicheskata Politika, p. 441.

86. Marcheva, "Problemi na Modernizatsiia͡ta," p. 204.

87. Petŭr Petrov, "55 Godini Avtomatika, Kibernetika I Robotika v BAN" (Chronological piece available at the website of one of ITKR's successors at http://css.iict.bas.bg/history.html; last accessed October 23, 2016).

88. Petrov, "Angel Simeonov Angelov."

89. Slava Gerovitch, "Mathematical Machines of the Cold War: Soviet Computing, American Cybernetics and Ideological Disputes in the Early 1950s," *Social Studies of Science* vol. 31, no. 2 (April 2001), pp. 256–257.

90. Loren R. Graham, *Science, Philosophy and Human Behaviour in the Soviet Union* (New York: Columbia University Press, 1987), p. 273. Norbert Weiner is considered the father of cybernetics, introducing and elaborating the term in his 1948 seminal work, *Cyberetics: Or Control and Communications in the Animal and the Machine*) (New York: The Technology Press, John Wiley & Sons, Inc., 1948); Claude Shannon was also a key figure in the discipline with his 1949 *Mathematical Theory of Communicaiton,* a foundational work in information theory, introducing the concept of noise in communication channels. These philosophical works impacted discussions that are elaborated in chapter 6; for more on the cybernetic moment, see Robert Kline, *The Cybernetics Moment: Or Why We Call Our Age the Information Age* (Baltimore: Johns Hopkins University Press, 2017).

91. David Mitchell, Jerome Segal, and Slava Gerovitch, "From Communications Engineering to Communications Science: Cybernetics and Information Theory in the United States, France, and the Soviet Union," in Mark Walker (ed.), *Science and Ideology: A Comparative History* (London: Routledge, 2003), p. 81.

92. Igor Poletaev, *Signal: O Nekatoriyh Ponyatiyh Kibernetiki* (Moscow: Izdatelstvo Sovetskoe Radio, 1958), p. 23.

93. *Kratka Bŭlgarska Ent͡siklopediia͡ Tom 3 Kvant-Opere* (Sofia: Izdatelstvo na BAN, 1966), p. 19.

94. The history of Soviet cybernetics is beyond the scope of this book, so for an overview of the topic, see Slava Gerovitch's enthralling *From Newspeak to Cyberspeak: A History of Soviet Cybernetics* (Cambridge, MA: The MIT Press, 2002).

95. Slava Gerovitch, "InterNyet: Why the Soviet Union Did Not Build a Nationwide Computer Network," *History and Technology,* vol. 24, no. 4 (December 2008), p. 337; for an entertaining and more general look, see Francis Spufford, *Red Plenty* (New York: Graywolf Press, 2012).

96. Spufford, *Red Plenty;* for an excellent overview of the attempts to network Soviet society to achieve this control, see Benjamin Peters, *How Not to Network a Nation: The*

Uneasy History of the Soviet Internet (Cambridge, MA: The MIT Press, 2016), a work that will return in later chapters.

97. Gerovitch, "InterNyet," p. 340.

98. Gerovitch, "Mathematical Machines of the Cold War," pp. 265–266.

99. Boris Malinovsky, *Pioneers of Soviet Computing* (Published electronically, 2010; http://www.sigcis.org/files/SIGCISMC2010_001.pdf; last accessed October 24, 2016), p. 23.

100. Gerovitch, "Mathematical Machines of the Cold War," p. 268.

101. James W. Cortada, "Information Technologies in the German Democratic Republic (GDR), 1949–1989," *IEEE Annals of the History of Computing,* vol. 34, no. 2 (February 2012), p. 37.

102. Petri Paju and Helena Durnova, "Computing Close to the Iron Curtain: Inter/national Computing Practices in Czechoslovakia and Finland 1945–1970," *Comparative Technology Transfer and Society,* vol. 7, no. 3 (December 2009), p. 311.

103. Anonymous, "History of Computing Developments in Romania," *IEEE Annals of the History of Computing,* vol. 21, no. 3 (1999), p. 58.

104. Zsuzsa Szentgyorgi, "A Short History of Computing in Hungary," *IEEE Annals of the History of Computing,* vol. 21, no. 3 (1999), p. 51.

105. Andrew Targowski, *The History, Present State, and Future of Information Technology* (Santa Rosa, CA: Informing Science Press, 2016), pp. 123–126.

106. Cortada, "Information Technologies in the GDR," p. 37.

107. Here I call on the work of Alexander Gerschrenkron, *Economic Backwardness in Historical Perspective: A Book of Essays* (Cambridge, MA: Belknap Press, 1962).

CHAPTER 2

1. A popular slogan of supporters of Levski-Sofia, one of the country's most popular football clubs.

2. Kiril Boi͡anov, *Shtrihi ot Razvitieto na Izchislitelnata Tehnika v Bŭlgarii͡a* (Sofia: Akademichno Izdatelstvo Prof Marin Drinov, 2010), p. 22.

3. TsDA f. 1B op. 68 a.e. 1836 l. 201 (Politburo Discussions on Electronic Development, 1986).

4. For an argument about this Soviet amalgamation, see Vladislav Zubok and Constantine Pleshakov, *Inside the Kremlin's Cold War: From Stalin to Khrushchev* (Cambridge, MA: Harvard University Press, 1996).

5. Elidor Mehilli, *From Stalin to Mao: Albania and the Socialist World* (Ithaca, NY: Cornell University Press, 2017), p. 5.

6. Quoted in Ivan Chalŭkov's *Da Napravish Holograma: Kniga za Uchenite, Svetlinata I Vsichko Ostanalo* (Sofia: AI Prof. Marin Drinov, 1998) p.17; the provocative joke of Zhivkov as venture capitalist was first uttered by Quinn Slobodian at a workshop at Columbia University in April 2016.

7. Boi͡anov, *Shtrihi,* p. 16.

8. Boĭanov, *Shtrihi,* pp. 17–18.

9. Boĭanov, *Shtrihi,* p. 19.

10. Boĭanov, *Shtrihi,* p. 22.

11. Boĭanov, *Shtrihi,* p. 24.

12. TsDA f. 1B op. 6 a.e. 5513 l. 150–153 (Politburo Discussions 1964).

13. TsDA f. 1B op. 6 a.e. 5513 l. 152.

14. TsDA f. 1B op. 6 a.e. 5513, l. 136.

15. TsDA f. 1B op. 6 a.e. 5513 l. 138–144.

16. I thank Michel Christian for drawing my attention to UNIDO's importance in his paper, "Developing a National-Based Electronics Industry. UNIDO's Cooperation with the Bulgarian Research Institute for Instrument Design in the 1970s and 1980s," presented at the Pan-European Economic Spaces in the Cold War Conference in Geneva on June 2022. UNIDO is curiously absent from the Bulgarian archives, and this archival silence probably speaks volumes about how national documents reflect a certain technological nationalism. So I thank Michel very warmly for bringing in this angle from the UN in the eleventh hour! A mention of UNIDO can be seen in passing in Kiril Boĭanov's memoir *Istinata e Kladenets: Zhivotŭt Mi v Kompyutŭrnata Era* (Sofia: AI Akad Marin Drinov, 2018) but without much detail—again, demonstrating how different the focus is if it is local!

17. *Higher Mechanical and Electrical Engineering Institute Varna* (Sofia: Septemvri Publishing House, 1987), pp. 6–7, 13.

18. *45 Godini Katedra Elektronika*—available at https://electronica-tugab.eu/images/stories/Documents/knijka_45.pdf (last accessed: August 1 2022).

19. TsDA f. 1B op. 6 a.e. 5513 l. 142 (Politburo Discussions).

20. Postanovlenie 26 in *Dŭrzhaven Vestnik,* issue 36, 1965.

21. Li͡ubomir Antonov, *Kakvi Sum Gi Vurshil* (unpublished memoir, available at https://6593fa9ac5.cbaul-cdnwnd.com/ea204c52c98c1613523a2e268ca812d5/200000313-d338ed432c/LAntonovAvtobio0512.pdf (last accessed August 1, 2022), pp. 86–87.

22. Antonov, *Kakvi Sum Gi Vurshil,* p. 88.

23. Dimitŭr Shishkov, *Zvezdnite Migove na Bŭlgarskata Elektroika* (Sofia: TangraTanNakRA IK 2002), p. 352.

24. Evgeniĭ Kandilarov, "Elektronikata v Ikonomicheskata Politika na Bŭlgarii͡a prez 60te–80te Godini na XX Vek," in *GSU-IF,* vol. 96/7 (2003/2004), pp. 445–446.

25. Antonov, *Kakvi Sum Gi Vurshil,* pp. 95–97.

26. Antonov, *Kakvi Sum Gi Vurshil,* pp. 95–97.

27. TsDA f. 517 op. 2 a.e. 169 l. 45–7 (CSTP Discussion on Soviet Co-Operation 1966–7).

28. TsDA f. 517 op. 2 a.e. 169 l. 45.

29. TsDA f. 517 op. 2 a.e. 169 l. 47.

30. TsDA f. 830 op. 1 a.e. 88 l. 1–2 (Izotimpex Exports 1968).

31. TsDA f. 830 op. 1 a.e. 90 l. 1–2 (Izotimpex Exports 1970).

32. TsDA f. 830 op. 1 a.e. 90 l. 5.

33. TsDA f. 830 op. 1 a.e. 91 l. 3 (Izotimpex Statistical Report 1971).

34. TsDA f. 830 op. 1 a.e. 92 l. 2 (Izotimpex Statistical Report 1972).

35. TsDA f. 830 op. 1 a.e. 93 l. 3 (Izotimpex Statistical Report 1973).

36. Antonov, *Kakvi Sŭm Gi Vŭrshil,* p. 104.

37. Antonov, *Kakvi Sŭm Gi Vŭrshil,* p. 100.

38. Milena Dimitrova, *Zlatnite Desetiletii͡a na Bŭlgarskata Elektronika* (Sofia: IK Trud, 2008), p. 216.

39. Jouko Nikula and Ivan Tchalakov, *Innovations and Entrepreneurs in Socialist and Post-Socialist Societies* (Newcastle: Cambridge Scholars Publishing, 2013), pp. 95–96.

40. Sayuri Guthrie-Shimizu, "Japan, the United States, and the Cold War 1945–1960," in Odd Arne Westad and M. Leffler (eds.), *Cambridge History of the Cold War, Vol. 1—Origins* (Cambridge: Cambridge University Press, 2010), p. 257.

41. Guthrie-Shimizu, "Japan, the United States, and the Cold War 1945–1960," p. 258.

42. Michael Schaller, "Japan and the Cold War 1960–1991," in Westad and Leffler (eds.), *Cambridge History of the Cold War, Vol. 3—Endings*, pp. 161–162.

43. Evgeniĭ Kandilarov, *Bŭlgarii͡a I I͡Aponii͡a: Ot Studenata Voina kum XXI Vek* (Sofia: Izdatelstvo Damyan Yankov, 2009), p. 42; the only monograph on Bulgarian-Japanese Cold War relations is also a great overview of the country's importance for Bulgarian technological policy.

44. Kandilarov, *Bŭlgarii͡a I I͡Aponiia,* p. 63.

45. Kandilarov, *Bŭlgarii͡a I I͡Aponii͡a,* p. 47.

46. Kandilarov, *Bŭlgarii͡a I I͡Aponiia,* p. 71.

47. Interview with Dimo Dimov, in Dimitrova, *Zlatnite Desitiletii͡a,* p. 79.

48. Dimitrova, *Zlatnite Desitiletii͡a,* p. 82.

49. Dimitrova, *Zlatnite Desitiletii͡a,* p. 84; Marangozov was the future father of the Bulgarian personal computer.

50. Kandilarov, *Bŭlgarii͡a i I͡Aponii͡a,* p. 86.

51. Dimitrova, *Zlatnite Desitiletii͡a,* p. 86.

52. Dimitrova, *Zlatnite Desitiletii͡a,* p. 87.

53. Dimitrova, *Zlatnite Desitiletii͡a,* p. 88.

54. Vera Vutova-Stefanova and Evgeniĭ Kandlilarov, *Bŭlgarii͡a I I͡Aponii͡a: Politika, Diplomatsiya, Lichnosti I Subitiya* (Sofia: Iztok-Zapad, 2019), p. 341.

55. Both Papazov and Doĭnov's experiences will be discussed more fully in chapter 7.

56. The Bulgarians took EXPO'70 very seriously, as a chance to boost relations with this important country. Preparation started in 1967, and the innovative pavilion design was approved in 1968, granting the project to the team of architects led by

Todor Kozhuharov and Evlogi T͡Svetkov, who aimed to symbolize the Balkan mountains through the four glass peaks. In six months, it was visited by over 9 million people and was the subject of 1,200 publications, a real surprise hit at the exposition. It was decided to bring the pavilion back to Bulgaria and make it permanent, either in Sofia or on the premises of the Plovdiv Trade Fair. The project, however, never saw the light of day due to ever-rising costs.

57. Todor Zhivkov, *Memoari* (Sofia: IK Trud I Pravo, 1996), pp. 512–513.

58. Zhivkov, *Memoari,* p. 513.

59. Zhivkov, *Memoari,* p. 514.

60. TsDA f. 1B op. 35 a.e. 1457 l. 11–13 (Politburo Discussions 1970).

61. TsDA f. 1B op. 35 a.e. 1457 l. 17.

62. TsDA f. 1B op. 35 a.e. 1457 l. 19–21.

63. TsDA f. 1B op. 35 a.e. 1457 l. 19–21.

64. The ITCR's story is key to chapters 5 and 6 and will be picked up in more detail there.

65. Li͡ubomir Iliev, *Matematikata v Narodna Republika Bŭlgarii͡a* (Sofia: Sayuz na Matematitsite v Bŭlgarii͡a, 1984).

66. Snezhana Khristova, *40 Godini Tzentralen Institut po Izhislitelna Tehnika* (unpublished memoir, available at http://bbaeii.webnode.com/bylg-electronica-i-inormatika/; last accessed: November 4, 2016).

67. TsDA f. 37A op. 1 a.e. 1 l. 7–32 (CICT Annual Plans 1966–8).

68. TsDA f. 37A op. 1 a.e. 2 l. 46 (Plans & Reports 1969–70).

69. Shishkov, *Zvezdnite Migove*, p. 225.

70. Kiril Boi͡anov, Speech Commemorating Angel Angelov's 80th Birthday, February 12, 2009, Sofia (text accessible at http://bbaeii.webnode.com/bylg-electronica-i-inormatika, last accessed August 1, 2022).

71. Interview with Angel Angelov conducted by e-mail through his daughter Sonia Angelova Hirt, June 29, 2016.

72. TsDA f. 136 op. 44 a.e. 10 l. 27–33 (Council of Ministers 1966

73. TsDA f. 136 op. 44 a.e. 10 l. 27–33 l.1.

74. TsDA f. 1B op. 35 a.e. 381 l. 41–2 (Politburo Discussions on Electronic Siting 1968).

75. TsDA f. 1003 op. 1 a.e. 1 l. 2 (IZOT Decisions 1969–78).

76. TsDA f. 1003 op. 1 a.e. 1 l. 1.

77. The Bulgarian Agrarian National Union was a coalition partner of the BCP throughout the regime's history, a curiosity that had practical benefits, such as using it to negotiate with regimes and organizations that would not do so with the Communist Party.

78. TsDA f. 1B op. 35 a.e. 54 l. 24–5 (Politburo Seminar on Technical Progress 1968).

79. The integrated circuit, or the microchip, was the revolution in electronics that allowed for the fast development of computers after the 1960s and their entry into

everyday life. The first such circuit was created in 1958, with mass production starting in the 1960s and increased miniaturization throughout the following decades.

80. TsDA f. 136 op. 48 a.e. 434 (Council of Ministers 1969).

81. The semiconductor became the basis of one of Bulgarians' favorite Zhivkov gaffes, commonly cited as proof of his low education and unclear understanding of what he was championing: "This year we have built a factory for semi-conductors, next—for full conductors!"

82. TsDA f. 136 op. 48 a.e. 422 (Council of Ministers 1975).

83. *Doklad vurhu Proekta za Ikonomicheski Rastezh i Prehod kum Pazarnata Ikonomika v Bŭlgarii͡a,* prepared for the United States Chamber of Commerce by Richard Rahn and Ronald Utt, October 31, 1990, chapter 22, p. 16.

84. TsDA f. 1B op. 35 a.e. 1063 (Politburo Decision 1969).

85. TsDA f. 1B op. 35 a.e. 1172 (Politburo Discussion on Territorial Distribution of Factories 1970).

86. Evgeniĭ Kandilarov, "Elektronikata v Ikonomicheskata Politika na Bŭlgarii͡a prez 60te-80te Godini na XX Vek," *GSU-IF*, vol. 96/97 (2003/2004), p. 464.

87. TsDA f. 136 op. 51 a.e. 158 l. 23 (Council of Minister **Minutes**

88. RGAE f. 9480 op. 9 a.e. 217 l. 3 (5th Session of Bulgarian-Soviet Co-Operation Commission 1966).

89. RGAE f. 9480 op. 9 a.e. 217 l. 5, l. 21.

90. RGAE f. 9480 op. 9 a.e. 218 l. 11 (GKNT Data on Co-Operation 1966).

91. RGAE f. 9480 op. 9 a.e. 218 l. 1.

92. RGAE f. 9480 op. 9 a.e. 462 l. 12 (6th Session of Co-Operation Commission 1967).

93. Benjamin Peters, *How Not to Network a Nation: The Uneasy History of the Soviet Internet* (Cambridge, MA: The MIT Press, 2016), p. 163.

94. RGAE f. 9480 op. 9 a.e. 880 l. 12 (GKNT Report on Computers in USSR—1969).

95. RGAE f. 9480 op. 9 a.e. 880 l. 150.

96. RGAE f. 9480 op. 9 a.e. 880 l. 151–154.

97. RGAE f. 9480 op. 9 a.e. 880 l. 221.

98. Vasil Nedev, *Hronika na Bulgarskata Kompi͡utŭrna Tehnika* (unpublished, available at http://bbaeii.webnode.com/bylg-electronica-i-inormatika/; last accessed August 1, 2022).

99. Stoi͡an Shalamanov, *NRB V Sotsialisticheskata Ikonomicheska Integratsiya: Strukturna Politika* (Sofia: Partizdat, 1981), p. 140.

100. Boi͡anov, *Shtrihi,* p. 185.

101. "Western Computer Companies Step Up Sales Drive," *Financial Times,* February 12, 1969, p. 7.

102. TsDA f. 37A op. 1 a.e. 7 l. 16 (CICT On ES Elements 1968).

103. TsDA f. 37A op.1 a.e. 17 l. 7–11 (CICT Documents for ES Meetings 1968).

104. TsDA f. 37A op. 1 a.e. 22 l. 2–10 (CICT On ES Input-Output Meetings 1969).

105. Kandilarov, "Elektronikata v Ikonomicheskata Politika," p. 437.

106. Interview with Alexander T͡Svetkov, April 6, 2015.

107. Interview with Angel Angelov in Dimitrova, *Zlatnite Desiteletii͡a,* p. 130.

108. TsDA f. 136 op. 49 a.e. 243 l. 77–8 (Council of Ministers).

109. TsDA f. 1B op. 35 a.e. 493 l. 5–6 (Politburo on Comecon Integration 1968–9).

110. TsDA f. 1B op. 35 a.e. 493 l. 10.

111. TsDA f. 1B op. 35 a.e. 493 l. 11.

112. TsDA f. 1B op. 35 a.e. 493 l. 14.

113. TsDA f. 1B op. 35 a.e. 493 l. 16–17.

114. TsDA f. 1B op. 35 a.e. 493 l. 22, l. 39–41.

115. TsDA f. 517 op. 2 a.e. 173 l. 62 (CSTP on Trade Problems 1969).

116. William Butler (ed.), *A Source Book on Socialist International Organizations* (Alphen aan den Rijn, The Netherlands: Sijthoff & Noordhoff, 1978).

117. Elitza Stanoeva, "Balancing between Socialist Internationalism and Economic Internationalisation: Bulgaria's Economic Contacts with the EEC," in Angela Romano and Federico Romero (eds.), *European Socialist Regimes' Fateful Engagement with The West: National Strategies in the Long 1970s* (London: Routledge 2020).

118. TsDA f. 1B op. 35 a.e. 12 l. 18–23 (Politburo on Relations with EC 1967–8).

119. RGAE f. 9480 op. 9 a.e. 939 l. 3, 49 (11th Session of Commission 1969).

120. RGAE f. 9480 op. 9 a.e. 938 l. 19–20 (10th Session of Commission 1969).

121. RGAE f. 9480 op. 9 a.e. 1198 l. 29 (13th Session of Commission 1970).

122. RGAE f. 9480 op. 9 a.e. 1199 l. 17–19 (Session Participation 1970).

123. RGAE f. 9480 op. 9 a.e. 1471 l. 12–18; RGAE f. 9480 op. 9 a.e. 1725 l. 12, 20, 38; RGAE f. 9480 op. 9 a.e. 1727 l. 24 (14th, 16th, and 17th Sessions—1971–2).

124. RGAE f. 9480 op. 9 a.e. 1474 l. 20 (Report on Activity 1971).

125. RGAE f. 9480 op. 9 a.e. 1473 l. 32 (14th and 15th Sessions 1971).

126. RGAE f. 9480 op. 9 a.e. 1474 l. 21 (Report on Activity 1971).

127. RGAE f. 9480 op. 9 a.e. 1726 l. 22 (17th Session 1972).

128. RGAE f. 9480 op. 9 a.e. 2250 l. 6 (GKNT Reports 1974).

129. TsDA f. 378B a.e. 360 (Minutes of Zhivkov-Brezhnev Meeting 1973).

130. Khristo Khristov, *Tainite Faliti Na Komunizma* (Sofia: Ciela, 2007), p. 104.

131. RGAE f. 9480 op. 9 a.e. 1726 l. 12–23 (17th Session 1972).

132. RGAE f. 9480 op. 9 a.e. 1728 l. 8–10 (GKNT Ukrainian Co-operation 1972).

133. RGAE f. 9480 op. 9 a.e. 1197 l. 33 (12th Session 1970).

134. RGAE f. 9480 op. 9 a.e. 1474 l. 6 (GKNT Report 1972).

135. RGAE f. 9480 op. 9 a.e. 1727 l. 99–104 (16th and 17th Sessions 1972).

136. RGAE f. 9480 op. 9 a.e. 1987 l. 1 (GKNT Negotiations 1973–4).

137. RGAE f. 9480 op. 12 a.e. 585 l. 1 (GKNT Co-Operation in Information 1977).

138. TsDA f. 517 op. 4 a.e. 24 l. 79 (CSTP Licenses 1975).

139. TsDA f. 517 op. 4 a.e. 15 l. 8 (CSTP On Soviet Integration 1973).

140. Shishkov, *Zvezdnite Migove*, p. 226.

141. Angel Angelov et al., *Elektronikata v Bŭlgarii͡a: Minalo, Nastoi͡ashte, Bŭdishte* (Sofia: Tekhnika, 1983), pp. 44–46.

142. RGAE f. 9480 op. 9 a.e. 1471 l. 33 (14th Session 1971).

143. Interview with Nedko Botev, Boyan T͡Sonev, and Koĭcho Dragostinov, June 23, 2015.

144. TsDA f. 136 op. 51 a.e. 158 l. 15–18 (Council of Ministers).

145. Interview with Stoi͡an Markov, July 28, 2015; Interview with Plamen Vachkov, June 30, 2015.

146. TsDA f. 1B op. 35 a.e. 4189 l. 5–6 (Popov Report to Zhivkov 1973).

147. TsDA f. 1B op. 35 a.e. 4189 l. 89.

148. *Dŭrzhaven Vestnik,* issue 56, 1973.

149. He had also started his rise while Papazov was ambassador to Japan; from 1971 Papazov had replaced Popov as head of CSTP.

150. Ogni͡an Doĭnov, *Spomeni* (Sofia: Trud, 2002).

151. As will be seen in chapter 5, this did not stop him from retaining an active and constructive role in scientific policy, turning the organization into a space of innovation.

152. TsDA f. 1B op. 35 a.e. 5176 l. 1–20 (Central Committee Plenum 1975).

153. TsDA f. 1B op. 66 a.e. 1303 l. 34 (Doĭnov Politburo Report on Accelerated Development 1977).

154. TsDA f. 1B op. 66 a.e. 1303 l. 23.

155. Boi͡anov, *Shtrihi,* p. 213; the years of introductions below are taken from the list the author presents between pages 212 and 230.

156. Ognemir Genchev, *Panorama na Elektronnata Promishlenost na Bŭlgarii͡a: Fakti I Dokumenti* (Sofia: Ciela, 2003), p. 72.

157. Shalamanov, *NRB V Sotsialisticheskata Ikonomicheska Integratsiya,* p. 147.

158. TsDA f. 517 op. 5 a.e. 15 l. 119 (CSTP Development Strategy 1979–1980).

159. TsDA f. 136 op. 68 a.e. 107 l. 720 (Council of Ministers); the issue of debt and capitalist currency for electronics will be discussed in the conclusion.

160. The standard in hard drive technology from the 1970s until the 2010s, the Winchester drive was introduced by IBM in 1973. Its low mass and low-load heads allowed for much greater recording density and faster read times, as well as reduced complexity and costs—and thus higher profits.

161. TsDA f. 517 op. 5 a.e. 38 l. 67 (CSTP on Permanent National Exhbit 1979).

162. TsDA f. 517 op. 5 a.e. 38 l. 13.

163. TsDA f.1B op. 59 a.e. 42 l. 9 (Politburo on Socio-Economic Plan 1979).

164. TsDA f. 517 op. 6 a.e. 108 l. 198 (CSTP Automation-8 Program 1982).

165. TsDA f. 517 op. 7 a.e. 86 l. 1 (CSTP Automation-8 Program 1980).

166. Dimitrova, *Zlatnite Desitiletii͡a*, pp. 192–194.

167. Interview with Krasimir Markov, February 4, 2016.

168. See Appendix B for a list of Bulgarian electronics production.

169. TsDA f. 1003 op. 1 a.e. 22 l. 5 (IZOT Cost Reports 1981).

170. Boi͡anov, *Shtrihi*, p. 231.

171. TsDA f. 517 op. 6 a.e. 21 l. 42 (CSTP On Memory Devices 1981).

172. TsDA f. 830 op. 1 a.e. 89 l. 2 (Izotimpex Statistical Report 1969).

173. TsDA f. 830 op. 1 a.e. 90 l. 6 (Izotimpex Statistical Report 1970).

174. TsDA f. 830 op. 1 a.e. 90 l. 2.

175. TsDA f. 830 op. 2 a.e. 25 l. 13 (Izotimpex Report 1984).

176. TsDA f. 830 op. 2 a.e. 26 l. 6 (Izotimpex Report 1985).

177. TsDA f. 830 op. 2 a.e. 34 l. 11 (Izotimpex Export Bulletin 1987).

178. TsDA f. 830 op. 2 a.e. 28 l. 1 (Izotimpex Report 1987).

179. TsDA f. 1003 op. 1 a.e. 71 l. 12 (IZOT Specialists Council Meetings 1978–80).

180. "Milliarden Dollar Schulden in Moskau," *Der Spiegel*, no. 46 (1982).

181. Quoted in Ivailo Znepolski (ed.), *NRB: Ot Nachaloto do Kraya* (Sofia: Ciela, 2011), p. 314.

CHAPTER 3

1. Miroslaw Sikora, "Cooperating with Moscow, Stealing in California: Poland's Legal and Illicit Acquisition of Microelectronic Knowhow from 1960 to 1990," in C. Leslie and M. Schmitt (eds.), *Histories of Computing in Eastern Europe. HC 2018. IFIP Advances in Information and Communication Technology*, vol. 549 (Cham, Switzerland: Springer, 2019), p. 171.

2. Sikora, "Cooperating with Moscow, Stealing in California," p.166; see also M. Sikora, "Clandestine Acquisition of Microelectronics and Information Technology by the Scientific-Technical Intelligence of Polish People's Republic in 1970–1990," 2017 Fourth International Conference on Computer Technology in Russia and in the Former Soviet Union (SORUCOM), (IEEE 2017) (Zelenograd, 2017), pp. 200–212.

3. For his ideas on the closed-world metaphor, see Paul Edwards, *The Closed World: Computers and the Politics of Discourse in Cold War America* (Cambridge, MA: MIT Press, 1996).

4. Yoko Yasuhara, "The Myth of Free Trade: The Origins of CoCom 1945–1950," *Japanese Journal of American Studies*, vol. 4 (1991), p. 128. For more on different

perspectives on CoCom, see Gary K. Bertsch, Heinrich Vogel, and Jan Zielonka, *After the Revolutions: East-West Trade and Technology Transfer in the 1990s* (New York: Routledge, 2019).

5. Bertsch, Vogel, and Zielonka, *After the Revolutions*, pp. 143–144.

6. C. Leslie, "From CoCom to Dot-Com: Technological Determinisms in Computing Blockades, 1949 to 1994," in C. Leslie and M. Schmitt M. (eds.), *Histories of Computing in Eastern Europe. HC 2018. IFIP Advances in Information and Communication Technology*, vol. 549 (Cham, Switzerland: Springer, 2018).

7. Interview with Alexander T͡Svetkov, April 6, 2015.

8. Interview with Krasimir Markov, February 4, 2016.

9. R. Bergien, "Programmieren mit dem Klassenfeind," *Vierteljahrshefte für Zeitgeschichte*, vol. 67, no. 1 (2019), pp. 1–30.

10. Commission on the Dossiers, *Dŭrzhavna Sigurnost i Nauchno-Tehnicheskoto Razuznavane* (Sofia: KRDOPBGDSRSBNA, 2013), p. 6.

11. Momchil Metodiev, *Mashina za Legitimnost: Rolyata na Dŭrzhavna Sigurnost v Komunisticheskata Durzhava* (Sofia: Ciela, 2008), p. 18.

12. Khristo Khristov, "Dŭrzhavna Sigurnost. Chast 2.1: Nauchno-Tehnicheskoto Razuznavane," *Kapital*, August 29, 2010.

13. For a great work on this issue, see Jordan Baev, *KGB v Bŭlgarii͡a* (Sofia: Voenno Izdatelstvo, 2009).

14. Metodiev, *Mashina za Legitimnost*, p. 113.

15. "Information" was the catch-all term that STI used to designate any document or item that was acquired; it could thus be anything from a material good to a research plan. The acquisition and implementation of "informations" was at the heart of its activities.

16. Khristov, "Dŭrzhavna Sigurnost. Chast 2.1."

17. TsDA f. 1B op. 64 a.e. 313 l. 30 (Politburo Reports 1964).

18. AKRDOPBGDSRSNBA-M/R f. 66 op. 1 a.e. 70 l. 1 (Central Committee Decisions 1966).

19. Metodiev, *Mashina za Legitimnost*, p. 117.

20. Metodiev, *Mashina za Legitimnost*, p. 117.

21. AKRDOPBGDSRSNBA-R, f. 9 op. 4 a.e. 589 l. 21 (Council of Ministers Order 1980).

22. AKRDOPBGDSRSNBA-R, f. 9 op. 4 a.e. 466 l. 35 (Politburo Decision 1979).

23. AKRDOPBGDSRSNBA-R, f. 9 op. 2 a.e. 412 l. 13 (Organizational Structure 1967).

24. AKRDOPBGDSRSNBA-R, f. 9 op. 2 a.e. 412 l. 69 (Organizational Structure 1967).

25. AKRDOPBGDSRSNBA-R, f. 9 op. 2 a.e. 367 l. 1–2 (Organizational Instructions 1976).

26. AKRDOPBGDSRSNBA-R, f. 9 op. 2 a.e. 418 l. 20 (Suggestions to First Directorate 1974).

27. AKRDOPBGDSRSNBA-R, f. 9 op. 2 a.e. 418 l. 21 (Suggestions to First Directorate 1974).

28. AKRDOPBGDSRSNBA-R, f. 9 op. 2 a.e. 367 l. 2 (Temporary Instructions 1976).

29. AKRDOPBGDSRSNBA-R, f. 9 op. 4 a.e. 466 l. 38 (Politburo Decision 1979).

30. AKRDOPBGDSRSNBA-R, f. 9 op. 3 a.e. 149A l. 67 (First Directorate Order 1977).

31. AKRDOPBGDSRSNBA-R, f. 9 op. 4 a.e. 432 l. 125–7 (Internal Report 1971).

32. AKRDOPBGDSRSNBA-R, f. 9 op. 4 a.e. 462 l. 124–6 (Instructions on ISKRA Usage 1981).

33. Joshua Sanborn is currently working on a study of the KGB and computers, and I thank him for sharing some of his preliminary insights with me.

34. AKRDOPBGDSRSNBA-R, f. 9 op. 4 a.e. 453 l. 305–7 (Report on Trip to KGB 1980).

35. AKRDOPBGDSRSNBA-M, f. 1 op. 11 a.e. 445 l. 12–13 (Information Plan of Directorate 1980).

36. AKRDOPBGDSRSNBA-R, f. 9 op. 4 a.e. 527 l. 1 (Meeting with Mongolian Representative 1983).

37. AKRDOPBGDSRSNBA-R, f. 9 op. 4 a.e. 476 l. 9 (Trip to Havana 1981).

38. AKRDOPBGDSRSNBA-R, f. 9 op. 4 a.e. 528 l. 1 (Joint Meeting with Vietnam 1983).

39. Charles Tilly, "War Making and State Making as Organized Crime," in P. B. Evans, D. Rueschemeyer, and T. Skocpol (eds.), *Bringing the State Back In* (Cambridge: Cambridge University Press 1985), p. 184.

40. Evgeniĭ Kandilarov, "Elektronikata v Ikonomicheskata Politika na Bŭlgarii͡a prez 60te-80te Godini na XX Vek," *GSU-IF,* vol. 96/97 (2003/2004), p. 461.

41. AKRDOPBGDSRSNBA-R, f. 9 op. 2 a.e. 408 l. 38–9 (Activity Report 1972).

42. AKRDOPBGDSRSNBA-R, f. 9 op. 2 a.e. 408 l. 40 (Activity Report 1972).

43. AKRDOPBGDSRSNBA-R, f. 9 op. 2 a.e. 408 l. 38 (Activity Report 1972).

44. AKRDOPBGDSRSNBA-R, f. 9 op. 2 a.e. 408 l. 14 (Activity Report 1972).

45. AKRDOPBGDSRSNBA-R, f. 9 op. 2 a.e. 378 l. 67 (Activity Report 1976).

46. AKRDOPBGDSRSNBA-R, f. 9 op. 2 a.e. 375 l. 54–6 (Activity Report 1973).

47. AKRDOPBGDSRSNBA-R, f. 9 op. 2 a.e. 376 l. 57 & f. 9 op. 2 a.e. 366 l. 56 (Activity Reports 1974 & 1975).

48. AKRDOPBGDSRSNBA-R, f. 9 op. 2 a.e. 378 l. 67 (Activity Report 1976).

49. AKRDOPBGDSRSNBA-R, f. 9 op. 2 a.e. 390 l. 19 (Post-Helsinki Situation 1976).

50. AKRDOPBGDSRSNBA-R, f. 9 op. 2 a.e. 390 l. 20 (Post-Helsinki Situation 1976).

51. AKRDOPBGDSRSNBA-R, f. 9 op. 4 a.e. 433B l. 61 (Exchanges with KGB 1976).

52. Khristov, "Dŭrzhavna Sigurnost. Chast 2.1."

53. AKRDOPBGDSRSNBA-R, f. 9 op. 3 a.e. 171 l. 46 (Exchanges with KGB 1978).

54. AKRDOPBGDSRSNBA-R, f. 9 op. 2 a.e. 441 l. 1 (Report on Allied Services Meeting 1970).

55. AKRDOPBGDSRSNBA-R, f. 9 op. 2 a.e. 441 l. 1–3 (Report on Allied Services Meeting 1970).

56. AKRDOPBGDSRSNBA-R, f. 9 op. 2 a.e. 382 l. 20 (Received Materials 1968).

57. AKRDOPBGDSRSNBA-R, f. 9 op. 2 a.e. 398 l. 181 (Soviet Materials 1969).

58. AKRDOPBGDSRSNBA-R, f. 9 op. 3 a.e. 171 l. 44 (Exchnges with KGB 1978).

59. AKRDOPBGDSRSNBA-R, f. 9 op. 4 a.e. 540 l. 205–6 (Exchanges with KGB 1985).

60. AKRDOPBGDSRSNBA-R, f. 9 op. 4 a.e. 540 l. 204 (Exchanges with KGB 1985).

61. AKRDOPBGDSRSNBA-R, f. 9 op. 4 a.e. 556 l. 166–172 (KGB Exchanges 1986).

62. AKRDOPBGDSRSNBA-R, f. 9 op. 4 a.e. 576 l. 166 (KGB Exchanges 1987).

63. Which Charles Maier argues is a key part in the GDR's collapse—see chapter 2 of his *Dissolution: The Crisis of Communism and the End of East Germany* (Princeton, NJ: Princeton University Press, 1999).

64. AKRDOPBGDSRSNBA-R, f. 9 op. 2 a.e. 449 l. 227 (Report on Berlin Talks 1972).

65. AKRDOPBGDSRSNBA-R, f. 9 op. 3 a.e. 150B l. 27 (Talks with Germans 1975).

66. AKRDOPBGDSRSNBA-R, f. 9 op. 4 a.e. 433B l. 76 (Talks with Germans 1976).

67. AKRDOPBGDSRSNBA-R, f. 9 op. 4 a.e. 451 l. 101 & f. 9 op. 4 a.e. 487 l. 36 (Talks with Germans 1980 & 1982).

68. AKRDOPBGDSRSNBA-R, f. 9 op. 4 a.e. 451 l. 101 & f. 9 op. 4 a.e. 487 l. 150 (Talks with Germans 1980 & 1982).

69. AKRDOPBGDSRSNBA-R, f. 9 op. 4 a.e. 543 l. 153 (Exchange with GDR 1985).

70. AKRDOPBGDSRSNBA-R, f. 9 op. 3 a.e. 181 l. 42 (Exchange with Czechoslovakia 1977).

71. AKRDOPBGDSRSNBA-R, f. 9 op. 4 a.e. 489 l. 33–4 (Exchange with Czechoslovakia 1982).

72. AKRDOPBGDSRSNBA-R, f. 9 op. 4 a.e. 508 l. 62–71 (Exchange with Czechosovakia 1983).

73. AKRDOPBGDSRSNBA-R, f. 9 op. 4 a.e. 476 l. 5–8 (Havana Trip 1981).

74. AKRDOPBGDSRSNBA-R, f. 9 op. 4 a.e. 508 l. 31 (Exchange with Czechoslovakia 1983).

75. AKRDOPBGDSRSNBA-R, f. 9 op. 4 a.e. 535 l. 60 (Exchanges with Czechoslovakia 1985).

76. AKRDOPBGDSRSNBA-R, f. 9 op. 3 a.e. 150V l. 4 & f. 9 op. 4 a.e. 473 l. 32 (Exchanges with Hungary 1976 & 1981).

77. AKRDOPBGDSRSNBA-R, f. 9 op. 3 a.e. 150V l. 4 & f. 9 op. 4 a.e. 473 l. 129 (Exchanges with Hungary 1976 & 1981).

78. AKRDOPBGDSRSNBA-R, f. 9 op. 4 a.e. 507 l. 41 (Exchange with Hungary 1983).

79. AKRDOPBGDSRSNBA-R, f. 9 op. 3 a.e. 156 l. 11 (Consultations with Poles 1968); see also the articles quoted at the beginning of this chapter.

80. AKRDOPBGDSRSNBA-R, f. 9 op. 3 a.e. 156 l. 125 (Consultations with Poles 1968).

81. AKRDOPBGDSRSNBA-R, f. 9 op. 3 a.e. 156 l. 193 (Consultations with Poles 1968).

82. AKRDOPBGDSRSNBA-R, f. 9 op. 4 a.e. 525 l. 87 (Exchange with Poland 1984).

83. Khristov, "Dŭrzhavna Sigurnost. Chast 2.1."

84. AKRDOPBGDSRSNBA-R, f. 9 op. 2 a.e. 368 l. 26 (Activity Report 1965).

85. AKRDOPBGDSRSNBA-R, f. 9 op. 2 a.e. 449 l. 228 (Berlin Talks 1972).

86. AKRDOPBGDSRSNBA-R, f. 9 op. 2 a.e. 371 l. 2–3 (Activity Report 1968).

87. Interviews regarding topics that touch on espionage are always delicate affairs. Getting people to talk more openly about the topic usually involves a measure of trust built over a few interviews, which is the case with Peter Petrov. Several interviews, in increasingly friendly circumstances, predisposed the interviewee to such stories. The question of his involvement with the intelligence services might always remain open, but his name has not been found on lists of agents or informants published by the Bulgarian Commission on Dossiers.

88. Interview with Petŭr Petrov, March 19, 2015.

89. Interview with Petŭr Petrov, December 11, 2015.

90. Interview with Petŭr Petrov, December 11, 2015.

91. Interview with Petŭr Petrov, December 11, 2015.

92. TsDA f. 37A op. 1 a.e. 2 l. 31 (CICT Report 1969–70).

93. AKRDOPBGDSRSNBA-R, f. 9 op. 2 a.e. 373 l. 33 (Activity Report 1970).

94. AKRDOPBGDSRSNBA-R, f. 1B op. 64 a.e. 438 l. 5 (Politburo Decision on DS 1974).

95. AKRDOPBGDSRSNBA-R, f. 9 op. 3 a.e. 139 l. 24 & f. 9 op. 3 a.e. 137 l. 130 (Activity Reports 1978 & 1979).

96. The company name was changed to Thomson-CSF after 1968.

97. AKRDOPBGDSRSNBA-R, f. 9 op. 2 a.e. 379 l. 9 (Work Evaluation 1965).

98. AKRDOPBGDSRSNBA-R, f. 9 op. 2 a.e. 370 l. 32 (Activity Report 1967).

99. AKRDOPBGDSRSNBA-R, f. 9 op. 2 a.e. 373 l. 25 (Activity Report 1970).

100. AKRDOPBGDSRSNBA-R, f. 9 op. 2 a.e. 373 l. 25 (Activity Report 1970).

101. AKRDOPBGDSRSNBA-R, f. 9 op. 2 a.e. 408 l. 5 (Activity Report 1972).

102. AKRDOPBGDSRSNBA-R, f. 9 op. 2 a.e. 408 l. 7 (Activity Report 1972).

103. AKRDOPBGDSRSNBA-R, f. 9 op. 2 a.e. 408 l. 12 (Activity Report 1972).

104. AKRDOPBGDSRSNBA-R, f. 9 op. 2 a.e. 408 l. 13 (Activity Report 1972).

105. The Eastern Bloc term for "station," a base of intelligence operations in a foreign country.

106. AKRDOPBGDSRSNBA-R, f. 9 op. 4 a.e. 443 l. 271 (Plan Fulfilment 1980).

107. AKRDOPBGDSRSNBA-R, f. 9 op. 2 a.e. 376 l. 56 (Activity Report 1974).

108. AKRDOPBGDSRSNBA-R, f. 9 op. 3 a.e. 140A l. 79–80 (Note on Lisbon Purchase 1977).

109. "Portugal Country Reader," interviews at *Association for Diplomatic Studies and Training Foreign Officers Oral History Project*, pp. 102–104 (https://adst.org/Readers/Portugal.pdf; last accessed: August 2, 2022).

110. AKRDOPBGDSRSNBA-R, f. 9 op. 3 a.e. 139 l. 108 (Report on STI Contribution to Economy 1978).

111. AKRDOPBGDSRSNBA-R, f. 9 op. 3 a.e. 140V l. 8 (Report on Elektronika S Program 1979).

112. AKRDOPBGDSRSNBA-R, f. 9 op. 3 a.e. 139 l. 131 (Ministerial Report to Zhivkov 1979).

113. AKRDOPBGDSRSNBA-R, f. 9 op. 4 a.e. 449 l. 1–4 (Economic Effect 1980).

114. AKRDOPBGDSRSNBA-R, f. 9 op. 4 a.e. 449 l. 12, l. 102 (Economic Effect 1980).

115. AKRDOPBGDSRSNBA-R, f. 9 op. 4 a.e. 482 l. 143–151 (Report on Magnetic Discs 1982).

116. AKRDOPBGDSRSNBA-R, f. 9 op. 4 a.e. 555 l. 48–9 (Doĭnov, Stoi͡an ov, Markov Note to Zhivkov 1986).

117. AKRDOPBGDSRSNBA-R, f. 9 op. 2 a.e. 418 l. 41 (Canadian Line 1974).

118. AKRDOPBGDSRSNBA-R, f. 9 op. 4 a.e. 442 l. 77 (Activity Report 1980).

119. "Bulgarian Seized in New York and Charged as Spy," *New York Times*, September 24, 1983.

120. AKRDOPBGDSRSNBA-R, f. 9 op. 4 a.e. 483 l. 75 & f. 9 op.3 a.e. 139 l. 26 (Orders & Reports 1978 & 1982).

121. AKRDOPBGDSRSNBA-R, f. 9 op. 4 a.e. 548 l. 38 (Activity Report 1986).

122. AKRDOPBGDSRSNBA-R, f. 9 op. 2 a.e. 388 l. 32 (Work Evaluation 1974).

123. Momchil Metodiev and Mariya Dermendzhieva, *Dŭrzhavna Sigurnost—Predimstvo Po Nasledstvo: Profesionalni Biografii na Vodeshti Ofitseri* (Sofia: Ciela 2015), p. 15.

124. Metodiev and Dermendzhieva, *Dŭrzhavna Sigurnost*, p. 62.

125. Metodiev and Dermendzhieva, *Dŭrzhavna Sigurnost*, p. 63.

126. AKRDOPBGDSRSNBA-R, f. 9 op. 2 a.e. 413, l. 2–9 (Note on Foreign Lines 1968).

127. Metodiev and Dermendzhieva, *Dŭrzhavna Sigurnost*, pp. 733–737.

128. AKRDOPBGDSRSNBA-R f. 9 op. 2 a.e. 418 l. 5–6 (Servicing the Economy Report 1974).

129. AKRDOPBGDSRSNBA-R f. 9 op. 2 a.e. 418 l. 7 (Servicing the Economy Report 1974).

130. Interview with Peter Petrov, December 11, 2015.

131. AKRDOPBGDSRSNBA-R, f. 9 op. 2 a.e. 376 l. 58 & f. 9 op. 2 a.e. 378 l. 70 (Activity Reports 1974 & 1976).

132. AKRDOPBGDSRSNBA-R, f. 9 op. 4 a.e. 518 l. 28 (Implementing Acquisitions 1984).

133. *Statisticheski Godishnik na Narodna Republika Bŭlgariia za 1984* (Sofia: KESSI, 1984), p. 106.

134. AKRDOPBGDSRSNBA-R, f. 9 op. 2 a.e. 375 l. 53 (Activity Report 1973).

135. AKRDOPBGDSRSNBA-R, f. 9 op. 3 a.e. 137 l. 127 (Plan Fulfilment 1979).

136. AKRDOPBGDSRSNBA-R, f. 9 op. 4 a.e. 518 l. 30 (Implementation 1984).

137. AKRDOPBGDSRSNBA-R, f. 9 op. 4 a.e. 554 l. 20–21 (CAI Cooperation 1986).

138. AKRDOPBGDSRSNBA-R, f. 9 op. 4 a.e. 572 l. 41 (Internal Historical Note 1987).

139. TsDA f. 37A op. 10 a.e. 12 l. 122 (CICT Correspondence 1982).

140. TsDA f. 37A op. 10 a.e. 13 l. 30–226 (CICT Correspondence 1983).

141. TsDA f. 37A op. 10 a.e. 16 l. 313–314 (CICT Correspondence 1986).

142. TsDA f. 37A op. 10 a.e. 16 l. 59 (CICT Correspondence 1986).

143. TsDA f. 37A op. 10 a.e. 17 l. 19, 200 (CICT Correspondence 1987).

144. TsDA f. 37A op. 10 a.e. 18 l. 29–30, 187 (CICT Correspondence 1988).

145. AKRDOPBGDSRSNBA-R, f. 9 op. 3 a.e. 139 l. 100 (Co-operation Report 1978).

146. AKRDOPBGDSRSNBA-R, f. 9 op. 4 a.e. 473 l. 32–33 (Cooperation with Hungary 1981).

147. AKRDOPBGDSRSNBA-R, f. 9 op. 3 a.e. 150D l. 11 (Talks with Poland 1979).

148. AKRDOPBGDSRSNBA-R, f. 9 op. 2 a.e. 412 l. 63 (Report on Enterprises as Covers 1967).

149. AKRDOPBGDSRSNBA-R, f. 9 op. 2 a.e. 364 l. 15 (CSTP Co-operartion 1970).

150. AKRDOPBGDSRSNBA-R, f. 9 op. 2 a.e. 415 l. 14 (Undercover in Bulgaria 1971).

151. AKRDOPBGDSRSNBA-R, f. 9 op. 3 a.e. 140A l. 74 (CSTP Co-operation 1977).

152. AKRDOPBGDSRSNBA-R, f. 9 op. 2 a.e. 410 l. 15 (Activity Report 1975).

153. AKRDOPBGDSRSNBA-R, f. 9 op. 4 a.e. 438 l. 6 (Work Evaluation 1969).

154. AKRDOPBGDSRSNBA-R, f. 9 op. 2 a.e. 366 l. 19–20 (Gen. Kotsev Report 1975).

155. AKRDOPBGDSRSNBA-R, f. 9 op. 2 a.e. 366 l. 1–2 (Gen. Kotsev Report 1975).

156. AKRDOPBGDSRSNBA-R, f. 9 op. 2 a.e. 410 l. 11 (Meetings Minutes 1975).

157. Khristo Khristov, *Imperiyata na Zadgranichnite Firmi: Suzdavane, Deynost, I Iztochvane na Druzhestvata s Bulgarsko Uchastie zad Granitsa 1961–2007* (Sofia: Ciela, 2009), p. 15.

158. While some use this term for current geopolitical tensions, here I use it in its original meaning to denote the changes in US-Soviet relations after the Afghanistan invasion, used at the time by authors such as Frank Halliday in *The Making of the Second Cold War* (London: Verso, 1983).

159. Gus W. Weiss, "The Farewell Dossier: Duping the Soviets," *Studies in Intelligence*, vol. 39, no. 5 (1996), pp. 124–125.

160. This story, most likely false, can be found in Thomas C. Reed, *At the Abyss: An Insider's History of the Cold War* (New York: Presidio Press, 2004).

161. Bruce B. Weyhrauch "Operation Exodus: The United States Government's Program to Intercept Illegal Exports of High Technology," *Computer/Law Journal,* vol. 7, no. 2 (1986), p. 212.

162. OSWR Report CIA/SW 89–10023X, June 1989, pp. iii–v (Available at the CIA FOIA Electronic Reading Room, https://www.cia.gov/library/readingroom/; last accessed November 14, 2020).

163. OSWR Report CIA/SW 89–10023X, June 1989, p. 1.

164. OSWR Report CIA/SW 89–10023X, June 1989, p. 6.

165. Metodiev, *Mashina za Legitimnost,* p. 114.

166. Interview with Alexander T͡Svetkov, April 6, 2015.

167. Interview with Stoi͡an Markov, July 28, 2015.

168. Interview with Nedko Botev and Boyan T͡Svonev, June 23, 2015.

169. AKRDOPBGDSRSNBA-R, f. 9 op. 4 a.e. 596 (Report to Minister 1987).

CHAPTER 4

1. AMVnR f. 20 op. 38 a.e. 1171 l. 45–6 (Gandhi Visit 1981).

2. The daughter of Todor Zhivkov and the minister of culture during the late 1970s and early 1980s, Liudmila Zhivkova became a towering figure in both the domestic cultural sphere and cultural diplomacy. For more on her ideas and development, see Veneta Ivanova, "Occult Communism: Culture, Science and Spirituality in Late Socialist Bulgaria," PhD dissertation, University of Illinois, 2017. For Zhivkova, Bulgaria, and India more widely, see V. Atanasova, "Aktzenti na Bulgarskata Kulturna Politika po Otnoshenie na Indiya (60-te I 70-te Godini na XX Vek)," *Istoricheski Pregled,* vol. LXVII, no. 1–2 (2011), pp. 174–193, and Bŭlgarskoto Ogledalo, *Obrazut na Indiya v Bŭlgarii͡a—Krayat na XIX Vek-Krayat na XX Vek* (Sofia: Akad. Izd. Prof. Marin Drinov, 2015).

3. The term I use is differs from how the Bulgarian documents often address the region—"Third World" and "developing world" were terms often employed by the analysts. However, "Global South" captures a wider sense of the newly liberated countries, which differed hugely in their development. It also allows me to include the less-developed socialist countries, or those which had socialist regimes for a time, which were not strictly speaking "Third World" in the geopolitical sense but battled many of the same development problems.

4. For the role of Bulgarian architecture globally during this period, see parts of Lukasz Stanek's *Architecture in Global Socialism: Eastern Europe, West Africa, and the Middle East in the Cold War* (Princeton, NJ: Princeton University Press, 2020).

5. For an overview, see David Engerman, "The Second World's Third World," *Kritika: Explorations in Russian and Eurasian History,* vol. 12, no. 1 (January 20, 2011), pp. 183–211. For India specifically, see his "Learning from the East: Soviet Experts and India in the Era of Competitive Coexistence," *Comparative Studies of South Asia, Africa and the Middle East,* vol. 33, no. 2 (August 28, 2013), pp. 227–238. For an

example of another socialist state's efforts, this time in the Middle East, see M. Trentin, *Engineers of Modern Development: East German Experts in Ba'thist Syria, 1965–1972* (Padua: CLEUP 2010).

6. For a very good overview, see C. R. Unger, "Histories of Development and Modernization: Findings, Reflections, Future Research," H-Soz-Kult (December 9, 2010); available at: https://www.hsozkult.de/literaturereview/id/forschungsberichte-1130 (accessed May 16, 2018); for an exploration of the US efforts in India in such spheres, see C. Unger, 'Towards Global Equilibrium: American Foundations and Indian Modernization 1950s–1970s," *Journal of Global History*, vol. 6, no. 1 (March 2011), pp. 121–142.

7. Nick Cullather, *The Hungry World: America's Cold War Battle against Poverty in Asia* (Cambridge, MA: Harvard University Press, 2010), p. 5.

8. C. Unger and D. Engerman, "Introduction: Towards a Global History of Modernization," *Diplomatic History*, vol. 33, no. 3 (April 2009), 381–383.

9. A. Hilger, "GDR and Soviet Bloc Policy towards India 1971–1989," Parallel History Project on Cooperative Security (February 2009), available at: http://www.php.isn.ethz.ch/lory1.ethz.ch/collections/coll_india/documents/intro_Hilger3ee91.html?navinfo=56154 (accessed May 16, 2018).

10. C. R. Unger, "Industrialization vs Agrarian Reform: West German Modernization Policies in India in the 1950s and 1960s," *Journal of Modern European History*, vol. 8, no. 1 (2010), p. 51.

11. For great work on how the developing world impacted Eastern European intellectuals and development ideas, the new project "Reconfiguring Backwardness: Polish Social Scientists and the Making of the Third World" by Malgorzata Mazurek is particularly illuminating.

12. For more on Indian modernization and computing's role in it, see N. Menon, "'Fancy Calculating Machine': Computers and Planning in Independent India," *Modern Asian Studies*, vol. 52, no. 2 (March 2018), pp. 421–457.

13. TsDA f. 1B op. 64 a.e. 268 l. 3–5 (Politburo Protocol 1960).

14. The history of Texim is worthy of a work on its own. Headed by Georgi Naidenov, a trade representative in Turkey and Egypt, as well as a State Security agent, it became a veritable empire of trade and logistics, worth hundreds of millions of dollars by 1969, when it was closed under pressure from Moscow. Its assets became the core of the Bulgarian Merchant Fleet, while Naidenov was tried and found guilty of embezzlement, serving five years out of a 20-year sentence. In 1992, he refounded the organization as Texim Bank, branches of which dot Bulgaria. In many ways, Texim was a capitalist company par excellence operating in a planned economy—precisely the reason Moscow was opposed to it.

15. TsDA f. 136 op. 86 a.e. 523 l. 1–2 (Council of Ministers Resolution 1961).

16. TsDA f. 259 op. 17 a.e. 80 l. 1–2 (Foreign Trade Ministry—MFT—on Egypt 1965–6).

17. TsDA f. 259 op. 17 a.e. 81 l. 38–43 (Syria, Tunisia, Algeria 1965–6).

18. TsDA f. 259 op. 19 a.e. 382 l. 1–4 (Kabul Report 1971).

19. TsDA f. 259 op. 44 a.e. 408 l. 2–7 (Angolan Report 1977).

20. TsDA f. 1B op. 66 a.e. 1950 l. 59 (Politburo on Tanzania 1979).

21. TsDA f. 259 op. 44 a.e. 16 l. 115–123 (MFT Fairs 1978).

22. TsDA f. 1B op. 35 a.e. 704 l. 59 (Politburo Minutes 1969).

23. TsDA f. 1B op. 35 a.e. 106 l. 2 (Politburo on Mongolia 1968).

24. TsDA f. 1B op. 35 a.e. 3010 l. 45 (Politburo on Tanzania 1972).

25. TsDA f. 259 op. 20 a.e. 501 l. 4 (MFT India 1967).

26. TsDA f. 259 op. 44 a.e. 129 l. 44 (MFT Contracts 1977).

27. TsDA f. 517 op. 2 a.e. 74 l. 28 (CSTP Report 1972).

28. TsDA f. 517 op. 2 a.e. 74 l. 27 (CSTP Report 1972)

29. TsDA f. 517 op. 2 a.e. 74 l. 38 (CSTP Report 1972)

30. TsDA f. 1B op. 35 a.e. 4459 l. 52 (Politburo Reports 1973).

31. TsDA f. 259 op. 45 a.e. 351 l. 86 (MFT Exports 1977–81).

32. TsDA f. 259 op. 45 a.e. 351 l. 139–140 (MFT Exports 1977–81).

33. TsDA f. 259 op. 45 a.e. 353 l. 44–46 (MFT Exports 1985–7).

34. TsDA f. 1003 op. 1 a.e. 28 (IZOT Report 1986).

35. TsDA f. 259 op. 44 a.e. 131 (MFT Imports 1977).

36. TsDA f. 259 op. 44 l. 44.

37. TsDA f. 259 op. 44 l. 78.

38. TsDA f. 37A op. 10 a.e. 11 l. 8 (CICT Correspondence 1981).

39. TsDA f. 259 op. 45 a.e. 848 l. 6 (MFT on Mozambique 1981–7).

40. TsDA f. 259 op. 45 a.e. 203 l. 12 & f. 37A op. 10 a.e. 11 l. 8 (MFT & CICT Reports 1982).

41. TsDA f. 830 op. 2 a.e. 3 l. 4 (Izotimpex on Zimbabwe 1983).

42. TsDA f. 830 op. 2 a.e. 27 l. 28 (Izotimpex Report 1986).

43. TsDA f. 830 op. 2 a.e. 28 l. 6–7 (Report 1987).

44. TsDA f. 830 op. 2 a.e. 87 l. 5–6 (Market Program 1982).

45. TsDA f. 378 op. 1 a.e. 1101 (Foreign Ministry Report 1976).

46. TsDA f. 1003 op. 1 a.e. 14 l. 34 (Izotimpex Plan 1974).

47. TsDA f. 259 op. 45 a.e. 832 l. 2 (MFT on Joint Firms 1987–8).

48. TsDA f. 259 op. 45 a.e. 832 l. 25 (MFT on Joint Firms 1987–8).

49. TsDA f. 37A op. 9 a.e. 8 l. 102 (CICT Trips 1985).

50. Albena Shkodrova, *Sots-Gurme: Kurioznata Istorii͡a na Kuhnyata v NRB* (Sofia: Zhanet 45, 2014), p. 176.

51. TsDA f. 259 op. 44 a.e. 449 (MFT Libya 1975).

52. TsDA f. 37A op. 4 a.e. 11 l. 10 (CICT Progress Report 1971).

53. AMVnR f. 20 op. 43–5 a.e. 136 l. 7–17 (Hanoi Report 1986–7).

54. AMVnR f. 20 op. 43–5 a.e. 136 l. 19 (Hanoi Report 1986–7).

55. TsDA f. 830 op. 2 a.e. 90 l. 23 (Izotimpex Market Program 1985).

56. TsDA f. 1B op. 35 a.e. 4459 l. 34 (Politburo Minutes 1973).

57. TsDA f. 830 op. 2 a.e. 21 l. 5 (Izotimpex Report 1979).

58. TsDA f. 37A op. 9 a.e. 8 l. 48 (CICT Trips 1985).

59. TsDA f. 830 op. 2 a.e. 91 l. 78 (Izotimpex Market Program 1986).

60. TsDA f. 259 op. 45 a.e. 677 l. 3–31 (MFT China 1985–9).

61. TsDA f. 259 op. 45 a.e. 677 l. 53 (MFT China 1985–9).

62. TsDA f. 259 op. 45 a.e. 677 l. 68 (MFT China 1985–9).

63. TsDA f. 830 op. 2 a.e. 91 l. 69–71 (Izotimpex Market 1986).

64. TsDA f. 259 op. 20 a.e. 501 l.95; Indira Gandhi generally expressed more pro-Soviet and anti-US sentiments than her predecessors, and the Indian alignment with the Eastern Bloc was helped by the early-mid 1960s wars with China and Pakistan, as well as the US intervention in Vietnam. See R. J. McMahon, "On the Periphery of a Global Conflict: India and the Cold War 1947–1991," in A. Hilger and C. Unger (eds.), *India in the World since 1947: National and Transnational Perspectives* (Frankfurt am Main: Peter Lang GmbH, 2012), especially pp. 290–294.

65. TsDA f. 259 op. 20 a.e. 501 l. 98 (Indian Cooperation Agreement 1967).

66. TsDA f. 259 op. 20 a.e. 501 l. 7 (Indian Cooperation Agreement 1967).

67. TsDA f. 259 op. 20 a.e. 501 l. 96 (Indian Cooperation Agreement 1967).

68. TsDA f. 1B op. 35 a.e. 138 l. 16 (Politburo Meeting 1968).

69. TsDA f. 378 op. 30 a.e. 1205 l. 9 (Todorov Visit 1974).

70. Sugata Bose, "Instruments and Idioms of Colonial and National Development: India's Historical Experience in Comparative Perspective," in Frederick Cooper and Randall Packard (eds.), *International Development and the Social Sciences: Essays on the History and Politics of Knowledge* (Berkeley: University of California Press, 1997), pp. 45–63.

71. TsDA f. 259 op. 20 a.e. 501 l. 60 (Co-Operation 1967).

72. TsDA f. 259 op. 20 a.e. 501 l. 18 (Co-Operation 1967).

73. TsDA f. 259 op. 19 a.e. 385 l. 7 (MFT Tractors 1969–70).

74. A Bulgarian import-export company set up in 1966, tasked with exporting a variety of industrial goods. A company run by state security, it often took the lead in breaking into new markets, but its biggest purview was weapon sales.

75. TsDA f.259 op. 19 a.e. 341 l. 134 (MFT Protocol 1970).

76. TsDA f. 1B op. 35 a.e. 1202 l. 204 (Politburo Discussion 1970).

77. Washington National Records Office, OSD Files; FRC 76–0197, Box 74, Pakistan 092 (Aug–Dec 1971), Memorandum for the Secretary of Defense.

78. TsDA f. 259 op. 36 a.e. 520 l. 14 (Todorov & Gandhi Meeting 1974).

79. TsDA f. 1B op. 35 a.e. 3079 l. 99 (Politburo Plan 1972).

80. TsDA f. 1B op. 35 a.e. 3079 l. 139 (Politburo Plan 1972).

81. TsDA f. 259 op. 36 a.e. 522 l. 41 (MFT Trips 1971–4).

82. TsDA f. 259 op. 36 a.e. 524 l. 345 (MFT Trips 1974).

83. Menon was the premier scientific administrator in India during the 1970s, the perfect counterpart to Popov. He was a protégé of Homi Bhabha, the founder of atomic research in India, and succeeded him as the head of the Tata Institute for Fundamental Research on his mentor's death in 1966. From 1971, he was the head of the Department of Electronics, and eventually head of CSIR itself. His last administrative position was in 1990 as Minister of Science. He was the man who took on IBM in India, and the one who set up the foundations of the Indian electronics industry. At the start of 2016, his personal archive was handed over to the Nehru Memorial Museum and Library, but sadly, it has not yet been processed.

84. TsDA f. 1477 op. 30 a.e. 1210 l. 14 (Ministry of Foreign Affairs—MFA—Note on Menon 1974).

85. NAI, List 193; 17/855/1974 PMS, pp. 8–9 (Computer Industry Report 1974).

86. NAI, List 193; 17/635/74 PMS, p. 1 (Economic Report 1974).

87. NAI, List 193; 17/635/70 PMS, pp. 1–4 (Economic Report 1970).

88. NAI, List 193; 17/802/71 PMS, p. 1 (CSIR Enquiry 1971).

89. NAI, List 193; 17/3/71 PMS, p. 17 (CSIR Correspondence 1971).

90. NAI, List 193; 17/1702/1972 PMS, p. 3 (Ministerial Meeting 1972).

91. NAI, List 193; 17/855/74 PMS, p. 10 (Industrial Report 1974).

92. K. R. Bhandarkar and Raja Kulkarni, *Computer and Labour Problems in India* (Bombay: United Asia Publications, 1971).

93. Bhandarkar and Kulkarni, *Computer and Labour Problems in India,* p. III.

94. Bhandarkar and Kulkarni, *Computer and Labour Problems in India,* p. 61.

95. Bhandarkar and Kulkarni, *Computer and Labour Problems in India,* p. 102.

96. NAI, List 193; 17/1596/72 PMS, p. 6 (Investment Report 1978).

97. NAI, List 193; 17/1596/72 PMS, p. 4 (Investment Report 1978).

98. NAI, List 193; 17/1596/72 PMS, p. 1 (Investment Report 1978).

99. NAI, List 193; 17/855/77 PMS, pp. 2–3 (Industiral Report 1977).

100. NAI, List 193; 17/855/77 PMS, pp. 2–3 (Industrial Report 1977).

101. NAI, List 193; 17/855/77 PMS, p. 39-correspondence (Industrial Report 1977).

102. NAI, List 193; 17/855/74 PMS, pp. 1–3 (Industrial Report 1974).

103. NAI, List 193; 17/1533/74 PMS, p. 2 (ETTDC Report 1974).

104. AMVnR f. 20 op. 32 a.e. 1335 l. 63 (Zhivkov Visit 1976).

105. AMVnR f. 20 op. 32 a.e. 1335 l. 56 (Zhivkov Visit 1976).
106. AMVnR f. 20 op. 32 a.e. 1335 l. 63 (Zhivkov Visit 1976).
107. AMVnR f. 20 op. 32 a.e. 1338 l. 9 (Zhivkov Visit 1976).
108. AMVnR f. 20 op. 24 a.e. 1300 l. 6 (Program for India 1968).
109. AMVnR f. 20 op. 24 a.e. 1300 l. 4 (Program for India 1968).
110. AMVnR f. 20 op. 24 a.e. 1305 l. 27 (Economic Correspondence 1968–9).
111. AMVnR f. 20 op. 32 a.e. 1353 l. 10–11 (Cooperation Questions 1976).
112. TsDA f. 259 op. 45 a.e. 203 (MFT Mozambique 1981–7).
113. AMVnR f. 20 op. 32 a.e. 1388 l. 11 (Specialists Exchange 1975–6).
114. AMVnR f. 20 op. 32 a.e. 1375 "News from Bulgaria," Issue 11 of 1976, p. 13.
115. AMVnR f. 20 op. 32 a.e. 1417 l. 30–31 (Administrative Documents 1976).
116. AMVnR f. 20 op. 32 a.e. 1387 l. 15–27 (Session Minutes 1975–6).
117. TsDA f. 1003 op. 1 a.e. 128 l.2–11 (IZOT on India 1978).
118. TsDA f. 259 op. 39 a.e. 343 l. 41 (MFT Trade Report 1976–7).
119. TsDA f. 1003 op. 1 a.e. 128 l. 18 (IZOT 1978).
120. TsDA f. 1003 op. 1 a.e. 128 l. 19 (IZOT 1978).
121. TsDA f. 1003 op. 1 a.e. 128 l. 5 (IZOT 1978).
122. TsDA f. 1003 op. 1 a.e. 128 l. 16 (IZOT 1978).
123. TsDA f. 830 op. 2 a.e. 84 l. 9 (Izotimpex Market 1976).
124. AMVnR f. 20 op. 32 a.e. 1397 l. 15 (MFA Economic Report 1975–6).
125. AKRDOPBGDSRSNBA f. 9 op. 4 a.e. 548 l. 38 (Activity Report 1986).
126. AKRDOPBGDSRSNBA f. 9 op. 4 a.e. 438 l. 5–7 (CSTP Work 1980).
127. AMVnR f. 20 op. 40 a.e. 1587 l. 34–6 (Embassy Evaluations 1983).
128. AMVnR f. 20 op. 43–5 a.e. 150 (Embassy Self-Evaluation 1986).
129. AMVnR f. 20 op. 33 a.e. 1277 l. 108 (Future Plans 1975–7).
130. TsDA f. 259 op. 36 a.e. 513 l. 10–14, 176 (MFT Delhi Report 1974).
131. TsDA f. 259 op. 36 a.e. 513 l. 80–81, 176 (MFT Delhi Report 1974).
132. RGAE f. 9480 op. 12 a.e. 244 l. 32 (GKNT Subcommittee 1976).
133. TsDA f. 259 op. 44 a.e. 284 l. 10 (MFT Report 1978).
134. TsDA f. 259 op. 44 a.e. 287 l. 55–57 (MFT Commission 1977).
135. AMVnR f. 20 op. 36 a.e. 1249 l. 18 (Embassy Letters 1980).
136. AMVnR f. 20 op. 34 a.e. 1210 (Administrative Use 1978).
137. AMVnR f. 20 op. 36 a.e. 1277 l. 15 (Report on Electronics 1980).
138. AMVnR f. 20 op. 36 a.e. 1277 l. 7 (Report on Electronics 1980).
139. AMVnR f. 20 op. 36 a.e. 1277 l. 16 (Report on Electronics 1980).

140. TsDA f. 259 op. 44 a.e. 292 l. 10 (MFT Meetings 1979).

141. AMVnR f. 20 op. 36 a.e. 1277 l. 17 (Report 1980).

142. AMVnR f. 20 op. 38 a.e. 1168 l. 1 (Visits 1981).

143. Violina Atanasova, "Bulgarskiya Kulturno-Informatzionen Tzentur v Delhi—Istoriîa I Deynost," *Svetilnik,* vol. 14 (2002), pp. 30–34.

144. For more on Bulgaria's cultural links with India, see Violina Atanasova, "Akcenti na Bulgarskata Kulturna Politika po Otnoshenie na India (60te I 70te godini na XX vek)," *Istoricheski Pregled,* vol. 1, no. 2 (2011), pp. 174–193.

145. AMVnR f. 20 op. 39 a.e. 1282 (Propaganda Report 1982).

146. AMVnR f. 20 op. 36 a.e. 1276 l. 83 (Future Plans 1978–80).

147. AMVnR f. 20 op. 39 a.e. 1282 l. 7 (Propaganda Report 1982).

148. AMVnR f. 20 op. 39 a.e. 1297 l. 9 (News from Bulgaria 1982).

149. AMVnR f. 20 op. 39 a.e. 1297 l. 28 (News from Bulgaria 1982).

150. AMVnR f. 20 op. 39 a.e. 1297 l. 30–31 (News from Bulgaria 1982).

151. AMVnR f. 20 op. 40 a.e. 1618 l. 31 (Doĭnov Visit 1983).

152. AMVnR f. 20 op. 40 a.e. 1618 l. 31 (Doĭnov Visit 1983).

153. NMML, Dr. Gopal Singh Papers, List 378; Series 2, Doc. 29, p. 304 (Ambassador Correspondence 1970–2).

154. AMVnR f. 20 op. 40 a.e. 1663, "News from Bulgaria," no. 11, 12, 1983, p. 9.

155. TsDA f. 259 op. 45 a.e. 630 l. 32 (MFT Sessions 1983–8).

156. Bipan Chandra, Mirdula Mukherjee, and Aditya Mukherjee, *India after Independence* (Delhi: Penguin, 2008), pp. 348–349.

157. Chandra, Mukherjee, and Mukherjee, *India after Independence,* p. 350.

158. Sumit Ganguly and Rahul Mukherjee, *India since 1980* (Cambridge: Cambridge University Press, 2011), p. 77.

159. AMVnR f. 20 op. 41 a.e. 1918 l. 82–4 (Economic Report 1984).

160. AMVnR f. 20 op. 44–5 a.e. 182 l. 8–9 (Press Overview 1987).

161. AMVnR f. 20 op. 44–5 a.e. 182 l. 10 (Press Overview 1987).

162. AMVnR f. 20 op. 44–5 a.e. 182 l. 11 (Press Overview 1987).

163. AMVnR f. 20 op. 44–5 a.e. 193 l. 11 (Trade Talks 1987).

164. *Kompîutŭr Za Vas,* no. 9, 1987, pp. 9–11.

165. It is worth noting that this computer was faster than the Elbrus series of Soviet supercomputers of the same generation, designed for similar usages. It was based on the massive parallel design concept of supercomputing.

166. Milena Dimitrova, *Zlatnite Desitiletiîa na Bulgarskata Elektronika* (Sofia: Trud, 2008), p. 192.

167. Dimitrova, *Zlatnite Desitiletiîa,* p. 8.

168. AMVnR f. 20 op. 44–5 a.e. 191 l. 11 (US-Indian Cooperation 1987).

169. Mani Ayar (ed.), *Rajiv Gandhi's India*, Vol. 3—*Foreign Policy* (New Delhi: UBS Publishers, 1998), p. 116.

170. Mani Ayar (ed), *Rajiv Gandhi's India*, Vol. 2—*Economics* (New Delhi: UBS Publishers ,1998), p. 108.

171. AMVnR f. 20 op. 46–5 a.e. 163 l. 33 (Activity Report 1989).

172. Interview with Krasimir Markov, February 4, 2016.

173. Dimitrova, *Zlatnite Desitiletiia*, pp. 200, 259.

174. TsDA f. 259 op. 45 a.e. 630 l. 116 (MFT Sessions 1983–8).

175. AMVnR f. 20 op. 46–5 a.e. 163 l. 31 (Activity Report 1989).

176. AMVnR f. 20 op. 46–5 a.e. 165 l. 3 (Joint Commission 1989).

177. TsDA f. 259 op. 36 a.e. 107 l. 337 (MFT Reports 1973).

178. TsDA f. 259 op. 36 a.e. 107 l. 339 (MFT Reports 1973).

179. TsDA f. 259 op. 36 a.e. 107 l. 340 (MFT Reports 1973).

180. TsDA f. 259 op. 44 a.e. 116 (MFT Advertising 1974).

181. TsDA f. 259 op. 44 a.e. 128 (MFT Reports 1977).

182. TsDA f. 259 op. 44 a.e. 128, l. 22–23.

183. TsDA f. 259 op. 44 a.e. 128, l. 25.

184. TsDA f. 259 op. 44 a.e. 128, l. 26–27.

185. Interview with Atanas Shopov, "Bez Emotsii za Starozagorskata Elektronika," in *Septemvri* newspaper, no. 52, 1989.

186. TsDA f. 830 op. 2 a.e. 99 l. 4 (Izotimpex Advertising 1982–3).

187. TsDA f. 830 op. 2 a.e. 99 l. 7 (Izotimpex Advertising 1982–3).

188. TsDA f. 830 op. 2 a.e. 99 l. 9 (Izotimpex Advertising 1982–3).

189. TsDA f. 830 op. 2 a.e. 99 l. 9–11 (Izotimpex Advertising 1982–3).

190. TsDA f. 830 op. 2 a.e. 100 l. 7 (Izotimpex Fairs 1981).

191. TsDA f. 830 op. 2 a.e. 56 l. 29 (Izotimpex Report 1982).

192. TsDA f. 830 op. 2 a.e. 56 l. 29 (Izotimpex Report 1982).

193. TsDA f. 830 op. 2 a.e. 99 l. 12 (Izotimpex Advertising 1982–3).

194. TsDA f. 830 op. 2 a.e. 99 l. 12 (Izotimpex Advertising 1982–3).

195. TsDA f. 830 op. 2 a.e. 100 l. 3 (Izotimpex Fairs 1981).

196. TsDA f. 259 op. 39 a.e. 392 l. 287 (MFT Iran 1976).

197. TsDA f. 259 op. 39 a.e. 343 l. 101 (MFT Reports 1976–7).

198. TsDA f. 37A op. 10 a.e. 16 l. 17 (CICT Correspondence 1986).

199. NMML, Dr. Gopal Singh Papers, List 378; Series 2, Doc. 28, p. 108 (Ambassadorial Correspondence 1965–75).

CHAPTER 5

1. Vladislav Zubok and Constantine Pleshakov, *Inside the Kremlin's Cold War: From Stalin to Khrushchev* (Cambridge, MA: Harvard University Press, 1996), p. 8

2. TsDA f. 517 op. 2 a.e. 277 l. 2 (CSTP Plan 1969).

3. For more on these policies, see Iliîana Marcheva, "Problemi na Modernizatsiyata pri Sotsializma: Industrializatsiyata v Bŭlgariîa," in E. Kandilarov and T. Turlakova (eds.), *Izsledvaniya po Istoriîa na Socializma v Bŭlgariîa 1944–1989* (Sofia: Grafimaks, 2010), pp. 207–208.

4. Martin Ivanov, "Ikonomikata na Komunisticheska Bulgaria (1963–1989)," in I. Znepolski (ed.), *Istoriîa na Narodna Republika Bŭlgariîa: Rezhimut I Obshtestvoto* (Sofia: Ciela, 2009), p. 306.

5. This paragraph draws heavily on Kevin Baker's innovative work on the GDR and cybernetics, "Red Helmsman: Cybernetics, Economics, and Philosophy in the German Democratic Republic," (MA thesis, Georgia State University, 2011), especially chapter 3: "Control: Cybernetics, Market Stimulation and State Planning."

6. TsDA f. 1B op. 35 a.e. 154 l. 10 (Politburo Discussions 1967–8).

7. This idea is most masterfully studied in the Soviet case by Benjamin Peters in *How Not to Network a Nation: The Uneasy History of the Soviet Internet* (Cambridge, MA: the MIT Press 2016).

8. TsDA f. 1B op. 35 a.e. 453 l. 8–9 (Politburo Commission 1968).

9. TsDA f. 1B op. 35 a.e. 453 l. 12–15 (Politburo Commission 1968).

10. TsDA f. 1B op. 35 a.e. 888 l. 8–6 (Central Committee Directives 1969).

11. TsDA f. 1B op. 35 a.e. 888 l. 8–6 (Central Committee Directives 1969).

12. TsDA f. 1B op. 35 a.e. 888 l. 8–7; l. 8–17 (Central Committee Directives 1969).

13. TsDA f. 1B op. 35 a.e. 888 l. 8–18/19 (Central Committee Directives 1969).

14. TsDA f. 1B op. 35 a.e. 888 l. 8–20 (Central Committee Directives 1969).

15. TsDA f. 1B op. 35 a.e. 888 l. 8–21 (Central Committee Directives 1969).

16. TsDA f. 1B op. 35 a.e. 888 l. 8–25 (Central Committee Directives 1969).

17. TsDA f. 1B op. 35 a.e. 888 l. 8–26 (Central Committee Directives 1969).

18. TsDA f. 1B op. 35 a.e. 888 l. 8–27 (Central Committee Directives 1969).

19. TsDA f. 1B op. 35 a.e. 888 l. 8–35 (Central Committee Directives 1969); interestingly, the Chilean Cybersyn system's sole practical application turned out to be in logistics.

20. TsDA f. 1B op. 35 a.e. 3079 l. 16 (Politburo Plan 1972).

21. TsDA f. 1B op. 35 a.e. 3079 l. 21 (Politburo Plan 1972).

22. See Slava Gerovitch, *From Newspeak to Cyberspeak: A History of Soviet Cybernetics* (Cambridge, MA: The MIT Press, 2004), especially chapter 6.

23. TsDA f. 1B op. 35 a.e. 4940 l. 28–33 (Politburo Plan 1974).

24. TsDA f. 1B op. 66 a.e. 1287 l. 67–1 (Politburo Report 1978).
25. TsDA f. 1B op. 66 a.e. 1287 l. 67–2 (Politburo Report 1978).
26. TsDA f. 1B op. 66 a.e. 1287 l. 68–9 (Politburo Report 1978).
27. TsDA f. 1B op. 67 a.e. 142 l. 48–9 (Politburo Congress 1981).
28. TsDA f. 1B op. 67 a.e. 142 l. 13 (Politburo Congress 1981).
29. TsDA f. 1B op. 67 a.e. 668 l. 26 (Politburo Plan 1981).
30. TsDA f. 1B op. 67 a.e. 668 l. 32 (Politburo Plan 1981).
31. TsDA f. 1B op. 67 a.e. 668 l. 32–3 (Politburo Plan 1981).
32. TsDA f. 1B op. 67 a.e. 2198 l. 158–160 (Politburo Discussions 1983).
33. RGAE f. 9480 op. 13 a.e. 2487 l. 10 (GKNT Tasks 1986–7).
34. RGAE f. 9480 op. 13 a.e. 2487 l. 23 & 26–7 (GKNT Tasks 1986–7).
35. TsDA f. 517 op. 2 a.e. 172 l. 15 (CSTP Co-operation Plans 1969).
36. TsDA f. 517 op. 2 a.e. 172 l. 14 (CSTP Co-operation Plans 1969).
37. TsDA f.1B op. 35 a.e. 1246 l. 87 (Politburo Plans 1970).
38. TsDA f.1B op. 35 a.e. 1246 l. 88 (Politburo Plans 1970).
39. There is no need to repeat here Stephen Kotkin's arguments in *Magnetic Mountain: Stalinism as a Civilization* (Berkeley: University of California Press, 1997), which puts steel front and center in Stalinist industrialization. The Bulgarian regime, too, built an outsized steelworks and metal work sector, centered around the huge Kremkovtsi plant outside Sofia.
40. TsDA f. 1B op. 35 a.e. 1990 l. 15–17 (Politburo Discussions 1970–1).
41. TsDA f. 1B op. 35 a.e. 1990 l. 21 (Politburo Discussions 1970–1).
42. TsDA f. 1B op. 35 a.e. 1990 l. 22 (Politburo Discussions 1970–1).
43. TsDA f. 1B op. 35 a.e. 1990 l. 24 (Politburo Discussions 1970–1).
44. TsDA f. 1B op. 35 a.e. 1990 l. 24 (Politburo Discussions 1970–1).
45. RGAE f. 9480 op. 9 a.e. 261 l. 51–68 (6th Bulgarian-Soviet Session 1966).
46. TsDA f. 1B op. 35 a.e. 1501 l. 121 (Politburo Discussions 1970).
47. TsDA f. 1B op. 35 a.e. 1501 l. 130–1 (Politburo Discussions 1970).
48. TsDA f. 517 op. 2 a.e. 115 l. 5 (CSTP Reports 1972).
49. TsDA f. 517 op. 2 a.e. 115 l. 9 (CSTP Reports 1972).
50. TsDA f. 517 op. 2 a.e. 115 l. 10 (CSTP Reports 1972).
51. TsDA f. 517 op. 2 a.e. 115 l. 10 (CSTP Reports 1972).
52. TsDA f. 517 op. 2 a.e. 115 l. 11 (CSTP Reports 1972).
53. TsDA f. 517 op. 2 a.e. 115 l. 12 (CSTP Reports 1972).
54. TsDA f. 517 op. 2 a.e. 115 l. 13–15 (CSTP Reports 1972).
55. TsDA f. 517 op. 4 a.e. 35 l. 64 (CSTP Automation-7 Program—1974).
56. TsDA f. 517 op. 4 a.e. 35 l. 53 (CSTP Automation-7 Program—1974).

57. TsDA f. 517 op. 4 a.e. 35 l. 52 (CSTP Automation-7 Program—1974).

58. TsDA f. 1B op. 35 a.e. 5368 l. 30 (Politburo Theses 1975).

59. TsDA f. 1B op. 35 a.e. 5368 l. 35 (Politburo Theses 1975).

60. TsDA f. 1B op. 35 a.e. 5368 l. 68–71 (Politburo Theses 1975).

61. TsDA f. 1B op. 35 a.e. 5368 l. 94 (Politburo Theses 1975).

62. TsDA f. 517 op. 5 a.e. 14 l. 63 (CSTP Agreements 1978).

63. TsDA f. 517 op. 6 a.e. 29 l. 7 (CSTP Automation Program 1981).

64. TsDA f. 517 op. 6 a.e. 29 l. 10 (CSTP Automation Program 1981).

65. TsDA f. 517 op. 6 a.e. 29 l. 27 (CSTP Automation Program 1981).

66. TsDA f. 517 op. 6 a.e. 29 l. 28 (CSTP Automation Program 1981).

67. This is best seen in the most (in)famous pet project—the giant Radomir machine-building plant (the "factory for factories"), which was to be based on the latest Japanese technologies. Clashes with the pro-Soviet party faction in the face of Lukanov, however, led to the factory being equipped to Soviet standards, losing its Japanese contracts (which its profits were predicated on), and Radomir became a black hole for over 1 billion levs of investments.

68. *Zhenata Dnes,* no. 2 (1987), p. 4.

69. TsDA f. 517 op. 6 a.e. 38 l. 11 (CSTP Report 1982).

70. TsDA f. 517 op. 6 a.e. 38 l. 12–13 (CSTP Report 1982).

71. TsDA f. 517 op. 6 a.e. 38 l. 14 (CSTP Report 1982).

72. TsDA f. 517 op. 6 a.e. 38 l. 15 (CSTP Report 1982).

73. TsDA f. 517 op. 6 a.e. 38 l. 16 (CSTP Report 1982).

74. TsDA f. 517 op. 6 a.e. 38 l. 37 (CSTP Report 1982).

75. TsDA f. 517 op. 6 a.e. 38 l. 45 (CSTP Report 1982).

76. TsDA f. 517 op. 6 a.e. 52 l. 24 (CSTP Report 1983).

77. TsDA f. 517 op. 6 a.e. 52 l. 23 (CSTP Report 1983).

78. TsDA f. 517 op. 6 a.e. 52 l. 20 (CSTP Report 1983).

79. TsDA f. 517 op. 6 a.e. 52 l. 23 (CSTP Report 1983).

80. TsDA f. 517 op. 6 a.e. 52 l. 22 (CSTP Report 1983).

81. TsDA f. 517 op. 6 a.e. 52 l. 26 (CSTP Report 1983).

82. TsDA f. 517 op. 6 a.e. 52 l. 27 (CSTP Report 1983).

83. TsDA f. 517 op. 6 a.e. 52 l. 29 (CSTP Report 1983).

84. TsDA f. 517 op. 6 a.e. 52 l. 31 (CSTP Report 1983).

85. Information can be found in both Bulgarian and Russian archives; see RGAE f. 9480 op. 13 a.e. 481; TsDA f. 517 op. 6 a.e. 52; and a Soviet dissertation: Yuri Polyakov, "Optimizatsiya Planov Proizvodstoa Predpriyateiy Mashinostroeniya (Na Primere ASU Sigma)," PhD dissertation, SAN-Novosibirsk, 1984.

86. DA-V; f. 1230 op. 1 a.e. 1 l. 1 (Ministerial-DSO Correspondence 1972).

87. DA-V; f. 1230 op. 1 a.e. 31 l. 1 (Catalogues 1969–78).

88. DA-V; f. 1230 op. 1 a.e. 33 (Centre Plan 1977–8).

89. DA-V; f. 1230 op. 1 a.e. 31 (Catalogues).

90. DA-V; f. 1230 op. 1 a.e. 15 l. 28–9; l. 54–5; l. 69 (Director's Reports 1972–3).

91. DA-V; f. 1230 op. 1 a.e. 10; a.e. 47; op. 2 a.e. 8; a.e. 14; a.e. 26 (Misc Correspondences 1976–83).

92. DA-V; f. 1230 op. 1 a.e. 30 l. 1–4 (FORAN Report 1976).

93. DA-V; f. 1230 op. 1 a.e. 61 l. 30 (Implementation Notes 1976).

94. DA-V f. 1230 op. 2 a.e. 48 (Qualifications Documents 1983).

95. DA-V f. 1230 op. 3 a.e. 54 (Staff Statistics 1983–93).

96. DA-V f. 1230 op. 2 a.e. 49 (Qualifications Documents 1981–3).

97. DA-V f. 1230 op. 2 a.e. 1 & 4 (Protocols 1979–83).

98. DA-V f. 1230 op. 1 a.e. 80 (Educationa Statistics 1981–3).

99. DA-V f. 1230 op. 2 a.e. 32 (Annual Plan 1982).

100. DA-V f. 1230 op. 2 a.e. 14 (ASU Program 1981).

101. DA-V f. 1230 op. 3 a.e. 2 a.e. 41–43 (Ministerial Orders 1981–9).

102. TsDA f. 517 op. 2 a.e. 102 l. 42 (CSTP Report 1967).

103. TsDA f. 517 op. 2 a.e. 95 l. 13–15 (CSTP Report 1965).

104. Alexander Petkov, "Razvitie na Informatsionnite Sistemi I Tehnologii v Ruse," *Izvestiya na Suyuza na Uchenite-Ruse,* no. 2 (2001), p. 1.

105. "U.K. Part in Bulgarian Contract," *Financial Times* (June 28, 1968), p. 6; and Adolf Herman, "Western Computer Companies Step Up Sales Drive," *Financial Times* (February 12, 1969), p. 7.

106. Petkov, "Razvitie na Informatsionnite Sistemi," pp. 1–2.

107. TsDA f. 517 op. 2 a.e. 106 l. 15 (CSTP Co-operation 1971).

108. Petkov, "Razvitie na Informatsionnite Sistemi," p. 4.

109. Vela Lazarova, *Istoriia͡ na Gabrovo* (Sofia: Otechestven Front, 1980), p. 398.

110. TsDA f. 517 op. 2 a.e. 115 l. 17 (CSTP Reports 1972).

111. TsDA f. 517 op. 2 a.e. 115 l. 18 (CSTP Reports 1972).

112. TsDA f. 517 op. 2 a.e. 115 l. 18 (CSTP Reports 1972).

113. TsDA f. 517 op. 2 a.e. 115 l. 19 (CSTP Reports 1972).

114. TsDA f. 517 op. 2 a.e. 115 l. 19 (CSTP Reports 1972).

115. TsDA f. 517 op. 2 a.e. 115 l. 12 (CSTP Reports 1972).

116. TsDA f. 517 op. 2 a.e. 115 l. 13 (CSTP Reports 1972).

117. TsDA f. 517 op. 2 a.e. 115 l. 14 (CSTP Reports 1972).

118. TsDA f. 517 op. 2 a.e. 115 l. 14 (CSTP Reports 1972).

119. TsDA f. 517 op. 2 a.e. 115 l. 15 (CSTP Reports 1972).

120. TsDA f. 1B op. 35 a.e. 5106 l. 194 (Politburo Plan 1974).

121. TsDA f. 1B op. 35 a.e. 5106 l. 192 (Politburo Plan 1974).

122. TsDA f. 1B op. 35 a.e. 5106 l. 196 (Politburo Plan 1974). The cross-point or cross-bar switch is a collection of switches arranged in a matrix; the electronic cross-point allowed for large-scale telephone exchanges and telephony systems after the mid-1960s.

123. TsDA f. 1B op. 35 a.e. 5106 l. 72–78 (Politburo Plan 1974).

124. TsDA f. 1B op. 35 a.e. 5106 l. 79 (Politburo Plan 1974).

125. Taken from the brief historical overview on the website of the private company Information Services AD, a successor of the state-owned ESSI systems under their various guises after 1970: https://www.is-bg.net/bg/about (last accessed: December 8, 2020).

126. TsDA f. 517 op. 4 a.e. 33 l. 44 (CSTP Report 1974).

127. TsDA f. 517 op. 4 a.e. 33 l. 44–5 (CSTP Report 1974).

128. TsDA f. 517 op. 4 a.e. 35 l. 66 (CSTP Automation-7 1974).

129. TsDA f. 1B op. 35 a.e. 5106 l. 265 (Politburo Plan 1974).

130. TsDA f. 1b op. 35 a.e. 5368 l. 70 (Politburo Theses 1975).

131. BAN-NA f. 20 op. 5 a.e. 113 l. 2 (BAS Paper 1976).

132. BAN-NA f. 20 op. 5 a.e. 113 l. 24 (BAS Paper 1976).

133. BAN-NA f. 20 op. 5 a.e. 113 l. 25 (BAS Paper 1976).

134. BAN-NA f. 20 op. 5 a.e. 113 l. 27 (BAS Paper 1976).

135. BAN-NA f. 20 op. 5 a.e. 113 l. 30–35 (BAS Paper 1976).

136. BAN-NA f. 20 op. 5 a.e. 113 l. 49–53 (BAS Paper 1976).

137. RGAE f. 9480 op. 12 a.e. 583 l. 1 (GKNT Co-Operation 1977); for more on the creation of the system and the assistance by Glushkov and the Soviets, and its technical specifications, see A. A. Morozov, V. V. Glushkova, and T. B. Korobkova, "Sozdanie Edinoĭ Sistemy So͡tsual'noĭ Informa͡TSii (ESSI)—Bolgarskoĭ OGAS," *Matematichni Mashini I Sistemi,* no. 3 (2013), pp. 3–21.

138. TsDA f. 517 op. 7 a.e. 58 l. 6 (CSTP Report 1979–80).

139. TsDA f. 517 op. 7 a.e. 58 l. 8 (CSTP Report 1979–80).

140. TsDA f. 1003 op. 1 a.e. 130 l. 57–8 (IZOT Reports 1980).

141. RGAE f. 9480 op. 12 a.e. 1777 l. 36 (GKNT Co-operation 1980).

142. RGAE f. 9480 op. 12 a.e. 1777 l. 37 (GKNT Co-operation 1980).

143. TsDA f. 517 op. 5 a.e. 20 l. 2–4 (CSTP Reports 1978).

144. TsDA f. 517 op. 5 a.e. 20 l. 6 (CSTP Reports 1978).

145. TsDA f. 1B op. 67 a.e. 2695 l. 104 (Politburo Discussions 1983).

146. TsDA f. 37A op. 9 a.e. 37 l. 7 (CICT Papers 1982).

147. TsDA f. 37A op. 9 a.e. 37 l. 8 (CICT Papers 1982).

148. For more, see Paul E. Ceruzzi, *A History of Modern Computing* (Cambridge, MA: The MIT Press, 1998), chapters 7 and 8.

149. TsDA f. 517 op. 6 a.e. 38 l. 66 (CSTP Report 1982).

150. TsDA f. 517 op. 6 a.e. 38 l. 56 (CSTP Report 1982).

151. TsDA f. 517 op. 6 a.e. 38 l. 62–3 (CSTP Report 1982).

152. TsDA f. 517 op. 6 a.e. 38 l. 66 (CSTP Report 1982).

153. TsDA f. 517 op. 6 a.e. 38 l. 67 (CSTP Report 1982).

154. TsDA f. 517 op. 6 a.e. 38 l. 69 (CSTP Report 1982).

155. TsDA f. 517 op. 6 a.e. 38 l. 70–71 (CSTP Report 1982).

156. TsDA f. 517 op. 6 a.e. 38 l. 79 (CSTP Report 1982).

157. TsDA f. 517 op. 6 a.e. 48 l. 34–6 (CSTP Reports 1983).

158. TsDA f. 517 op. 6 a.e. 98 l. 12 (CSTP Reports 1983).

159. TsDA f. 1B op. 67 a.e. 3090 l. 167; Milena Dimitrova, *Zlatnite Desitiletii͡a na Bulgarskata Elektronika* (Sofia: IK Trud, 2008), p. 221.

160. TsDA f. 1B op. 65 a.e. 24 l. 39–41 (Central Committee Plenum, 1978).

161. TsDA f. 1B op. 65 a.e. 24 l. 46 (Central Committee Plenum, 1978).

162. Evgeniĭ Kandilarov, "Elektronikata v Ikonomicheskata Politika na Bŭlgarii͡a prez 60te-80te Godini na XX Vek," *GSU-IF*, vol. 96/97 (2003/2004), p. 461.

163. TsDA f. 1B op. 65 a.e. 25 l. 79 (Closed Plenum 1978).

164. Vera Vutova-Stefanova and Evgeniĭ Kandilarov, *Bŭlgarii͡a I I͡Aponii͡a: Politika, Diplomatsiya, Lichnosti I Subitiya* (Sofia: Iztok-Zapad, 2019), p. 249.

165. TsDA f. 517 op. 5 a.e. 20 l. 81–3 (CSTP Report 1978).

166. TsDA f. 517 op. 6 a.e. 29 l. 10 & TsDA f. 1B op. 67 a.e. 1604 l. 61 (CSTP and Politburo Reports).

167. TsDA f. 517 op. 7 a.e. 55 l. 88B & l. 89; TsDA f. 1B op. 67 a.e. 668 l. 33 (CSTP and Politburo Reports).

168. RGAE f. 9480 op. 13 a.e. 875 l. 13 (Robotics Working Group 1982).

169. TsDA f. 1B op. 65 a.e. 24 l. 97–8 (Plenum 1978).

170. *Zhenata Dnes*, no. 2 (1987), p. 4.

171. TsDA f. 517 op. 6 a.e. 56 l. 122 (CSTP Report 1983).

172. TsDA f. 535 op. 3 a.e. 58 l. 69 (Automation Plan 1985).

173. TsDA f. 1B op. 66 a.e. 66 l. 7 (Plenum 1985).

174. DA-V f. 1230 op. 1 a.e. 54 l. 101 (Centre Plan 1971).

175. DA-V f. 1230 op. 1 a.e. 12 l. 6 (Activity Report 1975).

176. TsDA f. 830 op. 1 a.e. 11 l. 11 (Izotimpex Activity 1970).

177. Interview with Nedelcho Vichev, August 8, 2019.

178. Mar Hicks, *Programmed Inequality: How Britain Discarded Women Technologists and Lost Its Edge in Computing* (Cambridge, MA: The MIT Press, 2017).

179. *Zhenata Dnes,* no. 7 (1985), pp. 6–7.

180. *Zhenata Dnes,* no. 7 (1985), pp. 6–7.

181. *Zhenata Dnes,* no. 7 (1985), pp. 6–7.

182. *Zhenata Dnes,* no. 9 (1985).

183. *Zhenata Dnes,* no. 8 (1986), p. 33.

184. *Zhenata Dnes,* no. 1 (1986), pp. 15–16.

185. ARAN f. 579 op. 6 a.e. 380 l. 84–7 (Co-operation Agreement 1972).

186. BAN-NA f. 20 op. 5 a.e. 40 l. 21 & BAN-NA f. 20 op. 5 a.e. 65 l. 6 (BAS Reports).

187. BAN-NA f. 20 op. 5 a.e. 55 l. 7 (BAS Plan 1974–5).

188. Petŭr Petrov, *55 Godini Avtomatika, Kibernetika I Robotika v BAN* (Unpublished; shared with me by author), p. 10.

189. BAN-NA f. 20 op. 5 a.e. 93 l. 17–21 (BAS Report 1976).

190. Interview with Petŭr Petrov, December 11, 2015; curiously, we get a glimpse of the working day of an anonymous BAS member working on the "Astra" project in Petko Simeonov's *Individualna Deynost: Sotsiologichesko Znachenie I Ritum* (Sofia: Nauka I Izkustvo, 1982), a sociological study of individuals' days and work rhythms. On March 10, 1975, this person spent around 3 hours working on technical documentation related to the project in the afternoon and reported on recent trips to the mine to fix parts.

191. Interview with Petŭr Petrov, December 11, 2015; corroborated by Vasil Sgurev in an interview on July 7, 2016.

192. Albena Shkodrova, *Sots-Gurme: Kurioznata Istorii͡a na Kuhnyata v NRB* (Sofia: Zhanet'45, 2014).

193. Shkodrova, *Sots-Gurme,* p. 110.

194. Shkodrova, *Sots-Gurme,* p. 113.

195. TsDA f. 1B op. 67 a.e. 3792 l. 104–6 (Politburo Discussions 1984).

196. TsDA f. 517 op. 6 a.e. 56 l. 129 (CSTP Report 1983).

197. Quoted in Seymour Goodman, "Information Technologies and the Citizen: Towards a 'Soviet-Style Information Society'?" in Loren R. Graham (ed.), *Science and the Soviet Social Order* (Cambridge, MA: Harvard University Press, 1990), p. 60.

CHAPTER 6

1. Slava Gerovitch, *From Cyberspeak to Newspeak: A History of Soviet Cybernetics* (Cambridge, MA: The MIT Press, 2002). See also his "Mathematical Machines of the Cold War: Soviet Computing, American Cybernetics and Ideological Disputes in the Early 1950s," *Social Studies of Science,* vol. 31, no. 2 (April 2001), pp. 253–287; and "Russian Scandals: Soviet Readings of American Cybernetics in the Early Years of the Cold War," *The Russian Review* vol. 60 (October 2001), pp. 545–568.

2. See Ksenia Tatarchenko, "A House with the Window to the West: The Akademgorodok Computer Center (1958–1993)," PhD dissertation, Princeton University, 2013;

Ksenia Tatarchenko, "Thinking Algorithmically: From Cold War Computer Science to the Socialist Information Culture," *Historical Studies in the Natural Sciences,* vol. 49, no. 2 (2019), pp. 194–225; Peter Galison, *Image and Logic: A Material Culture of Microphysics* (Chicago: University of Chicago Press, 1997).

3. I use the term in the sense that James C. Scott does in his *Seeing Like a State: How Certain Schemes to Improve the Human Condition Have Failed* (New Haven, CT: Yale University Press, 1999), which discusses how high modernist states aim to make their societies "legible" to themselves through categorization and other tools that quantify populations and make them easily "read" by bureaucracies.

4. BAN-NA f. 20 op. 3 a.e. 77 l. 27 (Institute Council 1965).

5. Vasil Nedev, "Zapisi ot Khronikata na Bulgarskata Kompi͡utŭrna Tehnika" (available at https://bbaeii.webnode.page/bylg-electronica-i-inormatika/; last accessed August 3, 2022); and Evgeniĭ Kandilarov, "Elektronikata v Ikonomicheskata Politika na Bŭlgarii͡a prez 60te-80te Godini na XX Vek," *GSU-IF,* vol. 96/97 (2003/2004), p. 444.

6. TsDA f. 517 op. 2 a.e. 97 l. 42–46 (CSTP Report 1966).

7. For more on this, see Georgi Konstantinov, *Tom III, Chast 1—Napred I Ako Putyat Vodi Kum Golgota* (Sofia: Shrapnel, 2009), pp. 15–20; and Georgi Konstantinov, *Tom III. Chast 2—Svobodata, Sancho, E Veliko Neshto!* (Sofia: Shrapnel, 2009), pp. 6–7.

8. Tzvetana Dzerhmanova, *Spomeni ot Lagerite* (Sofia: Farago, 2011), pp. 169–170.

9. TsDA f. 517 op. 2 a.e. 105 l. 19 (CSTP Progress Report 1970).

10. TsDA f. 517 op. 2 a.e. 89 l. 44–45 (CSTP Protocols 1972).

11. TsDA f. 517 op. 2 a.e. 113 l. 100 (CSTP Reports 1972).

12. TsDA f. 517 op. 2 a.e. 74 l. 5 (CSTP-BAS Co-operation 1972).

13. TsDA f. 517 op. 2 a.e. 111 l. 21–22 (University Meetings 1972).

14. Interview with Petŭr Petrov, March 19, 2015; the author's father himself spent his service in the Navy in radio decryption.

15. TsDA f. 1B op. 67 a.e. 3517 l. 24 (Politburo Discussions 1984).

16. TsDA f. 1B op. 66 a.e. 1731 l. 7 (Politburo Discussions 1979).

17. TsDA f. 1B op. 66 a.e. 1731 l. 22, 37, 57 (Politburo Discussions 1979).

18. *Kompi͡utŭr za Vas,* no. 1 (1984), p. 36.

19. *Kompi͡utŭr za Vas,* no. 2 (1985), p. 9.

20. TsDA f. 517 op. 6 a.e. 89 l. 5–15 (CSTP Reports 1985).

21. *Kompi͡utŭr za Vas,* no. 3 (1985), p. 6.

22. ARAN f. 2061 op. 1 a.e. 18 l. 7 (Soviet Academy Trips 1985).

23. TsDA f. 1B op. 59 a.e. 153 l. 57 (Politburo Programs 1987).

24. TsDA f. 1B op. 68 a.e. 3416 l. 38 (Politburo Reports 1988).

25. TsDA f. 1B op. 68 a.e. 3425 l. 68 (Politburo Projects 1988).

26. TsDA f. 1B op. 67 a.e. 3595 l. 15 (Politburo Discussions 1984).

27. The chronology draws on the historical note at the start of the BAN-NA fond 20, which is the archive of the institute; as well as Petŭr Petrov's "55 Godini Avtomatika, Kibernetika I Robotika v BAN" (available at http://css.iict.bas.bg/55yearsACRinBAS_red-2.pdf ; last accessed: August 3, 2022).

28. Interview with Petŭr Petrov, March 19, 2015.

29. BAN-NA f. 20 op. 3 a.e. 69 l. 1 (Institute Budget 1964).

30. BAN-NA f. 20 op. 3 a.e. 69 l. 35 (Institute Budget 1964).

31. BAN-NA f. 20 op. 3 a.e. 71 l. 3–4 (Scientific Councils 1964).

32. BAN-NA f. 20 op. 5 a.e. 3 l. 8–9 (Staff List 1972).

33. BAN-NA f. 20 op. 5 a.e. 143 l. 8 (Staff List 1978).

34. BAN-NA f. 20 op. 5 a.e. 69 l. 9 (Institute Plan 1975).

35. Interview with Angel Angelov conducted by e-mail through his daughter Sonia Angelova Hirt, June 29, 2016.

36. From Petŭr Petrov's draft of an article on Angel Angelov's life, for the BAS bulletin.

37. Petrov, "55 Godini Avtomatika, Kibernetika I Robotika v BAN."

38. *Zhenata Dnes,* no. 10 (1988), pp. 16–17.

39. BAN-NA, f. 20 op. 5 a.e. 28 l. 41 (BAS Paper 1972).

40. BAN-NA, f. 20 op. 5 a.e. 28 l. 89 (BAS Paper 1972).

41. BAN-NA f. 20 op. 5 a.e. 33 l. 7 (BAS Paper 1972).

42. For more on this individual, see Loren R. Graham, *The Ghost of the Executed Engineer: Technology and the Fall of the Soviet Union* (Cambridge, MA: Harvard University Press, 1993).

43. The first publicly available book on cybernetics in the Eastern Bloc, Soviet mathematician Igor Poletaev's *Signal: O Nekatoyh Ponatiyah Kibernetiki* (Moscow: Sovetskoe Radio, 1958), described cybernetics as follows on page 23: "The laws of existence and transformation of information are objective and accessible for study. The determination of these laws, their precise description, and the use of information-processing algorithms, especially control algorithms, together constitute the content of cybernetics."

44. BAN-NA, f. 20 op. 5 a.e. 8 l. 14–17 (Institute Plan 1972).

45. BAN-NA, f. 20 op. 5 a.e. 79 l.7 (BAS Paper 1975).

46. BAN-NA, f. 20 op. 5 a.e. 101 l. 133–136 (BAS Paper 1976).

47. BAN-NA, f. 20 op. 5 a.e. 101 l. 137 (BAS Paper 1976).

48. BAN-NA, f. 20 op. 5 a.e. 116 l. 19 (BAS Paper 1976).

49. BAN-NA, f. 20 op. 5 a.e. 116 l. 21 (BAS Paper 1976).

50. BAN-NA, f. 20 op. 5 a.e. 116 l. 32–4 (BAS Paper 1976).

51. BAN-NA f. 20 op. 5 a.e. 113 (BAS Paper 1976).

52. BAN-NA, f. 20 op. 5 a.e. 137 l. 21 (BAS Paper 1977).

53. BAN-NA, f. 20 op. 5 a.e. 137 l. 25 (BAS Paper 1977).

54. BAN-NA, f. 20 op. 5 a.e. 137 l. 26 (BAS Paper 1977).

55. ARAN f. 1807 op. 1 a.e. 423 l. 5–7 (Soviet Report on Meeting 1980).

56. BAN-NA f. 20 op. 5 a.e. 133 l. 73 (BAS Paper 1977).

57. BAN-NA f. 20 op. 5 a.e. 133 l. 77 (BAS Paper 1977).

58. BAN-NA f. 20 op. 5 a.e. 135 l. 10 (BAS Paper 1977).

59. BAN-NA f. 20 op. 5 a.e. 139 l. 40–41 (BAS Paper 1977).

60. ARAN f. 579 op. 9 a.e. 81 l. 22; l. 127 (Soviet Academy Co-Operation 1975).

61. Loren R. Graham, *Science, Philosophy and Human Behaviour in the Soviet Union* (New York: Columbia University Press, 1987), p. 278.

62. A. Berg, A. Kolman, and V. Pekelis, "Za Kibernetichnata Informatsiya I Choveshkovo Mislene," *Filosofska Misŭl,* no. 8 (1969), pp. 79–80.

63. Trifon Trifonov, "Psihologiyata na Truda u Nas," *Filosofska Misŭl,* no. 9 (1969), p. 74.

64. Trifonov, "Psihologiyata na Truda u Nas," p. 76.

65. Trifonov, "Psihologiyata na Truda u Nas," p. 78.

66. V. Buriev and V. Boev, "Opit za Teoretiko-Informatzionna Interpretatsiya na Protivorechieto Mezhdu Natrupanite Znaniya I Vuznikvashtite Problemi," *Filosofska Misŭl,* no. 8 (1969), p. 61.

67. Petko Ganchev, "Sotsialna Informatsiya I Prognozirane v Obshtestvoto," *Filosofska Misŭl,* no. 10, (1969), p. 23.

68. Ganchev, "Sotsialna Informatsiya I Prognozirane v Obshtestvoto," pp. 25–27.

69. Nikolaĭ Stanulov, "Kibernetika I Socialno Upravlenie," *Filosofska Misŭl,* no. 1 (1973), p. 43.

70. Stanulov, "Kibernetika I Socialno Upravlenie," p. 44.

71. Stanulov, "Kibernetika I Socialno Upravlenie," p. 45.

72. Stanulov, "Kibernetika I Socialno Upravlenie," p. 47.

73. Stanulov, "Kibernetika I Socialno Upravlenie," p. 48.

74. Stanulov, "Kibernetika I Socialno Upravlenie," p. 49.

75. Stanulov, "Kibernetika I Socialno Upravlenie," pp. 50–51.

76. Nikolaĭ Stanulov, "Maluk Kompendium po Filosofskite Vuprosi na Kibernetikata," *Filosofska Misŭl,* no. 11 (1973), pp. 69–71.

77. Stanulov, "Maluk Kompendium po Filosofskite Vuprosi na Kibernetikata," p. 72.

78. Ivan Popchev et al. "Kompiutŭrna Simulatsiya v Nalichieto na Model za Izbirane na Optimalna Alternativa," in Nikolaĭ Naplatanov, Deniu Bechev, and Nikolaĭ Stanulov (eds.), *Suvremmeni Problemi na Tehnicheskata Kibernetika* (Sofia: Izdatelstvo na BAN, 1975), p. 20.

79. Dimtŭr Georgiev, "Nauchno-Tehnicheskata Revolyutsiya I Vsestrannoto Razvitie na Lichnostta," *Filosofska Misŭl,* no. 10 (1973), pp. 63–64.

80. Mitiu Iankov, *Nauchno-Tehnicheskata Revolyutsiya I Problemut na Sotsialnoto I Biologichnoto u Choveka* (Sofia: Meditsina I Fizkultura, 1977), p. 7.

81. Iankov, *Nauchno-Tehnicheskata Revolyutsiya,* p. 52.

82. Iankov, *Nauchno-Tehnicheskata Revolyutsiya,* pp. 100–104.

83. Pavel Georgiev, "Kritika na burzhoazni filosofski spekulatsii s naukata na informaciyata," *Filosofska Misŭl,* no. 8 (1978), p. 105.

84. Zhivkova is a controversial figure, who has not yet been given her due in academic writing. On her policy toward children and her international initiative, however, there is an in-depth analysis by Yuliyana Gencheva, "The International Children's Assembly 'Banner of Peace': Performing the Child in Socialist Bulgaria," PhD dissertation, Indiana University, 2010.

85. V. Afanasiev, "Informatsiya, Chovek, Kompi͡utŭr," *Filosofska Misŭl,* no. 7 (1974), p. 115.

86. Dimitŭr Georgiev, "Nauchno-Tehnicheskata Revolyutsiya I Vsestrannoto Razvitie na Lichnosta," *Filosofska Misŭl,* no. 10 (1973), p. 64.

87. Interview with Plamen Vachkov, the director of the Pravetz factories, June 30, 2015.

88. *Kompi͡utŭr za Vas,* no. 4 (1987), p. 4; for similar efforts in the USSR, see Gregory Afinogenov, "Andrei Ershov and The Soviet Information Age," *Kritika: Explorations in Russian and Eurasian History,* vol. 14, no. 3 (2013): 561–584.

89. *Kompi͡utŭr za Vas,* no. 4 (1987), p. 7.

90. Ts. Kardashev et al., "Kompi͡utŭrizatsiyata v Obrazovanieto," *Filosofska Misŭl,* no. 6, (1985), p. 123.

91. Biliana Papazova, "Decata v Kompi͡utŭriziraniyat Svyat: Utreshnite Problem—Dnes," *Filosofska Misŭl,* no. 10 (1985), pp. 106–107.

92. Ana Krŭsteva, "Informatsiiata—Novo Svetovno Predizvikatelstvo?" *Filosofska Misŭl,* no. 3 (1985), pp. 46–47.

93. Vladimir Stoychev, "Filosofski Analiz na Kibernetichniya Podhod v Izkustvoto," *Filosofska Misŭl,* no. 3 (1985), pp. 101–103.

94. Stoychev, "Filosofski Analiz na Kibernetichniya Podhod v Izkustvoto," p. 103.

95. Nikolai Naplatanov, "Elektroizchislitelnite Mashini I Nauchno-Tehnicheskoto Tvorchestvo," *Filosofska Misŭl,* no. 3 (1986), pp. 86–87.

96. TsDA f. 1B op. 67 a.e. 3595 l. 15 (Politburo Discussion 1984).

97. Iskra Arseneva, "Sŭvremenni Burzhoazni Teorii za Informatizatsiiata na Obshestvoto," *Filosofska Misŭl,* no. 4 (1988), pp. 70–77.

98. Petŭr Mitov, "Infomatsiya, Ikonomika, Obshtestvo," *Filosofska Misŭl,* no. 9 (1989), pp. 4–14.

99. Valentin Korkinov, "Niakoi Sotsialno-Psihologicheski Problemi, Porazhdashti 'Psihicheskata Bariera' Spryamo Kompi͡utŭrizatsiyata," *Filosofska Misŭl,* no. 9 (1989), pp. 28–36.

100. DA-V, f. 1230 op. 3 a.e. 55 l. 31–5 (Qualifications 1985–92).

101. A. I. Rakitov, "Filosofiyata na Progresa I Informatsionnata Revolyutsiya," *Filosofska Misŭl,* no. 9 (1989), p. 60.

102. *Kompi͡utŭr za Vas,* no. 1 (1984), p. 1.

103. *Kompi͡utŭr za Vas,* no. 1 (1984), p. 6.

104. *Kompi͡utŭr za Vas,* no. 1 (1985), p. 9.

105. *Kompi͡utŭr za Vas,* no. 1 (1985), p. 11.

106. *Kompi͡utŭr za Vas,* no. 2–3 (1986), p. 3.

107. *Kompi͡utŭr za Vas,* no. 9–10 (1986), p. 4.

108. *Kompi͡utŭr za Vas,* no. 5 (1987), p. 2.

109. *Kompi͡utŭr za Vas,* no. 1–2 (1988), p. 3.

110. *Kompi͡utŭr za Vas,* no. 8, (1986), p. 4.

111. *Kompi͡utŭr za Vas,* no. 1 (1984), p. 38.

112. *Zhenata Dnes,* no. 2 (1985), p. 9.

113. *Zhenata Dnes,* no. 4 (1985), p. 16.

114. *Zhenata Dnes,* no. 1 (1986), p. 16.

115. "Sblizhavane s Kompi͡utŭra," *Zhenata Dnes,* no. 2 (1985), pp. 8–9.

116. "Kompi͡utŭr v Peleni," *Zhenata Dnes,* no. 5 (1985), p. 12.

117. *Kompi͡utŭr za Vas,* no. 5 (1985), p. 1.

118. *Kompi͡utŭr za Vas,* no. 3 (1988), p. 11.

119. *Kompi͡utŭr za Vas,* no. 3–4 (1989), p. 1.

120. This information is from the Bulgarian science fiction portal: http://bgf.zavinagi.org/ (last accessed: December 30, 2020).

121. Information from the Bulgarian science fiction portal (last accessed: December 30, 2020).

122. Information from https://esfs.info/esfs-awards/1970-1979/ (last accessed: January 23, 2017).

123. Li͡uben Dilov, *Putyat na Ikar* (Plovdiv: Hristo G Danov, 1984), pp. 166–168.

124. Dilov, *Putyat na Ikar,* p. 185.

125. Li͡uben Dilov, *Propusnatiyat Shans* (Plovdiv: Hristo G Danov, 1986), pp. 21–22.

126. Dilov, *Propusnatiyat Shans,* pp, 191–211.

127. *Kompi͡utŭr za Vas,* no. 4 (1985), pp. 2–3.

128. Anton Staykov, *Kratka Istorii͡a na Bulgarskiya Komiks* (Sofia: Kibea, 2013).

129. Nikola Kesarovski, *Petiyat Zakon* (Sofia: Otechestvo, 1983).

130. Li͡ubomir Nikolov, "Sto Perviy Zakon Robotehniki," in the special issue of *Integralniy Klub Fantastiki, Evristiki I Prognostiki Ivana Efremova* (Sofia: Izdatelstvo Gradski Mladezhki Dom Lilyana Dimitrova, 1989).

CHAPTER 7

1. DA-V, f. 1230 op. 3 a.e. 1 (Restructuring Documents 1991–3).

2. DA-V, f. 1230 op. 3 a.e. 6 (Privatizaion Reports 1994–6).

3. Nathan Ensmenger, *The Computer Boys Take Over: Computers, Programmers, and the Politics of Technical Expertise* (Cambridge, MA: MIT Press, 2010).

4. Gil Eyal, Ivan Szelenyi, and Eleanor Townsley, *Making Capitalism without Capitalists: Class Formation and Elite Struggles in Post-Communist Central Europe* (London: Verso, 1998); and Gil Eyal, *The Origins of Post-Communist Elites: From the Prague Spring to the Break-up of Czechoslovakia* (Minneapolis: University of Minnesota Press, 2003).

5. See Roy Rosenzweig, "Wizards, Bureaucrats, Warriors, and Hackers: Writing the History of the Internet," *The American Historical Review,* vol. 103, no. 5 (December 1998), pp. 1530–1552 for a blending of both stories, from Big Science to the hacker. Fred Turner, *From Counterculture to Cyberculture: Stewart Brand, The Whole Earth Network, and the Rise of Digital Utopianism* (Chicago: University of Chicago Press, 2008). For a critical view of the techno-utopianism that comes with the rise of information technologies and knowledge work, see the captivating work by Alan Liu, *The Laws of Cool: Knowledge Work and the Culture of Information* (Chicago: University of Chicago Press, 2004).

6. BAN-NA f. 20 op. 5 a.e. 90 l. 2 (Scientific Council 1976).

7. *Nov Tekhnicheski Avangard* 4 (69/457), September 2006.

8. AKRDOPBGDSRSNBA-R, f. 9 op. 2 a.e. 371 l. 3 (Activity Report 1986).

9. TsDA f. 517 op. 4 a.e. 40 l. 8–13 (CSTP Trips 1974).

10. TsDA f. 517 op. 4 a.e. 40 l. 8–13 (CSTP Trips 1974).

11. TsDA f. 517 op. 4 a.e. 40 l. 11 (CSTP Trips 1974).

12. TsDA f. 1B op. 35 a.e. 12 l. 32–33 (Politburo Meetings 1967–8).

13. BAN-NA f. 20 op. 5 a.e. 45 l. 17–21 (BAS Trips 1973).

14. BAN-NA f. 20 op. 5 a.e. 96 l. 17–27 (BAS Trips 1976).

15. BAN-NA f. 20 op. 5 a.e. 152 l. 54–9 (BAS Trips 1978).

16. BAN-NA f. 20 op. 5 a.e. 127 l. 13–17 (BAS Trips 1977).

17. BAN-NA f. 20 op. 5 a.e. 127 l. 24–26 (BAS Trips 1977).

18. TsDA f. 37A op. 9 a.e. 2 l. 7 (CICT Trips 1983).

19. TsDA f. 37A op. 9 a.e. 2 l. 60–62 (CICT Trips 1983).

20. TsDA f. 37A op. 9 a.e. 4 l. 23 (CICT Trips 1983).

21. TsDA f. 37A op. 9 a.e. 4 l. 26 (CICT Trips 1983), and TsDA f. 37A op. 9 a.e. 5 l. 88 (Trips 1983&4).

22. TsDA f. 37A op. 9 a.e. 8 l. 11, 92, 169 (CICT Trips 1985).

23. TsDA f. 830 op. 2 a.e. 97 l. 4 (Izotimpex Trips 1978–86).

24. TsDA f. 517 op. 2 a.e. 108 l. 128 (CSTP US Visit 1972).

25. Eglė Rindzevičiūtė, *The Power of Systems: How Policy Sciences Opened Up the Cold War World* (Ithaca, NY: Cornell University Press, 2016).

26. TsDA f. 517 op. 2 a.e. 109 l. 132 (CSTP Report 1972).

27. For more on Mateev's economics, see Georgi Burlakov, "Proektut na Akademik Evgeni Mateev za Avtomatizirana Sistema za Upravlenie v Konteksta na Gospodstvashti Ikonomicheski Teorii," *Ikonomicheski Alternativi*, no. 5 (2008), pp. 103–114.

28. TsDA f. 517 op. 6 a.e. 108 l. 73–75 (CSTP Automation-8 1982).

29. TsDA f. 517 op. 6 a.e. 113 l. 112 (CSTP Report 1982).

30. RGAE f. 9480 op. 12 a.e. 1348 l. 3–17 (GKNT Information 1979).

31. RGAE f. 9480 op. 12 a.e. 1348 l. 3–17 (GKNT Information 1979).

32. TsDA f. 1B op. 67 a.e. 1929 l. 342 (Politburo Reports 1983).

33. TsDA f. 1B op. 67 a.e. 1929 l. 350 (Politburo Reports 1983).

34. TsDA f. 517 op. 7 a.e. 94 l. 25 (CSTP Prognosis 1984).

35. Interview with Peter Petrov, June 10, 2016.

36. Interview of the author with Nedko Botev and Boyan T͡Sonev, June 23, 2015.

37. Interview of the author with Stoi͡an Markov, July 28, 2015.

38. Nikolai Stanoulov, "On the Concepts of Information and Control," in Norbert Wiener and J. P. Schade (eds.), *Progress in Biocybernetics*, vol. 3 (Amsterdam: Elsevier, 1966), pp. 97–104.

39. This section is based on Ivo Khristov's "Pravoto na Prehoda," in Ivan Chalakov et al. (eds.), *Mrezhite na Prehoda: Kakvo Vsushnost se Sluchi v Bŭlgarii͡a sled 1989 g.?* (Sofia: Iztok Zapad, 2008), pp. 64–89.

40. Khristo Khristov, *Imperiyata na Zadgranichnite Firmi: Suzdavane, Deynost i Iztochvane na Druzhestvata s Bulgarsko Uchastie zad Granitsa 1961–2007* (Sofia: Ciela, 2009), pp. 10–11.

41. Khristov, *Imperiyata na Zadgranichnite Firmi*, p. 12.

42. Khristov, *Imperiyata na Zadgranichnite Firmi*, p. 322.

43. Khristov, *Imperiyata na Zadgranichnite Firmi*, p. 13.

44. TsDA f. 1B op. 61 a.e. 6 l. 6–7 (Politburo Report 1975).

45. TsDA f. 1B op. 61 a.e. 6 l. 51–57 (Politburo Report 1975).

46. TsDA f. 1003 op. 1 a.e. 129 l. 3 (IZOT & Olivetti Correspondance 1978–80).

47. TsDA f. 1003 op. 1 a.e. 129 l. 30 (IZOT & Olivetti Correspondance 1978–80).

48. TsDA f. 1003 op. 1 a.e. 129 l. 45–7 (IZOT & Olivetti Correspondance 1978–80).

49. TsDA f. 1003 op. 1 a.e. 129 l. 35 (IZOT & Olivetti Correspondance 1978–80).

50. TsDA f. 1003 op. 1 a.e. 130 l. 2 (IZOT & Control Data Correspondance 1980).

51. TsDA f. 1003 op. 1 a.e. 130 l. 5–9 (IZOT & Control Data Correspondance 1980).

52. TsDA f. 1003 op. 1 a.e. 130 l. 11 (IZOT & Control Data Correspondance 1980).

53. TsDA f. 1003 op. 1 a.e. 130 l. 13 (IZOT & Control Data Correspondance 1980).

54. TsDA f. 1003 op. 1 a.e. 130 l. 31 (IZOT & Control Data Correspondance 1980).

55. TsDA f. 1003 op. 1 a.e. 130 l. 33 (IZOT & Control Data Correspondance 1980).

56. TsDA f. 1003 op. 1 a.e. 130 l. 35–6 (IZOT & Control Data Correspondance 1980).

57. TsDA f. 1003 op. 1 a.e. 130 l. 44 (IZOT & Control Data Correspondance 1980).

58. TsDA f. 1003 op. 1 a.e. 130 l. 52–4 (IZOT & Control Data Correspondance 1980).

59. TsDA f. 1003 op. 1 a.e. 130 l. 57–8 & 93–4 (IZOT & Control Data Correspondance 1980).

60. TsDA f. 259 op. 45 a.e. 212 l. 209 (MFT Report 1983).

61. TsDA f. 259 op. 45 a.e. 212 l. 209–210 (MFT Report 1983).

62. TsDA f. 259 op. 45 a.e. 212 l. 210 (MFT Report 1983).

63. TsDA f. 259 op. 45 a.e. 212 l. 220–226 (MFT Report 1983).

64. TsDA f. 830 op. 2 a.e. 89 l. 9 (Izotimpex Program 1984).

65. "Oral History of Bisser Dimitrov," interview by Jim Porter, July 7, 2005, Computer History Museum.

66. "Oral History of Bisser Dimitrov," p. 2.

67. TsDA f. 830 op. 2 a.e. 97 l. 69–72 (Izotimpex Trips 1978–86).

68. The timeline can be traced in the document TsDA f. 259 op. 45 a.e. 212 (MFT Reports 1983).

69. Khristov, *Imperiyata na Zadgranichnite Firmi*, pp. 191–192.

70. "Oral History of Bisser Dimitrov," p. 5.

71. "Oral History of Bisser Dimitrov," p. 3.

72. Ogni͡an Doĭnov, *Spomeni* (Sofia: IK Trud, 2002), pp. 32–43.

73. Doĭnov, *Spomeni*, p. 48.

74. Doĭnov, *Spomeni*, pp. 49–50.

75. Doĭnov, *Spomeni*, p. 50.

76. Doĭnov, *Spomeni*, p. 52.

77. Quoted in Doĭnov, *Spomeni*, p. 56.

78. Doĭnov, *Spomeni*, pp. 70–71.

79. Doĭnov, *Spomeni*, p. 123.

80. TsDA f. 1B op. 67 a.e. 2431 l. 130 (Politburo Discussions 1983).

81. Doĭnov, *Spomeni*, p. 124.

82. Doĭnov, *Spomeni*, pp. 127–128.

83. Doĭnov, *Spomeni*, p. 129.

84. DA-V, f. 1230 op. 3 a.e. 10 (Council Meeting 1985).

85. Nacho Papazov, *I͡Aponii͡a ot Samuraiskiya Mech do Izkustveniya Intelekt* (Sofia: Otechestven Front, 1989).

86. Chapter 12, "Upravlenie po I͡Aponski," in Papazov, *I͡Aponii͡a*.

87. Chapter 14, "MITI—Agentsiyata 'Kormchiya,'" in Papazov, *I͡Aponii͡a*.

88. "Posleslov," in Papazov, *I͡Aponii͡a*.

89. Interview with Krasimir Markov, February 4, 2016.

90. Interview with Stoi͡an Markov, July 28, 2015.

91. Interview with Plamen Vachkov, June 30, 2015.

92. "Denyat na Direktora, *Otechestvo,* vol. 247 (January 28, 1986).

93. Interview with Plamen Vachkov, June 30, 2015.

94. A point Ivailo Znepolski makes in his *Bulgarskiya Komunizum: Sotsiokulturni Cherti I Vlastova Traektoriya* (Sofia: Ciela, 2012).

95. Albena Shkodrova, *Sots-Gurmet: Kurioznata Istorii͡a na Kuhnyata v NRB* (Sofia: Zhanet 45, 2014).

96. AMVnR f. 20 op. 42 a.e. 1967 l. 15 (Indian Press Reports 1985).

97. Interview of Charles S. Kennedy with John D. Caswell in the *Association for Diplomatic Studies and Training Foreign Officers Oral History Project,* from August 2000, p. 60 (https://adst.org/OH%20TOCs/Caswell,%20John%20D.toc.pdf; last accessed: August 3, 2022)

98. TsDA f. 1B op. 67 a.e. 3595 l. 15 (Politburo Discussions 1984).

99. TsDA f. 1B op. 65 a.e. 64 l. 31 (Central Committee Plenum 1984).

100. TsDA f. 1B op. 65 a.e. 64 l. 59–61 (Central Committee Plenum 1984).

101. TsDA f. 1B op. 67 a.e. 3595 (Politburo Discussions 1984).

102. TsDA f. 1B op. 67 a.e. 3595 l. 5–8 (Politburo Discussions 1984).

103. TsDA f. 1B op. 67 a.e. 3595 l. 1/17 (Politburo Discussions 1984).

104. TsDA f. 1B op. 67 a.e. 3595 l. 15 (Politburo Discussions 1984).

105. TsDA f. 1B op. 67 a.e. 3595 l. 323–328 (Politburo Discussions 1984).

106. TsDA f. 1B op. 59 a.e. 153 l. 7–11 (Discussions with Zhivkov—1987).

107. TsDA f. 1B op. 59 a.e. 153 l. 12 (Discussions with Zhivkov—1987).

108. TsDA f. 1B op. 59 a.e. 153 l. 24 (Discussions with Zhivkov—1987).

109. TsDA f. 1B op. 59 a.e. 153 l. 30 (Discussions with Zhivkov—1987).

110. TsDA f. 1B op. 59 a.e. 153 l. 35–37 (Discussions with Zhivkov—1987).

111. TsDA f. 1B op. 59 a.e. 153 l. 48–52 (Discussions with Zhivkov—1987).

112. TsDA f. 1B op. 59 a.e. 153 l. 56 (Discussions with Zhivkov—1987).

113. TsDA f. 1B op. 59 a.e. 153 l. 72 (Discussions with Zhivkov—1987).

114. TsDA f. 1B op. 59 a.e. 153 l. 77 (Discussions with Zhivkov—1987).

115. TsDA f. 1B op. 59 a.e. 153 l. 82 (Discussions with Zhivkov—1987).

116. TsDA, f. 1B op. 65 a.e. 90 l. 68 (Central Committee 1988).

117. TsDA, f. 1B op. 65 a.e. 90 l. 88 (Central Committee 1988).

118. TsDA, f. 1B op. 65 a.e. 90 l. 103 (Central Committee 1988).

119. TsDA f. 1B op. 65 a.e. 92 l. 7–8 (Central Committee Plenum 1988).

120. TsDA f. 1B op. 65 a.e. 92 l. 45 (Central Committee Plenum 1988).

121. Chalakov et al., *Mrezhite na Prehoda,* p. 50.

122. Chalakov et al., *Mrezhite na Prehoda*, p. 48.

123. Doĭnov, *Spomeni*, pp. 218–222.

124. Khristov, *Imperiyata na Zadgranichnite Firmi*, p. 13.

125. AKRDOPBGDSRSNBA-R, f. 9 op. 4 a.e. 574 l. 44 (Activity Report 1987).

126. Khristo Khristov, "Dŭrzhavna Sigurnost. Chast 2.4: NTR I Proektut 'Neva'," *Kapital*, September 12, 2010.

127. Khristov, "Dŭrzhavna Sigurnost. Chast 2.4."

128. Khristo Khristov, "Dŭrzhavna Sigurnost. Chast 2.3: Proektut 'Monblan' I 'Neyavnite Firmi' na DZU," *Kapital*, September 6, 2010.

129. AKRDOPBGDSRSNBA-R, f. 9 op. 630 l. 24 (Incoming Cypher-gram 1990).

130. AKRDOPBGDSRSNBA-R, f. 9 op. 630 l. 18 (Incoming Cypher-gram 1990).

131. AKRDOPBGDSRSNBA-R, f. 9 op. 630 l. 12 (Incoming Cypher-gram 1990).

132. *Dnevnik*, August 15, 2010.

133. Interview with Plamen Vachkov, June 30, 2015.

134. "Plamen Vachkov Oglavi Novata IT Agentsiya," Vesti.bg, September 15, 2005; https://www.vesti.bg/novini/s-tazi-promiana-syobshteniiata-se-izvazhdat-ot-strukturata-na-ministerstvoto-na-transporta-778202 (last accessed: January 7, 2021).

135. *Kompi͡utŭr za Vas*, no. 1 (1985), pp. 10–11.

136. *Kompi͡utŭr za Vas*, no. 3 (1985), pp. 58–59, and no. 4 (1986), p. 3.

137. *Kompi͡utŭr za Vas*, no. 2–3 (1986), p. 3, and no. 9 (1987), p. 3.

138. *Kompi͡utŭr za Vas*, no. 12 (1986), p. 5.

139. *Kompi͡utŭr za Vas*, no. 3–4 (1989), pp. 8–15.

140. *Kompi͡utŭr za Vas*, no. 1–2 (1990), p. 7.

141. *Kompi͡utŭr za Vas*, no. 3–4 (1990), pp. 9–10.

142. *Kompi͡utŭr za Vas*, no. 5–6 (1990), p. 1.

143. *Kompi͡utŭr za Vas*, no. 5–6 (1990), p. 1.

144. "Bulgarians Linked to Computer Virus," *New York Times*, December 21,1990.

145. David S. Bennahun, "Heart of Darkness," *Wired*, November 1, 1997.

146. "Bulgarian 'Dark Avenger' Part of East-Bloc Legacy," *Christian Science Monitor*, May 19, 1992.

147. "How Eastern Europe's Villains Changed Sides in the Malware War—and Made You Protect Your PC," ZDNet; http://www.zdnet.com/article/how-eastern-europes-villains-changed-sides-and-made-you-protect-your-computer/ (last accessed: March 5, 2017).

148. *Zhenata Dnes*, no. 6 (1989), pp. 4–5.

149. *Zhenata Dnes*, no. 7 (1989), pp. 6–7.

150. Manuel Castells, *The Rise of the Network Society* (Chichester, UK: Wiley-Blackwell, 2010), p. xviii–xxx.

151. Manuel Castells, *End of Millennium* (Malden, MA: Wiley-Blackwell, 2000), pp. 8–28.

152. Manuel Castells, *Communication Power* (New York: Oxford University Press, 2009), pp. 18–26.

153. Robert Castle, "Bulgaria's Delayed Transition: An Analysis of the Delays in Bulgaria's Political and Economic Transition from Socialism to Liberal Democracy," PhD dissertation, City University of New York, 2013.

154. "Elektrifitzirat selo Plochnik" in Plovdiv24.bg; https://www.plovdiv24.bg/novini/regionalni/Elektrificirat-selo-Plochnik-62488 (last accessed: August 3, 2022).

155. Interview with Vasil Sgurev, July 7, 2016.

156. Interview with Nedko Botev and Boyan T͡Sonev, June 23, 2015.

CONCLUSION

1. TsDA, f. 863, op. 3, a.e. 16, l. 3 (Currency Income Report for Electronic Association 1989).

2. TsDA, f. 863, op. 3, a.e. 16, l. 42.

3. TsDA, f. 863, op. 3, a.e. 16, l. 45.

4. In the author's eyes, the classic remains worth reading: Alexander Gerschenkron, *Economic Backwardness in Historical Perspective* (Cambridge, MA: Harvard University Press, 1962).

5. See Ivan Chalakov, Andrei Bundzhulov, Ivo Khristov, et al. (eds.), *Mrezhinte Na Prehoda: Kakvo Vsushnost Se Sluchi v Bŭlgarii͡a Sled 1989ta* (Sofia: Iztok-Zapad, 2008), especially section one (pp. 17–107), on which the following paragraphs draw.

6. "Absolyutno Globalni," *Forbes Bulgaria,* December 21, 2012.

7. "How These Communist-Era Apple II Clones Helped Shape Central Europe's IT Sector," ZDNet; http://www.zdnet.com/article/how-these-communist-era-apple-ii-clones-helped-shape-central-europes-it-sector/ (last accessed: April 12, 2021).

8. Statistics from Eurostat: https://appsso.eurostat.ec.europa.eu/nui/submitViewTableAction.do (last accessed: December 21, 2021).

9. "Bulgaria Strives to Become Tech Capital of the Balkans," *Financial Times,* October 17, 2016.

10. Interview with Krasimir Markov, February 4, 2016.

11. The "thickening" of connections is a term employed by C. A. Bayly in *The Birth of the Modern World 1780–1914* (London: Wiley-Blackwell, 2004).

12. Interview with Plamen Vachkov, June 30, 2015.

13. See https://www.moreto.net/novini.php?n=355885 (last accessed: January 30, 2021).

14. Information on the Olympiad in Pravetz can be found in Union of Mathematicians in Bulgaria, "Proceedings of the International Olympiad in Informatics, Pravetz, May 16–19, 1989;" http://www.math.bas.bg/talents/en/inf/International_Olimpiad_in_Informatics.pdf (last accessed: August 3, 2022).

15. Interview with Koĭcho Dragostinov, April 6, 2015.

INDEX

Note: Photos and Tables are indicated by italicized page numbers.

History of Computing

William Aspray and Thomas J. Misa, editors

Janet Abbate, *Recoding Gender: Women's Changing Participation in Computing*

John Agar, *The Government Machine: A Revolutionary History of the Computer*

William Aspray and Paul E. Ceruzzi, *The Internet and American Business*

William Aspray, *John von Neumann and the Origins of Modern Computing*

Charles J. Bashe, Lyle R. Johnson, John H. Palmer, and Emerson W. Pugh, *IBM's Early Computers*

Martin Campbell-Kelly, *From Airline Reservations to Sonic the Hedgehog: A History of the Software Industry*

Paul E. Ceruzzi, *A History of Modern Computing*

I. Bernard Cohen, *Howard Aiken: Portrait of a Computer Pioneer*

I. Bernard Cohen and Gregory W. Welch, editors, *Makin' Numbers: Howard Aiken and the Computer*

James Cortada, *IBM: The Rise and Fall and Reinvention of a Global Icon*

Nathan Ensmenger, *The Computer Boys Take Over: Computers, Programmers, and the Politics of Technical Expertise*

Thomas Haigh, Mark Priestley, and Crispin Rope, *ENIAC in Action: Making and Remaking the Modern Computer*

John Hendry, *Innovating for Failure: Government Policy and the Early British Computer Industry*

Mar Hicks, *Programmed Inequality: How Britain Discarded Women Technologists and Lost Its Edge in Computing*

Michael Lindgren, *Glory and Failure: The Difference Engines of Johann Müller, Charles Babbage, and Georg and Edvard Scheutz*

David E. Lundstrom, *A Few Good Men from Univac*

René Moreau, *The Computer Comes of Age: The People, the Hardware, and the Software*

Arthur L. Norberg, *Computers and Commerce: A Study of Technology and Management at Eckert-Mauchly Computer Company, Engineering Research Associates, and Remington Rand, 1946–1957*

Emerson W. Pugh *Building IBM: Shaping an Industry and Its Technology*

Emerson W. Pugh, *Memories That Shaped an Industry*

Emerson W. Pugh, Lyle R. Johnson, and John H. Palmer, *IBM's Early Computers: A Technical History*

Kent C. Redmond and Thomas M. Smith, *From Whirlwind to MITRE: The R&D Story of the SAGE Air Defense Computer*

Alex Roland with Philip Shiman, *Strategic Computing: DARPA and the Quest for Machine Intelligence, 1983–1993*

Raúl Rojas and Ulf Hashagen, editors, *The First Computers—History and Architectures*

Corinna Schlombs, *Productivity Machines: German Appropriations of American Technology from Mass Production to Computer Automation*

Dinesh C. Sharma, *The Outsourcer: A Comprehensive History of India's IT Revolution*

Dorothy Stein, *Ada: A Life and a Legacy*

Christopher Tozzi, *For Fun and Profit: A History of the Free and Open Source Software Revolution*

John Vardalas, *The Computer Revolution in Canada: Building National Technological Competence, 1945–1980*

Maurice V. Wilkes, *Memoirs of a Computer Pioneer*

Jeffrey R. Yost, *Making IT Work: A History of the Computer Services Industry*

Thomas Haigh and Paul E. Ceruzzi, *A New History of Modern Computing*

Daniel D. Garcia-Swartz and Martin Campbell-Kelly, *Cellular: An Economic and Business History of the International Mobile-Phone Industry*

Victor Petrov, *Balkan Cyberia: Cold War Computing, Bulgarian Modernization, and the Information Age behind the Iron Curtain*